Plant Biology: Theory and Applied Principles

Plant Biology: Theory and Applied Principles

Edited by Cristiano Shepherd

New York

Published by Syrawood Publishing House,
750 Third Avenue, 9th Floor,
New York, NY 10017, USA
www.syrawoodpublishinghouse.com

Plant Biology: Theory and Applied Principles
Edited by Cristiano Shepherd

International Standard Book Number: 978-1-68286-741-9 (Hardback)

Cataloging-in-Publication Data

Plant biology : theory and applied principles / edited by Cristiano Shepherd.
p. cm.
Includes bibliographical references and index.
ISBN 978-1-68286-741-9
1. Botany. 2. Plants. I. Shepherd, Cristiano.
QK45.2 .P53 2019
580--dc23

TABLE OF CONTENTS

PREFACE

Plant biology or botany is a branch of biology that is concerned with the study of plants. It studies plant structure, growth, metabolism, diseases, systematics and taxonomy. The study of gene expression mechanisms and control in plants, using molecular genetics and epigenetics is also under the scope of this field. Research in botany is significant in the fields of horticulture, agriculture and forestry for the provision of staple foods and materials like oil, rubber, fiber, timber and medicine. It is also applied in the synthesis of chemicals and raw materials for energy production and construction. The study of plant biology also helps in addressing the global environmental issues of resource management, conservation, food security, carbon sequestration, climate change and sustainability. This book includes some of the vital pieces of work being conducted across the world, on various topics related to plant biology. It provides significant information of this discipline to help develop a good understanding of botany and related fields. It aims to serve as a resource guide for students and experts alike and contribute to the growth of the discipline.

The information shared in this book is based on empirical researches made by veterans in this field of study. The elaborative information provided in this book will help the readers further their scope of knowledge leading to advancements in this field.

Finally, I would like to thank my fellow researchers who gave constructive feedback and my family members who supported me at every step of my research.

Editor

1

Evaluation of seasonal antioxidant activity and total phenolic compounds in stems and leaves of some almond (*Prunus amygdalus* L.) varieties

Aysel Sivaci[1*] and Sevcan Duman[2]

Abstract

Background: This study aimed to determine the seasonal changes of total antioxidant activity and phenolic compounds in samples taken from leaves (April, July, October) and stems (April, July, October, January) of some almond (*Prunus amygdalus* L.) varieties (Nonpareil, Ferragnes and Texas).

Results: It was indicated that antioxidant activity and phenolic compounds in leaves and stems of Nonpareil, Ferragnes and Texas showed seasonal differences. Antioxidant activity IC_{50} of these varieties reached the highest value in April for leaves whereas in October for stems. The highest level of total phenolic compounds was in January for stems while in October for leaves.

Conclusions: These results showed that total antioxidant activity and phenolics in leaves and stems of almond varieties changed according to season and plant organ.

Keywords: Almond, Antioxidant activity, Phenolics, *Prunus amygdalus*, Seasonal changes

Background

Climate is a factor which affects agricultural production. Increase of temperature or variations in precipitation ratio affect physiological events in plants [1–3].

Almond belongs to *Rosaceae* family and is an important product due to high commercial value of its fruits. Its fruits are nutritious due to their protein, fat, mineral substance, fibre and vitamin E content [4–10].

Natural products derived from plants are used for health supplements [11]. Antioxidants are compounds which prevent or delay the oxidation of lipids or other molecules by inhibiting the initiation or propagation of oxidative chain reactions have positive effects on human health [12,13]. Phenolic substances are one of the most widely known substances with their antioxidant characteristics [14,15]. Phenolic substances are metabolites with different structure and functions, having an aromatic ring containing generally one or more hydroxyl group [16,17]. Antioxidant effects of phenolic compounds are explained by bonding free radicals, forming chelate with metals and inactivating some enzymes [18]. Various studies carried out on almond cultivars showed that almond fruit and sections have phenolic compounds and antioxidant activity [19–23].

Analysis of previous research on almonds focused on investigating the antioxidant activity and phenolic compounds mostly in fruits, and the changes in stem and leaves have not been studied on seasonal basis. This study will be significant for determining beneficial compounds in different organs of almond varieties, on seasonal basis, the possibility of making use of these organs and explaining the variations in this plant under different climatic conditions. Therefore, this study investigated seasonal total antioxidant activity and total phenolic compounds in leaves and stems of some almond varieties (Nonpareil, Ferragnes and Texas) which are distributed in Adiyaman province of Turkey.

Results

Total antioxidant activity

It was found that total antioxidant activity varied according to season, plant organs and varieties (Figures 1 and 2). Total antioxidant capacity in the leaves of almond varieties (IC_{50}) was low in April in Texas, Ferragnes and Nonpareil (high antioxidant activity) (Texas, 88.67 μg mL^{-1};

* Correspondence: asivaci@gmail.com
[1]Department of Biology, Art and Science Faculty, Adiyaman University, Adiyaman, Turkey
Full list of author information is available at the end of the article

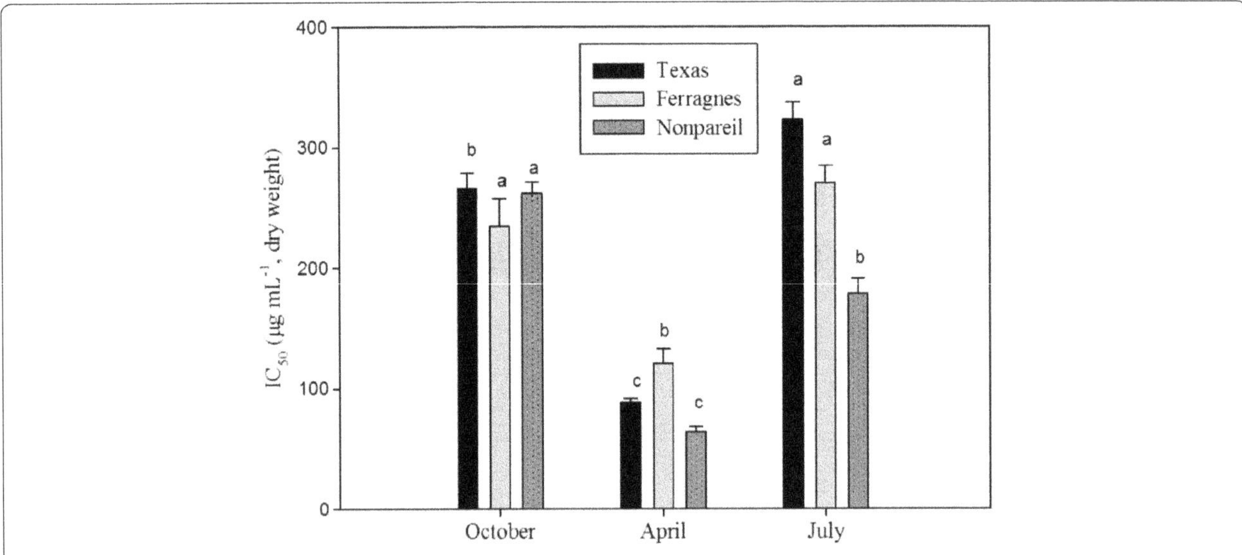

Figure 1 Seasonal total IC_{50} changes in leaves of Nonpareil, Texas and Ferragnes in DPPH. (Data followed by different letters are significantly different from each other ($p < 0.05$) according to Duncan's test).

Ferragnes, 121 µg mL^{-1}; Nonpareil, 64 µg mL^{-1}) ($p < 0.05$). The highest IC_{50} value (low antioxidant activity) was found in July for Texas, Ferragnes and in October for Nonpareil (Figure 1). It was determined that antioxidant capacity was the lowest for Nonpareil (high antioxidant activity) and high (low antioxidant activity) for Ferragnes in April (Figure 1) ($p < 0.05$).

IC_{50} values in the stems of almond varieties were low in October (high antioxidant activity) (Texas, 79.16 µg mL^{-1}; Ferragnes, 174.46 µg mL^{-1}; Nonpareil, 73.50 µg mL^{-1}); and high in April (low antioxidant activity) (Texas, 207.79 µg mL^{-1}; Ferragnes, 200.67 µg mL^{-1}; Nonpareil, 137.67 µg mL^{-1}) (Figure 2). The variation in antioxidant activity was significant in other varieties excluding Ferragnes ($p < 0.05$). IC_{50} values of Ferragnes and Texas varieties were similar in July and January. On the other hand, it was found that IC_{50} values were at the lowest level in Nonpareil and Texas (high antioxidant activity) and high in Ferragnes (low antioxidant activity) in October (Figure 2).

Total phenolic compounds

Phenolic compounds in the leaves of Nonpareil, Texas and Ferragnes varieties were high in October (Figure 3) ($p < 0.05$). In this month, values of phenolic compounds

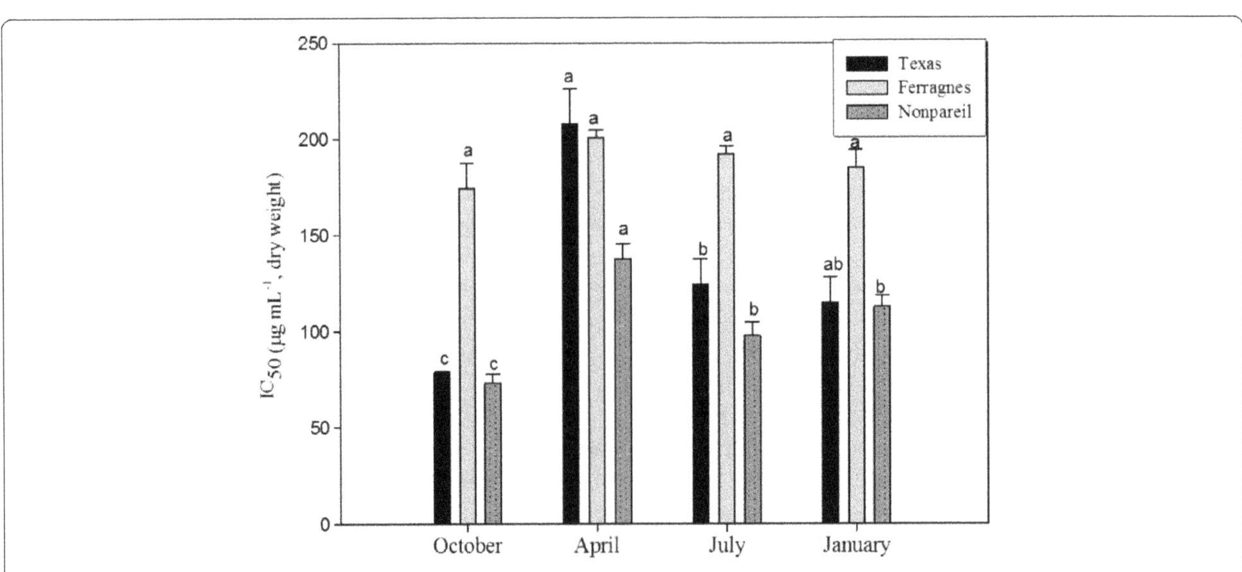

Figure 2 Seasonal total IC_{50} changes in stems of Nonpareil, Texas and Ferragnes in DPPH. (Data followed by different letters are significantly different from each other ($p < 0.05$) according to Duncan's test).

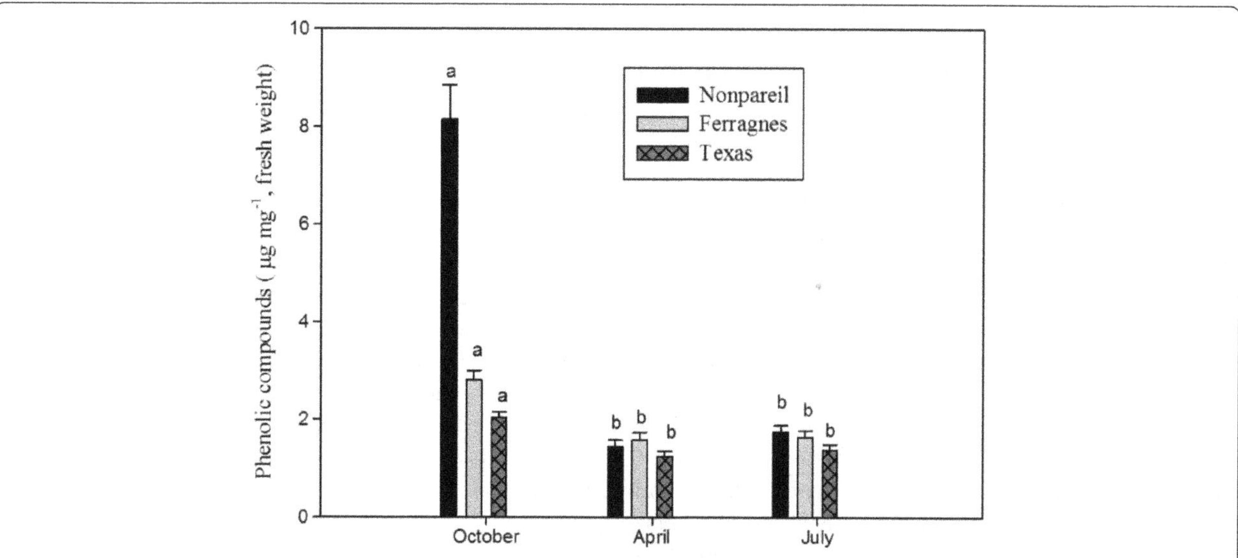

Figure 3 Seasonal total phenolic compounds in leaves of Nonpareil, Texas and Ferragnes. (Data followed by different letters are significantly different from each other ($p < 0.05$) according to Duncan's test).

of Texas, Ferragnes and Nonpareil were 2.03 µg mg^{-1}, 2.82 µg mg^{-1} and 8.15 µg mg^{-1} respectively. In all varieties, phenolic compounds were low in April and July and the variations observing in April and July were not significant statistically (Figure 3) ($p > 0.05$).

It was found that phenolic compounds in the stems of almond varieties also varied according to months. In all varieties, phenolic compounds were the highest in January (Teksas, 2.08 µg mg^{-1}; Ferragnes, 1.85 µg mg^{-1}; Nonpareil, 2.90 µg mg^{-1}) (Figure 4) ($p < 0.05$). The lowest phenolic compound contents were in October (0.95 µg mg^{-1}) and July (1.08 µg mg^{-1}) for Ferragnes and; in April for Texas (0.77 µg mg^{-1}). In Nonpareil, levels of phenolic compounds were higher than other two varieties in all months (Figure 4).

Discussion

Nunes et al. [24] carried out a study in red propolis and investigated the effect of season on antioxidant activity and total phenols. The researchers reported that there was a correlation between total antioxidant activity and season and that phenol content was high in hydra-

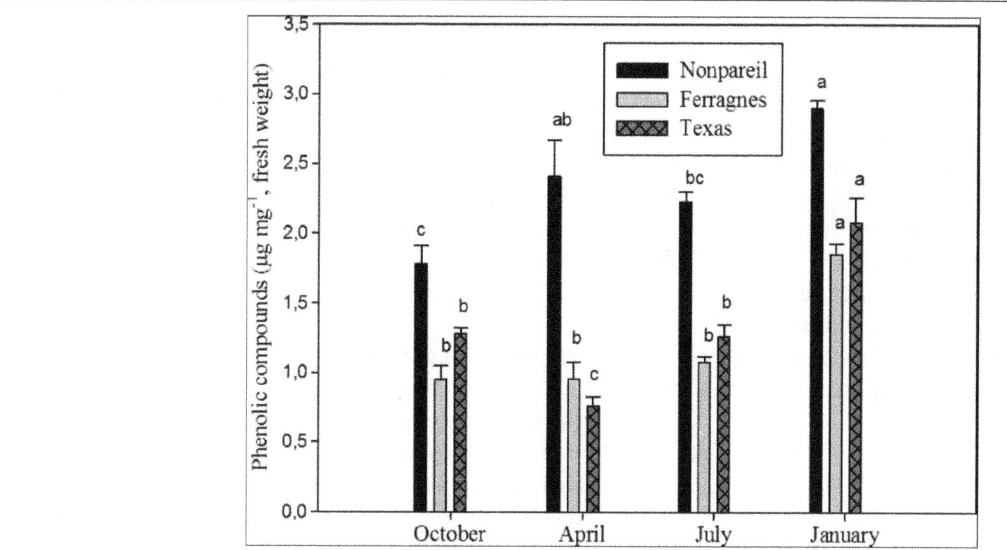

Figure 4 Seasonal total phenolic compounds in stems of Nonpareil, Texas and Ferragnes. (Data followed by different letters are significantly different from each other ($p < 0.05$) according to Duncan's test).

alcoholic (90%) concentration in October. Ignacio et al. [25] reported that photosynthetic pigment and antioxidant activity in *Fagus sylvetica* L. varied by sun and light conditions.

In another study, carried out on different cultivars of California almonds, it was determined that flavonoid content and antioxidant activity depended on the cultivar rather than season [20]. As indicated above, this study found that antioxidant activity showed seasonal variations in stem and leaves of almond varieties (Nonpareil, Texas and Ferragnes) (Figures 1 and 2). Esfahlan and Jamei [26] carried out a study in fruits of ten wild almond species and reported that there were variations in flavonoid, phenolic contents and antioxidant activities according to almond species. The present study found that antioxidant activity varied according to varieties and plant organs. In April, antioxidant activity was the highest in the leaves of Nonpareil variety and the lowest in Ferragnes (Figure 1). On the other hand, in stems, it was high in Nonpareil and Texas and low in Ferragnes in October (Figure 2).

Cosmulescu and Trandafır [27] investigated the seasonal variation of total phenols in the leaves of *Juglans regia* L. They found that total phenols increased in June and July; decreased in August and increased in early September. They reported that there could be a correlation between phenolic content, season, genetic and ecological factors in walnut leaves. Sivaci and Sökmen [28] carried out a study on stem cuttings of *Morus alba* and *Morus nigra* and found that antioxidant activity and phenolic compounds showed seasonal variation. The highest antioxidant activity in stems was found in October.

In another study, variation of some phenolic compounds (phenylpropane chlorogenic acid and flavonoids such as rutin, hyperoside, epigenin-7-O-glucoside, kaempherole, quercitrin, quercetin and amentoflavone) in four *Hypericum triquetrifolium* populations in Central Black Sea Region were explored. Chemical variation was identified between the populations and plant sections and it was reported that these variations could be a result from different genetic, environmental and morphological factors [29]. In our study, total phenolic compounds varied according to season, variety and plant parts. The highest phenolic compound content in all varieties was observed in October in leaves; and in January in stems. The highest phenolic compound contents belonged to Nonpareil when compared to other varieties (Figures 3 and 4).

Conclusions

It was found that total antioxidant activity and phenolic compounds in Nonpareil, Texas and Ferragnes varieties exhibited variations according to season, plant organ (leaf and stem) and variety. This could be result from ecological, genetic and metabolic differences as indicated other studies [27,29]. Also, in the period during almond tree has no fruit, the leaves and stems could be made use of due to their antioxidant activity.

Further studies should be conducted to investigate the total antioxidant activity and phenolic profiles of almonds in next seasons.

Methods

Plant materials

Almond varieties (Nonpareil, Ferragnes and Texas) were collected from Lokman village of Adiyaman/Turkey (37° 42′ 15″ N, 38° 19′ 11″ E, 1920 feet) in 2011-2012. Leaves (April, July, October) and stems (April, July, October, January) of the almonds were analyzed. No analysis was performed in January because the plants had no leaves.

Determination of antioxidant activity-DPPH

Leaf and stem samples collected from almond varieties were dried and grinded. Grinded samples were taken to methanol (MeOH) and extracted by shaking in water bath for 3 hours. Methanol extracts were then evaporated in evaporator under vacuum until they dried. Color of 2,2-diphenyl-1-picrylhydrazyl (DPPH) changes in the presence of antioxidant in the medium. Fifty microliters of various concentrations of almond variety extracts dissolved in methanol was added to in 5 mL of a 0.004% methanol solution of DPPH. The mixture was incubated at room temperature for 30 minutes and absorbance values were read at 517 nm [30]. Inhibition percent (I%) of DPPH was calculated according to the following equation:

$$I\% = \left(A_{blank}\text{-}A_{sample}/A_{blank}\right) \times 100$$

where A_{blank} is the absorbance of the control reaction (containing all reagents except the test compound) and A_{sample} is the absorbance of the test compound. Inhibition is concentration dependent, and extract concentration providing 50% inhibition (IC_{50}) is calculated from the graphplotted inhibition percentage against extract concentration. The assay was carried out in triplicate.

Determination of total phenolic compounds

The leaf and stem samples were homogenized in 2.5 ml ethanol and shaken in water bath at 25°C for 24 h. Homogenized samples were filtered. 1 ml ethanol, 5 ml distilled water and 1 ml Folin-Ciocalteu reagent were added to 1 ml of the filtered samples and shaken well. After 3 minutes, 3 ml of Na_2CO_3 (2%, w/v) was added and shaken in a dark medium at intervals for 2 hours. Absorbance values were read at 760 nm for phenolic compound amounts and amounts were determined according to standard gallic acid equivalence [31,32]. The assay was carried out in triplicate.

Statistical analysis

All analyses in this study were performed in three replicates. SPSS version 15.0 was used for statistical analyses. Duncan tests were used to determine the variations between the means. Differences at 5% ($p < 0.05$) level were considered as significant.

Competing interests

The authors declare that they have no competing interests.

Authors' contributions

AS carried out conception and design of the study, acquisition of data, /analysis and interpretation of data, drafting the manuscript and revising. SD carried out acquisition of data, analysis and interpretation of data, statistical analysis. Both authors read and approved the final manuscript.

Acknowledgements

This study was supported by Adıyaman University Scientific Research Projects Unit (SRP). I would like to thank Hüseyin Bereket from Lokman village, who is the grower of the almond varieties using in this study, and Dr. Rıza Binzet who helped field studies.

Author details

[1]Department of Biology, Art and Science Faculty, Adiyaman University, Adiyaman, Turkey. [2]Graduate School of Sciences, Adiyaman University, Adiyaman, Turkey.

References

1. Ahmed M, Fayyaz U-H, Aslamt M, Aslam MA: **Physiological attributes based resilience of wheat to climate change.** *Int J Agric Biol* 2012, **14:**407–412.
2. Olesen JE, Trnka M, Kersebaum KC, Skjelvag AO, Seguin B, Peltonen-Sainio P, Rossi F, Kozyra J, Micale F: **Impacts and adaptation of European crop production systems to climate change.** *European J Agron* 2011, **34:**96–112.
3. Wang X, Caı J, Jıang D, Lıu F, Daı T, Cao W: **Pre-anthesis high- temperature acclimation alleviates damage to the flag leaf caused by post-anthesis heat stress in wheat.** *J Plant Physiol* 2011, **168:**585–593.
4. Agunbıade SO, Olanlokun JO: **Evaluation of some nutritional characteristics of Indian almond (*Prunus amygdalus*) nut.** *Pakistan J Nutr* 2006, **5:**316–318.
5. Ahrens S, Venkatachalam M, Mıstry AM, Lapsley K, Sahte SK: **Almond (*Prunus dulcis* L.) protein quality.** *Plant Foods Hum Nutr* 2005, **60:**123–128.
6. Cordeıro V, Monteıro A: **Almond growing in Tras-os-Montes region (Portugal).** *Acta Hort* 2001, **591:**161–165.
7. Dokuzoguz M, Gulcan R: *Researchs on breeding of almond genotypes (Prunus amygdalus L.) by the selection in Eagean Region and they adaptation.* Turkey: Tübitak Toag; 1973:22.
8. Gulcan R: *Physiological and morphological studies on selected types of almond. E.Ü.* Bornova: Faculty of Agriculture Publications; 1976:72. No. 310.
9. Kuden A: **Almond germplasm and production in Turkey and the future of almonds in the GAP area.** *Acta Hort* 1997, **470:**29–33.
10. Mısırlı A, Gulcan R: **Almond growing in Turkey.** *Nucis* 2000, **9:**3–6.
11. Zafar Shoaıb M, Muhammad F, Javed I, Akhtar M, Khalıq T, Aslam B, Waheed A, Yasmın R, Zafar H: **White mulberry (*Morus alba*): a brief phytochemical and pharmacological evaluations account.** *Int J Agric Biol* 2013, **15:**612–620.
12. Exarchou V, Nenadıs N, Tsımıdou M, Gerothanasssıs IP, Troganıs A, Boskou D: **Antioxidant activities and phenolic composition of extracts from Greek Oregano, Greek Sage, and Summer Savory.** *J Agr Food Chem* 2002, **50:**5294–5299.
13. Velıoglu YS, Mazza G, Gao L, Oomah BD: **Antioxidant activity and total phenolics in selected fruits, vegetables, and grain products.** *J Agr Food Chem* 1998, **46:**4113–4117.
14. Madhavı DL, Deshpande SS, Salunkhe DK: *Food Antioxidants: Technological: Toxicological and Health Perspectives.* Newyork: Markel Dekker; 1996:41–50.
15. Shahıdı F, Wanasundara PKJPD: **Phenolic antioxidants.** *Critical Rev Food Sci and Nutr* 1992, **32:**67–103.
16. Naczk M, Shahıdı F: **Extraction and analysis of phenolics in food.** *J Chromatography A* 2004, **1054:**95–111.
17. Taiz L, Zeiger E: *Plant Physiology.* 4th edition. Inc: Sinauer Associates; 2006:764.
18. Yang R, Tsao R: **Optimization of a new mobile to know the complex and real polyphenolic composition: Towards a tool phenolic index using high performance liquid chromatography.** *J Chromatography A* 2003, **1018:**29–40.
19. Barreıra JCM, Ferreıra ICFR, Olıveıra MBPP, Pereıra JA: **Antioxidant activity and bioactive compounds of ten Portuguese regional and commercial almond cultivars.** *Food and Chem Toxicol* 2008, **46:**2230–2235.
20. Bollıng BW, Dolnikowskı G, Blumberg JB, Olıver Chen CY: **Polyphenol content and antioxidant activity of California almonds depend on cultivar and harvest year.** *Food Chem* 2010, **122:**819–825.
21. Esfahlan AJ, Jamei R, Esfahlan RJ: **The importance of almond (*Prunus amygdalus* L.) and its by-products.** *Food Chem* 2010, **120:**349–360.
22. Mılbury PE, Chen CY, Dolnıkowskı GG, Blumberg JB: **Determination of flavonoids and phenolics and their distribuition in almonds.** *J of Agr Food Chem* 2006, **54:**5027–5033.
23. Yıldırım AN, San B, Koyuncu F, Yıldırım F: **Variability of phenolics, α tocopherol and amygdalin contents of selected almond (*Prunus amygdalus* Batsch.) genotypes.** *J Food Agr Environ* 2010, **8:**76–79.
24. Nunes LCC, Galindo AB, Lustosa SR, Brasileiro MT, Do Egito AA, Freitas RM, Randau KP, Rolim Neto PJ: **Influence of seasonal variation on antioxidant and total phenol activity of red propolis extracts.** *Adv Studies Biol* 2013, **5:**119–133.
25. Ignacıo J, Plazaola G, Becerrıl JM: **Seasonal changes in photosynthetic pigments and antioxidants in beech (*Fagus sylvatica*) in a Mediterranean climate: implications for tree decline diagnosis.** *Aust J Plant Physiol* 2001, **28:**225–232.
26. Esfahlan AJ, Jamei R: **Properties of biological activity of ten wild almond (*Prunus amygdalus* L.) species.** *Turk J Biol* 2012, **36:**201–209.
27. Cosmulescu S, Trandafır I: **Seasonal variation of total phenols in leaves of walnut (*Juglans regia* L.) J.** *Med Plants Res* 2011, **5:**4938–4942.
28. Sivaci A, Sokmen M: **Seasonal changes in antioxidant activity, total phenolic and anthocyanin constituent of the stems of two *Morus* species (*Morus alba* L. and *Morus nigra* L.).** *Plant Growth Regul* 2004, **44:**251–256.
29. Çırak C, Radušienė J, Janulıs V, Ivanauskas L, Camaş N, Ayan AK: **Phenolic constituents of *Hypericum triquetrifolium* Turra (Guttiferae) growing in Turkey: variation among populations and plant parts.** *Turk J Biol* 2011, **35:**449–456.
30. Gulluce M, Sokmen M, Daferera D, Agar G, Ozkan H, Kartal N, Polıssıou M, Sokmen A, Sahın F: **In vitro antibacterial, antifungal, and antioxidant activities of the essential oil and methanol extracts of herbal parts and callus cultures of *Satureja hortensis* L.** *J Agr Food Chem* 2003, **51:**3958–3965.
31. Chandler SF, Dodds JH: **The effect of phosphate, nitrogen and sucrose on the production of phenolics and solasidine in callus cultures of *Solanum lacinitum*.** *Plant Cell Rep* 1983, **2:**205–208.
32. Slınkard K, Sıngleton VL: **Total phenol analyses: automation and comparison with manual methods.** *Am J Enol Viticult* 1977, **28:**49–55.

2

Sugars, organic acids, and phenolic compounds of ancient grape cultivars (*Vitis vinifera* L.) from Igdir province of Eastern Turkey

Sadiye Peral Eyduran[1], Meleksen Akin[2], Sezai Ercisli[3], Ecevit Eyduran[4] and David Maghradze[5*]

Abstract

Background: The Eurasian grapevine (*Vitis vinifera* L.) is the most widely cultivated and economically important horticultural crop in the world. As a one of the origin area, Anatolia played an important role in the diversification and spread of the cultivated form *V. vinifera* ssp. *vinifera* cultivars and also the wild form *V. vinifera* ssp. *sylvestris* ecotypes. Although several biodiversity studies have been conducted with local cultivars in different regions of Anatolia, no information has been reported so far on the biochemical (organic acids, sugars, phenolic acids, vitamin C) and antioxidant diversity of local historical table *V. vinifera* cultivars grown in Igdir province. In this work, we studied these traits in nine local table grape cultivars viz. 'Beyaz Kismis' (synonym name of Sultanina or Thompson seedless), 'Askeri', 'El Hakki', 'Kirmizi Kismis', 'Inek Emcegi', 'Hacabas', 'Kerim Gandi', 'Yazen Dayi', and 'Miskali' spread in the Igdir province of Eastern part of Turkey.

Results: Variability of all studied parameters is strongly influenced by cultivars ($P < 0.01$). Among the cultivars investigated, 'Miskali' showed the highest citric acid content (0.959 g/l) while 'Kirmizi Kismis' produced predominant contents in tartaric acid (12.71 g/l). The highest glucose (16.47 g/100 g) and fructose (15.55 g/100 g) contents were provided with 'Beyaz Kismis'. 'Kirmizi Kismis' cultivar had also the highest quercetin (0.55 mg/l), o-coumaric acid (1.90 mg/l), and caffeic acid (2.73 mg/l) content. The highest ferulic acid (0.94 mg/l), and syringic acid (2.00 mg/l) contents were observed with 'Beyaz Kismis' cultivar. The highest antioxidant capacity was obtained as 9.09 μmol TE g^{-1} from 'Inek Emcegi' in TEAC (Trolox equivalent Antioxidant Capacity) assay. 'Hacabas' cultivar had the highest vitamin C content of 35.74 mg/100 g.

Conclusions: Present results illustrated that the historical table grape cultivars grown in Igdir province of Eastern part of Turkey contained diverse and valuable sugars, organic acids, phenolic acids, Vitamin C values and demonstrated important antioxidant capacity for human health benefits. Further preservation and use of this gene pool will be helpful to avoid genetic erosion and to promote continued agriculture in the region.

Keywords: Table grape, Biochemical composition, HPLC, Spectrophotometer, Germplasm characterization

Background

Overwhelming evidence from *in vitro, in vivo*, epidemiological, and clinical trial data indicates that there are components in a plant-based diet, other than traditional nutrients, that can reduce cancer risk. More than a dozen classes of these biologically active plant chemicals, now known as 'phytochemicals', have been identified. The majority of naturally occurring health-enhancing substances appear to be of horticultural plant origin [1-5].

The grape, cultivated since ancient times is one of the most significant edible and processing for wine crop due to its beneficial influences on human health and economic significance on a large scale [6], and Turkey is one of the major players in this market: according to FAO records, Turkey placed 6th position in the World in terms of amount of grape production according to the data of 2012 [7].

There are vast germplasm resources available within the genus *V. vinifera* throughout grape growing areas in

* Correspondence: david.maghradze@gmail.com
[5]Scientific-Reaserch Center of Agriculture, Tbilisi, Georgia
Full list of author information is available at the end of the article

the world. Despite the existence of thousands of cultivars of *V. vinifera* in grape producer countries, only a few dozen cultivars account for the vast majority of world-wide production. Numerous local cultivars have regional importance and historically played a significant role in many viticulture regions [6]. In fact local cultivars show great variability in particular given biochemical traits and they can be use locally for wine, raisins, and fresh market (table grapes) [6,8,9]. Therefore it is crucial to determine how much variability can be found in local grape cultivars for the given biochemical traits to use them in breeding activities. This may also help the growers to pick the right cultivar according to their market opportunities.

Grape berry 'quality' is one of the cardinal variables that determine wine quality. Berry 'quality', however, is a generic term that refers to levels of a diverse range of berry chemical constituents including organic acids, sugars, phenolics, acidity etc. [9].

The presence of adequate levels of organic acids in the grape berries is one of the key factors to determine the quality of berries and wines. Conde et al. [8] underlined that in the general sense, organic acid contents of grape berries was an indicative in taste based on acid-sugar balance. Organic acids have favorable impacts on human metabolism as well [6]. The characterization of the phenolic compounds in grapes is of great importance in terms of positive contribution to human health and the organoleptic features of grape and wines. Polyphenols obtained from the grapes are reported to prevent cancer, cardiovascular, type-2 diabetes mellitus diseases and so forth [10].

Several variables viz. total soluble sugar, titratable acidity, nitrogen and phenolic compounds balance provide a major contribution in the description of the grape quality [11], which are very significant in case of table grape. Therefore, due to beneficial effects of the grape on human health, the researchers have more focused in particular on the identification of biochemical composition and bioactive compounds for grape cultivars grown in distinct geographic regions throughout the World. To exemplify, Topalovic et al. [12] evaluated the changes in sugar, organic acids, and phenolics at different dates (9, 16, 23, and 30 July) of the Cardinal table grape cultivar during ripening. Nile et al. [10] determined polyphenolic contents and antioxidant characteristics of different cultivars belonging to *Vitis vinifera*, *Vitis labrusca*, and *Vitis* hybrid. More comprehensively, Orak [13] investigated total antioxidant activities, phenolics, anthocyanins, and polyphenoloxidase activities for selected red grape cultivars and emphasized strong correlations among antioxidant capacity, total phenols, and anthocyanins. With the previous report by Sabir et al. [14], variation of several sugars, acids, and total phenols for juice of five grapevine (*Vitis spp.*) cultivars, depending on the various stages of the grape berry development, were scrutinized.

Considering all those former studies together, it is indispensible to conduct further studies on the identification of biochemical composition and phenolic contents that can influence the quality of the grape and its products, and notably, on the determination of the major factors (cultivar, year, location, harvest time and so on) linked to those contents under different climate conditions.

In Turkey, the native table grape cultivars are extensively cultivated in order of local usage fresh grape, raisin and grape juice. Therefore, the cultivation of some varieties has been recognized after the observations of plant and phenological characteristics in the arable soil of Igdır province located in the Eastern Anatolia region of Turkey. Igdir province includes well-known Agri Dagi - the same Ararat Mountain - that is recognized as a location where the Hoah's Ark stopped and the birthplace of viticulture, when Noah after the Flooding planted vineyard and made the wine according to Bible. In spite of the examination, the detection of organic acid, phenolic compounds, and antioxidant compositions of the regional grape cultivars from Igdir province have not been yet documented which may gain more significance in improving the grape industry of the province in the future. Hence, an attempt was made in the current investigation to identify the sugars, organic acids and phenolic compounds for the local nine grape cultivars from Igdır province. The region cultivates only those local grape 2cultivars and there are no well-known international cultivars in the region – in opposite: we have to take in consideration, that the local cultivar 'Sultanina' of Turkish origin is widely spread in America, Australia, Mediterranean, Northern Africa, Europe and other countries [15]; as well as the cultivar 'Askeri' is also spread in Eastern countries like Azerbaijan, Armenia, Iran and Middle Asia [16].

Results and discussion

Organic acids

Samples of nine native grape cultivars available in the single collection vineyard together in Necefali village of Igdır province have been evaluated to detect organic acids including citric, tartaric, malic, succinic, and fumaric acids at the harvest time. The statistical analysis results for organic acids are summarized in Table 1. The cultivars significantly affected levels of citric, tartaric, malic, succinic, and fumaric acids ($P < 0.01$), implying that the cultivars was a very significant source of the variation on the organic acids. Our finding is in agreement with those reported by previous authors [14,17,18]. In the statistical assessment of Table 1, it was determined that coefficients

Table 1 Content of organic acids in the table grape cultivars from Igdir province of Turkey

Cultivars	Citric acid g/l	Tartaric acid g/l	Malic acid g/l	Succinic acid g/l	Fumaric acid g/l
Askeri	0.642 ± 0.02b	4.71 ± 0.29f	2.27 ± 0.18cd	1.01 ± 0.06ab	0.0026 ± 0.0002b
Beyaz Kismis	0.263 ± 0.02f	4.92 ± 0.24ef	1.82 ± 0.10cde	0.34 ± 0.03de	0.0026 ± 0.0002b
El hakki	0.488 ± 0.05c	5.53 ± 0.28e	1.92 ± 0.13cd	0.57 + 0.01de	0.0012 ± 0.0001c
Hacabas	0.637 ± 0.01b	7.77 ± 0.12c	2.23 ± 0.21cd	1.34 ± 0.15a	0.0024 ± 0.0002b
Inek Emcegi	0.323 ± 0.03e	4.74 ± 0.12ef	3.29 ± 0.12ab	1.17 ± 0.10ab	0.0012 ± 0.0001c
Kerim Gandi	0.427 ± 0.02d	6.21 ± 0.09d	1.31 ± 0.16e	1.07 ± 0.03ab	0.0027 ± 0.0002b
Kırmızı Kismis	0.264 ± 0.02f	12.71 ± 0.12a	1.53 ± 0.16de	0.93 ± 0.01bc	0.0007 ± 0.0000c
Miskali	0.959 ± 0.04a	5.67 ± 0.09de	2.58 ± 0.16bc	0.25 ± 0.02e	0.0026 ± 0.0002b
Yazen Dayi	0.254 ± 0.03f	8.72 ± 0.13b	3.56 ± 0.29a	0.67 ± 0.01cd	0.0034 ± 0.0002a
CV (%)	1.09	4.90	13.44	14.33	12.38

**$P < 0.01$ (the significant cultivar effect) [a,b,c,d,f]. In each column means followed by different letter are statistically different each other at $P < 0.01$ (**), ND: Non determined.

of variation (%) of the organic acids ranged from 1.09 to 14.33.

In citric acid, the averages of the evaluated cultivars had a great range between 0.254 to 0.959 g/l. Among the available cultivars, 'Miskali' produced the highest average in citric acid (0.959 g/l) and was found significantly different from the others, statistically ($P < 0.01$).

A very desirable coefficient of variation with 1.09 (%) was found for citric acid (Table 1).

It could be inferred from the statistical results in Table 1 that tartaric acid varied from 9.43 to 25.43 g/l in the study. As seen from Table 1, it was understood markedly that the highest tartaric acid content was averagely obtained from 'Kırmızı Kismis' among the assessed cultivars. Additionally, it is apparent that 'Kırmızı Kismis' differed from other grape cultivars, statistically.

In relation to the results reported in the Table 1, the statistically significant influence of the cultivars on content of malic acid was determined ($P < 0.01$). Of those local grape cultivars, 'Yazen Dayi' recorded the highest average content for malic acid and was equal to 3.56 mg/l. Table 1 denotes that the very narrow range of 1.31 to 3.56 g/l was decided for malic acid.

When Table 1 was statistically assessed, the succinic acid significantly influenced by cultivar factor with a very narrow range changed between 0.25 and 1.34 g/l in the investigation. The highest average was numerically recorded in 'Hacabas' cultivar, whereas no differences between the 'Hacabas' and some other cultivars ('Kerim Gandi', 'Inek Emcegi', and 'Askeri') were detected.

The very narrow range described for fumaric acid was taken notice from Table 1. 'Yazen Dayi' was realized to have a statistically higher content in the organic acid compared with other varieties (Table 1). However, it was displayed that 'El Hakki', 'Kırmızı Kismis', and 'Inek Emcegi' cultivars were not different in the acid content from each other, depending upon the results of Tukey test. For the 'Cardinal' grape variety during ripening, Topalovic et al. [12] informed to be 1.90 g/kg for tartaric acid content and 0.97 g/kg for malic acid content, respectively. For the Alphonse Lavellee, Muscat of Hamburg, Isabella, Italia, and Muscat of Alexandria grape cultivars, Sabir et al. [14] reported tartaric acid (3.8, 4.2, 5.2, 4.8, and 5.0), malic acid (3.6, 2.8, 3.4, 3.1, and 3.0 g/l), and citric acid contents (0.4, 0.3, 0.3, 0.2, and 0.3 g/l), respectively. In our examination (Table 1), malic acid contents of the grape cultivars were found to be considerably lower compared with their tartaric acid contents, which was in agreement with those of Sabir et al. [14].

Antioxidant capacity, sugars, vitamin C

The results and statistical evaluations for the Trolox Equivalent Antioxidant Capacity (TEAC), vitamin C (ascorbic acid), and sugars (fructose and glucose) described in the present investigation are presented in the Table 2. As recognized in TEAC, vitamin C, fructose, and glucose produced the significantly wide variation due to the cultivar factor ($P < 0.01$). The coefficients of variation for TEAC, vitamin C, fructose, and glucose were very low, meaning that our investigation was reliable. For TEAC, with the range of 3.60 to 9.09 μmol TE g^{-1}, the highest average value was achieved for 'Inek emcegi' cultivar, which was ascertained to be significantly different from others ($P < 0.01$).

In the study, vitamin C varied between 11.21 and 35.74 mg per 100 g (Table 2). The results showed explicitly that the highest content identified on average for vitamin C as an indicator of the dietary of the foods was provided with 'Hacabas' cultivar, statistically ($P < 0.01$). On the other hand, 'El Hakki' and 'Yazen Dayi' cultivars were statistically similar and gave the lowest content for vitamin C (Table 2).

It was noted earlier by Sabir et al. [14] that the maturity degree of the grapes was strongly connected with sugar concentration, and principally was dependent

Table 2 Results for TEAC, vitamin C, fructose, and glucose in the table grape varieties from Igdir province of Turkey

Cultivars	TEAC (μmol TE g^{-1})	Vitamin C mg/100 g	Fructose g/100 g	Glucose g/100 g
Askeri	4.50 ± 0.19c	21.08 ± 0.44d	11.40 ± 0.14cd	13.88 ± 0.23c
Beyaz Kismis	4.07 ± 0.05c	30.61 ± 0.23c	15.55 ± 0.40a	16.47 ± 0.21a
El hakki	6.50 ± 0.25b	15.85 ± 0.32f	12.13 ± 0.09c	13.34 ± 0.07c
Hacabas	4.10 ± 0.20c	35.74 ± 0.33a	13.47 ± 0.22b	15.21 ± 0.08b
Inek Emcegi	9.09 ± 0.50a	11.21 ± 0.10 g	8.22 ± 0.08f	10.32 ± 0.18ef
Kerim Gandi	4.63 ± 0.17c	18.71 ± 0.49e	9.76 ± 0.07e	12.25 ± 0.16d
Kırmızı Kismis	3.65 ± 0.01c	33.55 ± 0.28b	9.33 ± 0.16e	10.49 ± 0.13e
Miskali	6.10 ± 0.15b	29.28 ± 0.20c	8.03 ± 0.37f	9.51 ± 0.27f
Yazen Dayi	3.60 ± 0.21c	15.73 ± 0.44f	10.32 ± 0.18de	11.39 ± 0.19d
CV (%)	7.71	2.50	3.50	2.50

**P < 0.01 (the significant cultivar effect) [a,b,c,d,e,f]. In each column means followed by different letter are statistically different each other at P < 0.01 (**), ND: Non determined.

upon quality criteria like sugar accumulation, phenolic compounds, and the ratio of sugar to acid. Rusjan and Korosec-Koruza [17] reported that sugar content of the grape was very significant to decide technological maturity and harvest time. Herewith, when fructose was taken into consideration in Table 2, the highest fructose average was appeared with 'Beyaz Kismis' cultivar, numerically and statistically ($P < 0.01$). The fructose gave the range of 8.03 to 15.55. The highest accumulation of glucose was defined in 'Beyaz Kismis' with the significant content of 16.47 g/100 g. Sabir et al. [14] reported glucose and fructose content between 86.4-107.0 g/l and 80.4-94.1 g/l, respectively.

As to Table 2, the accumulation of fructose contents for the present grape cultivars were found very similar to that of glucose contents with the verification of the very strong correlation of 0.957 (data not shown). The finding was consistent with those reported by some authors [12,13] previously. For 15 red wine cultivars, Rusjan and Korosec-Koruza [17] obtained the range of 3.0-16.8 g/l for sucrose content, the range of 50.9-89.9 g/l for glucose content, and for fructose content, the range of 54.8-83.9 g/l respectively. However, the total sugar accumulation ability in wine grapes is higher than of table grapes and it is depending also in place of cultivation.

Phenolic acids

Results of statistical evaluations for several phenolic compounds of the grape cultivars under the investigation are summarized in Tables 3 and 4, respectively. Cultivar ($P < 0.01$) had a significantly remarkable impact on phenolic compounds, which are very important in grape juice and wine processing industry [18]. For analysis of variance for only one factor, cultivar, coefficients of variation relevant to all the compounds identified in the grape were found to have a range of 0.46 to 12.4% as illustrated in Tables 3 and 4. As appeared obviously from Tables 3 and 4, ferrulic acid of the 'Yazen Dayi' cultivar, and vanillic acids of 'Hacabas' and 'Miskali' cultivars among the grape cultivars analyzed in the present investigation were unidentified.

As antioxidants, plant flavonoids are the most important in free radical scavenging. In a previous review on nutritional benefits of flavonoids, it was reported that catechin among those flavonoids had the highest antioxidant activity [19]. Among the grape cultivars, 'Yazen Dayi' contained predominant level of 1.82 mg/l in catechin, a polyphenolic antioxidant plant metabolite, which plays a major role in microbial defense of the grape berry, when compared with the catechin levels of the others (Table 3). The cultivar including the lowest catechin content was recorded as 'Kırmızı Kismis'. In a previous study, Breksa et al. [20] reported that the catechin contents identified for 16 raisin grape (*Vitis vinifera* L.) cultivars and selections had a very wide range of 1.8 to 209.1 μg/g. The present results demonstrated that the rutin content of the grape cultivars had a narrow range of 1.09 to 3.34 mg/l. The cultivar of the highest rutin content among the analyzed grape cultivars was verified to be 'Hacabas', numerically and statistically (Table 3), but the lowest content in rutin was obtained for the cultivars shown with letter "e" according to the results of Tukey test in Table 3. Breksa et al. [20] mentioned that the rutin contents with a very narrow range of 0.8 to 3.7 μg/g for 14 raisin grape (*Vitis vinifera* L.) cultivars were detected.

As to the statistical evaluation performed in quercetin, 'Kırmızı Kismis' was the grape cultivar that contained the utmost content in comparison to others ($P < 0.01$), whereas the lowest contents were established for the cultivars with letter "e", on the basis of Tukey test, as also depicted from Table 3. The maximal chlorogenic acid content was procured with 'Beyaz Kismis' and 'Inek

Table 3 Results for phenolic acids (Part 1) in the table grape cultivars from Igdir province of Turkey

Cultivars	Catechin (mg/l)	Rutin (mg/l)	Quercetin (mg/l)	Chlorogenic acid (mg/l)	Ferulic acid (mg/l)
Askeri	0.95 ± 0.01cd	1.14 ± 0.07e	0.31 ± 0.01d	1.28 ± 0.03e	0.11 ± 0.00d
Beyaz Kismis	1.26 ± 0.03b	2.22 ± 0.05c	0.36 ± 0.01cd	3.31 ± 0.04a	0.94 ± 0.01a
El Hakki	1.14 ± 0.04bc	2.25 ± 0.06c	0.43 ± 0.01b	2.36 ± 0.04c	0.08 ± 0.01d
Hacabas	0.95 ± 0.01cd	3.34 ± 0.02a	0.09 ± 0.00e	1.13 ± 0.02e	0.24 ± 0.01c
Inek Emcegi	0.76 ± 0.01d	2.33 ± 0.03c	0.14 ± 0.01e	3.48 ± 0.04a	0.22 ± 0.01c
Kerim Gandi	0.73 ± 0.02d	1.50 ± 0.04d	0.40 ± 0.02bc	1.22 ± 0.03e	0.12 ± 0.01d
Kırmızı Kismis	0.43 ± 0.01e	1.09 ± 0.03e	0.55 ± 0.01a	2.08 ± 0.03d	0.54 ± 0.01b
Miskali	1.26 ± 0.06b	1.18 ± 0.02e	0.17 ± 0.01d	2.80 ± 0.06b	0.23 ± 0.01c
Yazen Dayi	1.82 ± 0.13a	3.10 ± 0.02b	0.32 ± 0.01d	2.06 ± 0.03d	ND
CV (%)	0.86	3.57	6.16	3.03	4.93

**P < 0.01 (the significant cultivar effect) [a,b,c,d,e]. In each column means followed by different letter are statistically different each other at P < 0.01 (**), ND: Non determined.

Emcegi' cultivars, which were statistically found similar. Obtained from Tukey test, the letter "e" symbolized the lowest chlorogenic acid content (Table 3). In Ferulic acid, 'Beyaz Kismis' cultivar was superior to the others, statistically ($P < 0.01$), and the cultivars shown through the letter "d" produced the lowest contents (Table 3).

Of the probed local grapes, 'Kırmızı Kismis' had much more content in o-coumaric acid than the others, but 'Kerim Gandi' with the letter "f" produced much lower content as compared to the other cultivars as also understood from Table 4. 'Kirmizi Kismis' was found as the cultivars with the leading average caffeic acid content of 2.73 mg/l (Table 4). In terms of syringic acid, the highest content was proved with 'Beyaz Kismis' cultivar, which was different from the others, statistically ($P < 0.01$). With Table 4, it was obvious that, 'Askeri' was the cultivar with the most vanillic acid content when compared with the other cultivars. Amongst the cultivars under study, 'Beyaz Kismis', 'Inek Emcegi', and 'Miskali', statistically similar to each other, were determined to have the highest contents of gallic acid in reference to the results of Tukey test. Breksa et al. [20] reported that the gallic acid contents of only two genotypes (A95-15 and A95-27) among 16 raisin grape (*Vitis vinifera* L.) cultivars and selections were identified as 6.9 and 24.5 μg/g, respectively. In terms of the grape characterization, analyzing phenolic and aromatic compounds which can be varied to cultivar and *Terroir* of the grape is necessary for producing very high qualified wines.

Conclusions

The obtained results reflected that the historical nine table grape cultivars grown in the Igdir province of Eastern Turkey contained valuable sugars, organic acids, and phenolic compounds in health benefits. In the development of the grape industry of Igdır province, the present study will provide industrialists at first and then breeders to get baseline information for selection of grape cultivars for cultivation or breeding programs, which may be gained great importance with the support of the

Table 4 Results for phenolic acids (Part 2) in the table grape cultivars from Igdir province of Turkey

Cultivars	o-coumaric acid (mg/l)	p-coumaric acid (mg/l)	Caffeic acid (mg/l)	Syringic acid (mg/l)	Vanillic acid (mg/l)	Gallic acid (mg/l)
Askeri	1.69 ± 0.02b	0.06 ± 0.006cd	2.24 ± 0.05b	0.86 ± 0.02c	0.45 ± 0.01a	0.83 ± 0.06b
Beyaz Kismis	0.91 ± 0.05d	0.04 ± 0.003d	1.35 ± 0.06d	2.00 ± 0.03a	0.27 ± 0.01b	1.18 ± 0.03a
El Hakki	0.96 ± 0.01d	0.18 ± 0.006a	1.77 ± 0.04c	0.27 ± 0.02e	0.24 ± 0.02bc	0.27 ± 0.01c
Hacabas	1.38 ± 0.02c	0.19 ± 0.009a	0.65 ± 0.02e	0.31 ± 0.03de	ND	0.15 ± 0.02c
Inek Emcegi	0.67 ± 0.01e	0.13 ± 0.006b	1.15 ± 0.04d	0.50 ± 0.03d	0.06 ± 0.00d	1.27 ± 0.10a
Kerim Gandi	0.34 ± 0.02f	0.07 ± 0.003c	1.62 ± 0.05c	1.20 ± 0.09b	0.23 ± 0.01c	0.10 ± 0.01c
Kırmızı Kismis	1.90 ± 0.04a	0.01 ± 0.000e	2.73 ± 0.04a	0.16 ± 0.01e	0.02 ± 0.00e	0.67 ± 0.05b
Miskali	1.05 ± 0.04d	0.06 ± 0.003 cd	2.45 ± 0.08b	0.27 ± 0.05e	ND	1.27 ± 0.06a
Yazen Dayi	0.74 ± 0.02e	0.13 ± 0.010b	0.40 ± 0.04f	1.10 ± 0.07b	0.05 ± 0.00de	0.32 ± 0.03c
CV (%)	0.46	10.10	5.23	10.50	7.48	12.40

**P < 0.01 (the significant cultivar effect) [a,b,c,d,e,f]. In each column means followed by different letter are statistically different each other at P < 0.01 (**), ND: Non determined.

further studies to be carried out more comprehensively in future years.

Methods

Plant material

The present investigation was conducted on nine historical local table grape cultivars 'Beyaz Kismis', 'Askeri', 'El Hakki', 'Kirmizi Kismis', 'Inek Emcegi', 'Hacabas', 'Kerim Gandi', 'Yazen Dayi', and 'Miskali' grown in a single collection vineyard of Igdir province - located in the Eastern Anatolia Region of Turkey. The aim is to identify their biochemical content (sugars, organic acids, and phenolic acids, Vitamin C) and antioxidant capacity. In fact the prime name of 'Beyaz Kismis' is 'Sultanina' or Thompson seedless [21]. 'Hacabas' and 'Kirmizi Kismis = Kismis Kirmizi' are also mentioned in Armenia as a country of origin (19). 'Yazen Dayi' - No one variety with this name is mentioned yet neither in VIVC [21] nor in European Vitis Database [22]. Some basic characteristics of these cultivars are shown in Table 5.

The collection site is located in the village of Necefali (40°10′N latitude, 44°4′E longitude on the 865 meters above sea level) at a distance of 13 km to the city Igdir - the administrative center of the province.

Fruits (berries) for analyses were collected from bunch of 8 grapevine plants ensured good representative of the cultivars to make average sampling, and the berries of those grape cultivars were taken during commercial ripe stage of the grapes in the dates varying between 1 to 20 August of the year 2013. Thus the sampling structured according to differential ripening dates across cultivars.

The plants were established with planting scheme of 2.5 m × 2.5 m in distance. In winter months, plants were covered by soil to avoid winter injury. Pruning is applying in February and 3–4 buds left on shoots.

Extraction and detection of organic acids

Samples were stored at −20°C until processed. In the present paper succinic, citric, malic, fumaric and tartaric acids as organic acids were determined. Organic acids were extracted after the method suggested by Bevilacqua and Califano [23] was modified. In the study, 5 g from the whole fruit samples were transferred to centrifuge tubes. 20 ml of 0.009 N H_2SO_4 was pured on the samples and homogenized (Heidolph Silent Crusher M, Germany). Then, the samples were mixed for 1 hour on a shaker (Heidolph Unimax 1010, Germany) and centrifuged at 15000 g for 15 min. Their watery part extracted from the centrifuge was initially distilled from filter paper and then distilled twice from a membrane filter of 0.45 μm, and lastly, it was distilled from the cartridge SEP-PAK C_{18}. Organic acids were exposed to a HPLC device on the basis of the method identified by Bevilacqua and Califano [23].

In the HPLC system, Aminex HPX - 87 H, 300 mm × 7.8 mm column (Bio-Rad Laboratories, Richmond, CA, ABD), was used and the HPLC device was commanded by Agilent package program. In the system, DAD detector (Agilent, USA) was adjusted to the wavelengths of 214 and 280 nm.

Extraction and detection of phenolic acids

In the present paper, gallic acid, catechin, chlorogenic acid, caffeic acid, p-cumaric acid, o-cumaric acid, ferulic acid, vanillic acid, rutin, syringic acid and quercetin as phenolic acids were identified. The phenolic compounds were identified with HPLC using the modified method of Rodriguez-Delgado et al. [24].

The samples were mixed with distilled water in a ratio of 1:1, and centrifuged for 15 min at 15000 rpm. Afterwards, the upper part was filtered with 0.45 μm millipore filters, and enjected to HPLC. Chromatographic analysis was performed by Agilent 1100 (Agilent) HPLC system using DAD dedector (Agilent. USA) and 250*4.6 mm, 4 μm ODS column (HiChrom, USA). As a mobile phase, A methanol:acetic acid:water (10:2:28) and B methanol:acetic acid.water (90:2:8) were used. The extraction was made at 254 and 280 nm, 1 mL/min flow rate, and 20 μL injection volume.

Table 5 Basic morphological characteristics of studied cultivars

Cultivars	Bunch weight (g)	Bunch size	Berry color	Berry shape	pH	SSC (%)	Usage
Askeri	157	High	Light Yellow	Spherical	4.50	20.5	Fresh
Beyaz Kismis	129	Medium	Light Yellow	Spherical	3.34	22.5	Fresh
El Hakki	266	Medium	Red	Spherical	3.75	16.7	Fresh
Hacabas	338	High	Light Yellow	Spherical	3.71	18.3	Fresh
Inek Emcegi	431	High	White	Elipsoidal elongated	4.03	18.3	Fresh
Kerim Gandi	147	Medium	Light Green	Elipsoidal elongated	3.80	21.0	Fresh
Kırmızı Kismis	417	High	Pink Red	Spherical	3.54	22.4	Fresh
Miskali	515	High	Light Pink	Spherical	4.08	19.0	Fresh
Yezan Dayi	269	High	Light Yellow	Spherical	3.61	25.1	Fresh

Extraction and detection of sugars

The analysis of sugars was performed using the modified method of Melgarejo et al. [25]. The sugar analyses were done with the standards of fructose and glucose of fruit juice. The sample of 5 g was homogenized and centrifuged at 12000 rpm for 2 min, afterwhich run in SEP-PAK C18 column. The extraction was preserved at -20°C until analysis. The sugars from the samples to be filtered were determined using μbondapak-NH_2 column with 85% acetonitrile liquid phase in HPLC that has a refractive index detector. Calculation of the concentrations was done based on fruit juice standards.

Analysis of vitamin C (Ascorbic Acid)

The sample of 5 g from the whole fruits was transferred to test tubes and then 5 ml 2.5% M-phosphoric acid solution was poured on it. The mixture was centrifuged with 6500 g for 10 min at 4°C. From the clear part in the centrifuge tube 0.5 ml was taken and 2.5% M-phosphoric solution was poured until reaching 10 ml. The new mixture was filtered by 0.45 μm teflon filter and injected to HPLC. Ascorbic Acid was detected by the C18 column (Phenomenex Luna C18, 250 × 4.60 mm, 5 μ) in the HPLC. Ultra distilled water was used as a mobile phase with 1 ml/min flow rate and pH of 2.2 adjusted with H_2SO_4. The DAD detector with 254 nm wavelength was used for the readings. For determination of ascorbic acid different concentration levels of L-ascorbic acid (SigmaA5960) (50, 100, 500, 1000, and 2000 ppm) were used [26].

Detection of trolox equivalent antioxidant capacity (TEAC)

For a standard TEAC measurement, ABTS acetate was solved in a buffer and potassium persulphate was prepared [27]. To keep the stability of the mixture for a long time, 20 mM sodium acetate buffer solution was solved in an acidic pH of 4.5 and an absorbance of 734 nm, 0.700 ± 0.01. For spectrophotometric measurements, 3 ml $ABTS^+$ solution was mixed with 20 μl fruit extract, and incubated for 10 min. Absorbance values obtained for 734 nm.

Statistical analysis

Eight grapevine plants for each cultivar were used for sampling. Fruits (berries) were collected from different bunch in these 8 plants per cultivar. The descriptive statistics for all quantitative characteristics under investigation were expressed as Mean ± SE. The data were analyzed through One-Way ANOVA with four replications and mean separation was done by using Tukey multiple comparison test. In the present paper, a significance level of 1% was considered. All statistical computations were performed using SPSS 20 program.

Competing interests
The authors declare that they have no competing interests.

Authors' contributions
SPE, MA, and EE made a significant contribution to experiment design, acquisition of data, analysis and drafting of the manuscript. SE and DM have made a substantial contribution to interpretation of data, drafting and carefully revising the manuscript for intellectual content. All authors read and approved the final manuscript.

Author details
[1]Agricultural Faculty, Department of Horticulture, Igdır University, Igdir, Turkey. [2]Agricultural Faculty, Department of Horticulture, Oregon State University, Corvallis, Oregon, USA. [3]Agricultural Faculty, Department of Horticulture, Atatürk University, Erzurum, Turkey. [4]Agricultural Faculty, Department of Animal Science, Biometry Genetics Unit, Igdır University, Igdır, Turkey. [5]Scientific-Reaserch Center of Agriculture, Tbilisi, Georgia.

References

1. Hegedus A, Engel R, Abranko L, Balogh E, Blazovics A, Herman R, et al. Antioxidant and antiradical capacities in apricot (*Prunus armeniaca* L.) fruits: variations from genotypes, years, and analytical methods. J Food Sci. 2010;75(9):C722–30.
2. Jurikova T, Sochor J, Mlcek J, Balla S, Klejdus B, Baron M, et al. Polyphenolic profile of interspecific crosses of rowan (*Sorbus aucuparia* L.). Ital J Food Sci. 2014;26(3):317–25.
3. Piluzza G, Sulas L, Bullitta S. Dry matter yield, feeding value, and antioxidant activity in Mediterranean chicory (*Cichorium intybus* L.) germplasm. Turk J Agric For. 2014;38:506–14.
4. Rop O, Ercisli S, Mlcek J, Jurikova T, Hoza I. Antioxidant and radical scavenging activities in fruits of 6 sea buckthorn (*Hippophae rhamnoides* L.) cultivars. Turk J Agric For. 2014;38:224–32.
5. Baltacioglu H, Artik N. Study of postharvest changes in the chemical composition of persimmon by HPLC. Turk J Agric For. 2013;37:568–74.
6. Ivanova-Petropulos V, Ricci A, Nedelkovski D, Dimovska V, Parpinello GP, Versari A. Targeted analysis of bioactive phenolic compounds and antioxidant activity of Macedonian red wines. Food Chem. 2015;171:412–20.
7. Anonymous. 2014. http://faostat.fao.org/site/339/default.aspx (Access time: 23/10/2014).
8. Conde C, Silva P, Fontes N, Dias ACP, Tavares RM, Sousa MJ, et al. Biochemical changes throughout grape berry development and fruit and wine quality. Global Science Books, Food. 2007;1(1):1–22.
9. Zerihun A, McClymont L, Lanyon D, Goodwin I, Gibberd M. Deconvoluting effects of vine and soil properties on grape berry composition. J Sci Food Agric. 2015;95:193–203.
10. Nile SH, Kim SH, Young Ko E, Park SW. Polyphenolic contents and antioxidant properties of different grape (*V. vinifera*, *V. labrusca*, and *V. hybrid*) cultivars. BioMed Res Int. 2013:1–5.
11. Pereira GE, Gaudillere JP, Leeuwen CV, Hilbert G, Maucourt M, Deborde C, et al. ^{1}H NMR metabolite fingerprints of grape berry: comparison of vintage and soil effects in Bordeaux grapevine growing areas. Anal Chim Acta. 2006;563(1–2):346–52.
12. Topalovic A, Milukovic-Petkovsek M. Changes in sugars, organic acids and phenolics of grape berries of cultivar Cardinal during ripening. J Food Agric Environ. 2010;8(3&4):223–7.
13. Orak HH. Total antioxidant activities, phenolics, anthocyanins, polyphenoloxidase activities of selected red grape cultivars and their correlations. Sci Hortic. 2007;111:235–41.
14. Sabir A, Kafkas E, Tangolar S. Distribution of major sugars, acids and total phenols in juice of five grapevine (*Vitis* spp.) cultivars at different stages of berry development. Spanish J Agric Res. 2010;8(2):425–33.
15. Christersen LP. Raisin Grape Varieties. In: Raisin Production Manual. Oakland, CA: University of California, Agricultural and Natural Resources Publication 3393; 2000. p. 38–47.
16. Maghradze D, Rustioni L, Turok J, Scienza A, Failla O, editors. Caucasus and Northern Black Sea Region Ampelography. Vitis (special issue); 2012. p. 14.

17. Rusjan D, Korosec-Koruza Z. Morphometrical and biochemical characteristics of red grape varieties (*Vitis vinifera* L.) from collection vineyard. Acta Agric Slovenica. 2007;89(1):245–57.
18. Parpinello GP, Rombolà AD, Simoni M, Versari A. Chemical and sensory characterization of Sangiovese red wines: comparison between biodynamic and organic management. Food Chem. 2015;167:145–52.
19. Frankel EN. Food Factors for Cancer Prevention: Nutritional Benefits of Flavonoids. Tokyo: Springer-Verlag; 1997. p. 613–6.
20. Breksa AP, Takeoka GR, Hidalgo MB, Vilches A, Vasse J, Ramming DW. Antioxidant activity and phenolic content of 16 raisin grape (*Vitis vinifera* L.) cultivars and selections. Food Chem. 2010;121:740–5.
21. VIVC – Vitis International Variety Catalogue. http://www.vivc.de/ (Access time: 04.11.2014).
22. European Vitis Database. http://www.eu-vitis.de/ (Access time: 04.11.2014.
23. Bevilacqua AE, Califano AN. Determination of organic acids in dairy products by high performance liquid chromatography. J Food Sci. 1989;54:1076–9.
24. Rodriguez-Delgado MA, Malovana S, Perez JP, Borges T, Garcia-Montelongo FJ. Separation of phenolic compounds by high-performance liquid chromatography with absorbance and fluorimetric detection. J Chromatog. 2001;912:249–57.
25. Melgarejo P, Salazar DM, Artes F. Organic acids and sugars composition of harvested pomegranate fruits. Eur Food Res Technol. 2000;211:185–90.
26. Cemeroglu B. Food Analysis. Food Technology Association Publication. 2007. pp: 168–171. No: 34, Ankara, Turkey.
27. Ozgen M, Reese RN, Tulio AZ, Scheerens JC, Miller AR. Modified 2,2-Azino-Bis-3-Ethylbenzothiazoline-6-Sulfonic Acid (ABTS) method to measure antioxidant capacitiy of selected small fruits and a comparison to ferric reducing antioxidant power (FRAP) and 2,2-diphenyl-1-picrylhdrazyl (DPPH) methods. J Agric Food Chem. 2006;54(4):1151–7.

3

Compositional studies and biological activities of some mash bean (*Vigna mungo* (L.) Hepper) cultivars commonly consumed in Pakistan

Muhammad Zia-Ul-Haq[1*], Shakeel Ahmad[2], Shazia Anwer Bukhari[3], Ryszard Amarowicz[4], Sezai Ercisli[5*] and Hawa ZE Jaafar[6*]

Abstract

Background: In recent years, the desire to adopt a healthy diet has drawn attention to legume seeds and food products derived from them. Mash bean is an important legume crop used in Pakistan however a systematic mapping of the chemical composition of mash bean seeds is lacking. Therefore seeds of four mash bean (*Vigna mungo* (L.) Hepper, family *Leguminoseae*) cultivars (NARC-Mash-1, NARC-Mash-2, NARC-Mash-3, NARC-Mash-97) commonly consumed in Pakistan have been analyzed for their chemical composition, antioxidant potential and biological activities like inhibition of formation of advanced glycation end products (AGE) activity and tyrosinase inhibition activity.

Results: The investigated cultivars varied in terms of biochemical composition to various extents. Mineral composition indicated potassium and zinc in highest and lowest amounts respectively, in all cultivars. The amino acid profile in protein of these cultivars suggested cysteine is present in lowest quantity in all cultivars while fatty acid distribution pattern indicated unsaturated fatty acids as major fatty acids in all cultivars. All cultivars were found to be rich source of tocopherols and sterols. Fourier transform infrared spectroscopy (FTIR) fingerprints of seed flour and extracts indicated major functional groups such as polysaccharides, lipids, amides, amines and amino acids. Results indicated that all investigated cultivars possessed appreciable antioxidant potential.

Conclusions: All cultivars are rich source of protein and possess sufficient content of dietary fiber, a balanced amino acid profile, low saturated fatty acids and antioxidant capacity that rationalizes many traditional uses of seeds of this crop besides its nutritional importance. The collected data will be useful for academic and corporate researchers, nutritionists and clinical dieticians as well as consumers. If proper attention is paid, it may become an important export commodity and may fetch considerable foreign exchange for Pakistan.

Keywords: Nutrients, Antioxidant potential, Mash bean cultivar, Pakistan

Background

Mash bean (*Vigna mungo* (L.) Hepper) family *Leguminoseae* locally known as *sabut maash*, is a highly praised legume in Pakistan due to its dieto-therapeutic importance. Seeds are used in culinary dishes since primeval. The seeds are eaten after cooking. Seeds are the chief constituent of many traditional products like *wari, papad, idli, dosa, halwa* and *imrati* [1]. The seeds are well-known due to their therapeutic and nutritional potential.

The roots are narcotic and diuretic and are used for treating nostalgia, abscess, aching bones, dropsy, cephalgia and inflammation. The seeds are emollient, astringent, thermogenic, diuretic, aphrodisiac, nutritious, galactogauge, appetizer, laxative, styptic and nervine tonic. They are useful in treating scabies, leucoderma, gonorrhea, pains, epistaxis, piles, asthma, heart trouble, dyspepsia, anorexia, strangury, constipation, haemorrhoids, hepatopathy, neuropathy, agalactia, schizophrenia, hysteria, nervous debility, partial paralysis, facial paralysis and weakness of memory.

* Correspondence: ahirzia@gmail.com; sercisli@gmail.com; hawazej@gmail.com
[1]The Patent Office, Karachi, Pakistan
[5]Agricultural Faculty, Department of Horticulture, Ataturk University, Erzurum, Turkey
[6]Department of Crop Science, Faculty of Agriculture, 43400 UPM Serdang, Selangor, Malaysia
Full list of author information is available at the end of the article

Seeds are believed as spermatopoetic, and used for treating erectile dysfunction and premature ejaculation. Seeds are used for lengthening the hair, keeps them black and curing dandruff. Hot aqueous extracts of the leaves are used in the treatment of brain disorders, stomach, jaundice, rheumatic pain and inflammatory disorders. Seeds are considered fattening and flour made from seeds is excellent substitute for soap, leaving the skin soft and smooth and used in cosmetics in preparation of facial mask [2-6].

The mash bean occupies an important position in agriculture system of Pakistan and is grown annually on area of 27.6 thousand hectares with annual production of 13.6000 tonnes with 493 kg/ha as average yield [7]. It is grown all over the country, but its cultivation is concentrated mainly in Punjab, the major mash production province. It is the least researched crop among pulses in Pakistan as is apparent from scarcity of literature on it and as a result its area of cultivation and production are decreasing gradually [8].

The food industry globally is searching functional foods, nutraceuticals and botanicals to meet demand of consumers for natural, immunity-boosting and health-promoting plant based food products. To our knowledge, there is no study indicating chemical composition and antioxidant potential of seeds of mash bean cultivar indigenous to Pakistan. As part of our research studies to investigate the biochemical composition and antioxidant capacity of indigenous flora of Pakistan [9-13] this study has been conducted to determine the chemical composition, antioxidant activity and biological activities of seeds of mash bean cultivars.

Results and discussion

Composition and contents of various constituents and components like various bioactive constituents and secondary metabolites, fixed and essential oil, fatty acids, tocopherol and sterol profile, mineral, amino acid, vitamin, protein and carbohydrate contents present in a food commodity like seed, fruit, vegetable, spice, grain or any other product derived from them varies depending upon many factors like plant variety, agronomic practices utilized in cultivation, stage of collection and geological and climatic conditions of area from where that food commodity or plant part (seed or fruit) is collected, and the method employed for its determination. So there is need to establish food composition database on regional and country level for various food commodities for various regions and countries respectively. Previously our research group has compiled compositional and nutritional information on various other legumes like chickpea, pea, cowpea, lentil and mung bean. In current study we have determined biochemical composition, their impact on health as well biological activities of a less-researched legume crop i.e. mash bean.

The data on the proximate composition is summarized in Table 1. The observed range for protein was 24.62% for NARC-Mash-97 to 25.48% for NARC-Mash-2 Mash 2. The crude fiber content ranged from 4.25% to 5.09%. The range observed for fat content was between 1.80 and 2.25% while carbohydrates ranged from 53.43% to 55.55%. The high carbohydrate contents present in mash bean seeds indicate its potential use as a prime source of energy to prevent marsamus in infants especially. Like other legumes, its seeds are also rich in protein, contain sufficient amount of dietry fibre and lesser amount of oil. The results are in partial agreement to those reported earlier for mash bean [14,15] and other legumes [9-13]. Regular intake of dietry fibre is associated with low chances of cardiovascular disease, obesity, certain cancers and diabetes. High dietry fiber contents may be responsible for its traditional use as anti-cancer food. Since dietry fibre containing foods are used in bakery products, it also indicates its potential use in bakery and pastry products.

The data of vitamin contents is summarized in Figure 1. Niacin content was highest in NARC-Mash-1 (1.80 ± 0.07 mg/g) while NARC-Mash-3 had lowest content of riboflavin (0.19 ± 0.19 mg/g). Regarding vitamin contents of seeds of mash bean, niacin was present in higher concentration among all cultivars. As there is no report available on vitamin contents of mash bean, so vitamin contents cannot be compared to previous results. However the vitamin contents are in close proximity to that of *Pisum sativum* as per our previous studies [16]. High contents of niacin are good from medical point of view as this water-soluble vitamin is excreted by urine from human body and its continuous supply by eating mash bean seeds will complete its deficiency. Various agro-geo-climatological conditions affect vitamin contents in legume seeds.

Mineral contents (Table 2) indicated potassium as major mineral from 1599.82 ± 1.74 mg/100 g in NARC-Mash-97 to 1646.01 ± 0.92 mg/100 g in NARC-Mash-3. Phosphorus ranked second in quantity from 439.79 ± 0.42 and 500.17 ± 1.85 in same varieties. Zinc was present in lowest content (1.94 ± 0.76 mg/100 g) in NARC-Mash-97. All cultivars contained sufficient contents of potassium, phosphorus and copper. It is perhaps this high potassium content that makes it an aphrodisiac. The high content of potassium is useful for patients who use diuretics to manage hypertension and there is unnecessary seepage of potassium from their body fluids. The low content of sodium compared to potassium led to a low sodium: potassium ratio, which is favorable from nutritional point of view, as foods with low Na:K ratio are linked with lower frequency of blood hypertension. Na:K ratio is from 0.14 to 0.17 in NARC-Mash-1 and NARC-Mash-3 respectively. For prevention of high blood pressure, Na/K ratio of less than one is suggested. This may explain the

Table 1 Proximate composition (%) of seeds of mash bean cultivars

Component	NARC-Mash-1	NARC-Mash-2	NARC-Mash-3	NARC-Mash-97
Crude protein	27.91 ± 1.71ᵃ	26.48 ± 1.66ᵇ	25.07 ± 1.60ᶜ	28.60 ± 1.72ᵃ
Total lipids	5.13 ± 0.05ᵇ	6.00 ± 0.05ᵃ	5.80 ± 0.09ᵃ	6.22 ± 0.09ᵃ
Carbohydrates	56.55 ± 1.82ᵃ	54.81 ± 1.73ᵇ	58.13 ± 1.10ᵃ	54.81 ± 1.75ᵇ
Crude fiber	5.44 ± 1.7ᵇ	6.84 ± 1.60ᵃ	4.25 ± 1.20ᵇ	5.11 ± 1.60ᵇ
Ash	4.97 ± 0.19ᵇ	5.87 ± 0.18ᵇ	6.72 ± 0.19ᵃ	5.26 ± 0.18ᵇ

Values in the same row having different letters differ significantly with least significant difference (LSD) at probability ($p < 0.05$).

rationale behind the traditional use of its seeds in managing hypertension. Low Ca:P ratio leads to loss of Ca in the urine more than normal amount, so Ca concentration in bones is reduced. Food is considered "poor" if Ca:P ratio is less than 0.5 and "good" if it is above one. In present study, Ca:P ratio ranged from 0.78 to 1.00 in NARC-Mash-2 and NARC-Mash-1 respectively indicating regular consumption of mash bean seeds will serve as fine source of calcium for formation of bones. High levels of calcium are required during growth, gravidity and lactation of animals [13]. The results are in par to those already reported for mash bean elsewhere [17,18]. It is well-known that mineral contents of plant and crops parts like fruit and seeds depend on cultivars, collection time and maturity stage, climatological conditions, agronomic practices like type of fertilizer and water as well as selectivity, acceptability and intake of minerals by crops and plants. These results suggested that mash bean may provide adequate quantity of minerals to meet the mineral requirements of human body [19].

A protein-rich diet is not a guarantee to fulfill the requirements of the amino acids, a human body needs. A balanced protein diet should comprise all amino acids in sufficient amount and essential and non-essential amino acid ratio denotes the nutritional quality of protein. Glutamic acid (19.19 ± 0.62 to 21.49 ± 0.07 g/100 g) and aspartic acids (11.53 ± 0.11 to 13.20 ± 0.27 g/100 g) were present in highest amount in all cultivars. Except tryptophan and S-containing amino acids, all essential amino acids are present in sufficient amounts in all analyzed cultivars as is evident by data (Table 3). Most amino acids derived from plant sources are believed to possess antimicrobial, anti-inflammatory, immune-stimulating and antioxidant properties besides their role in nutrition. Results are comparable to those of previous studies on mash bean amino acids [17]. The deficient amino acids can be acquired by including large quantity of mash bean in diet, or by taking mash bean as well as other legumes.

Besides amino acid composition, protein digestibility is crucial for determining the protein quality. *In-vitro* protein digestibility data (Figure 2) suggested that values are lowest in NARC-Mash-1 (29.30 ± 0.82%) and highest in NARC-Mash-97 (38.53 ± 0.21%) while starch digestibility was 59.93 ± 0.17 to 67.09 ± 0.02 for same cultivars. Protein digestibility was below 50 percent while starch digestibility was above 50 percent in all analyzed cultivars. A significant variation has been observed for protein digestibility of legume seeds previously for mash bean and other legumes [16,17]. The sensory, textural and nutritional characteristics of products made from legumes are due to various functional properties of proteins. Anti-nutritional components like tannins, phytates and trypsin inhibitors, and structural distinctiveness of storage proteins slow down the digestibility of legume proteins. Treatments like roasting; autoclaving and cooking may be utilized to increase the legume proteins digestibility. In vitro starch digestibility values are close to those reported earlier [20]. Since legume starches generally contain more amylase, therefore these are less digestible. This low digestibility is useful as it decreases release of glucose in blood and so is helpful for patients suffering from diabetes. It may be reason of prescribed use of mash bean for diabetic patients by traditional healers. The low-digestibility however may be managed by utilization of legume seeds along with husk since dietary fibre present in husk will decrease the transit time in intestines and will help in bowel motility.

Fatty acids profile determines the oil quality of seeds or fruits or any other part of plant and products derived from them. Fatty acids profile of oil of seeds of investigated cultivars is summarized in Table 4. All cultivars were found to be rich source of α-linolenic acid (49.52 ± 0.09 to 51.80 ± 0.03%) and oleic acid (26.62 ± 0.07 to 27.34 ± 0.25%). Bulk of the oil consisted of unsaturated fatty acids for all cultivars. The results are comparable to previously published works for low-oil bearing legumes in general [21] and for mash bean in particular [22]. Saturated fatty acids were a small percentage of total fatty acids present. There is reduced risk of cholesterol-related heart diseases by consuming oils containing more unsaturated fatty acids. However since oil content is very low in seeds therefore it cannot be considered as commercial source of vegetable oil.

Data about tocopherol composition is summarized in Table 5. Despite differences, γ-tocopherol contents were present in highest quantity in all cultivars while considerable contents of δ-tocopherol followed α-tocopherol were also noted. Oil of seeds of all mash bean cultivars studied contained all major tocopherols. Like many other traits,

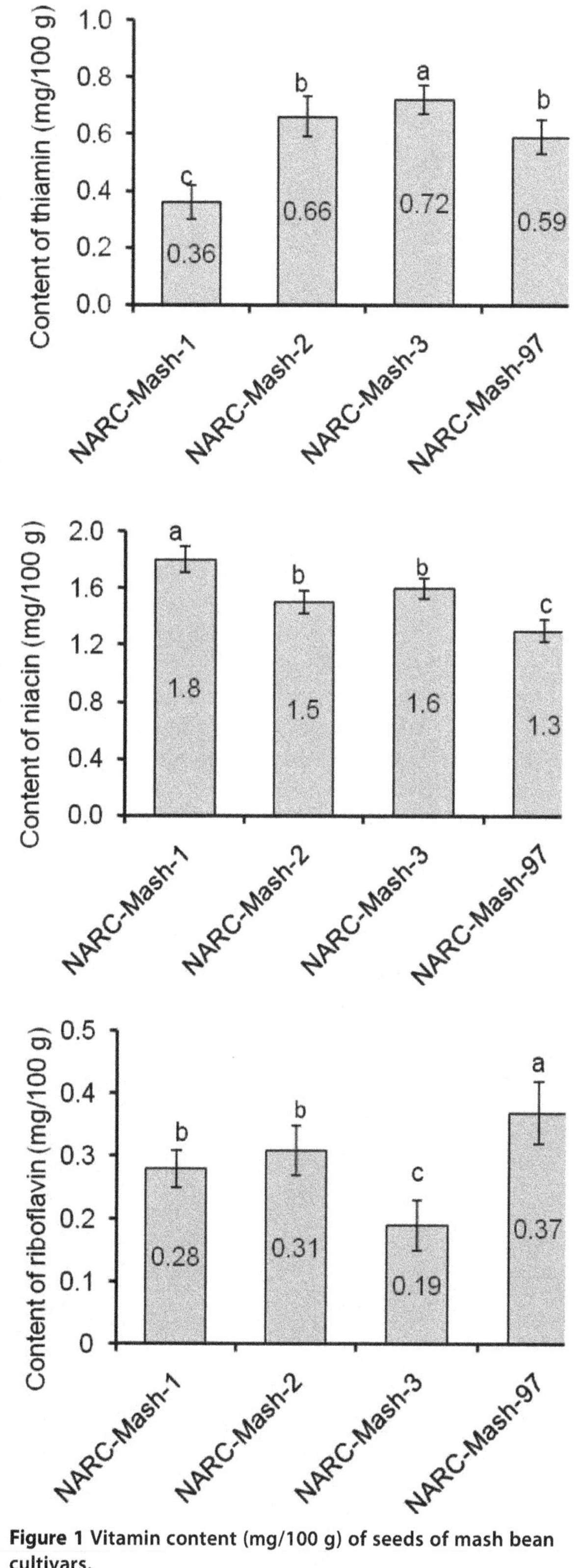

Figure 1 Vitamin content (mg/100 g) of seeds of mash bean cultivars.

no previous report is present regarding tocopherol and sterol contents of mash bean seeds. However all values are close to those reported for Indian mash bean seeds [22]. Since naturally occurring tocopherols are used for oils and fats stabilization against oxidative degradation, it suggests their usage in pharmaceutical, biomedical, and nutritional products.

Sterol profile is summarized in Table 6. Substantial amounts of campesterol, avenasterol and stigmasterol were found in oils of seeds of all four cultivars. The main sterol in oil of seeds of all investigated mash bean cultivars was β-sitosterol which is in agreement with previous studies for low-oil bearing legumes like chickpea, mungbean, cowpea, *Albizia lebbeck* and *Acacia leucophloea* in general [10,12,13,21] and for mash bean in particular [23]. Various agro-geo-climatological factors as well as solvent used for extraction of oil are believed to be responsible for the distribution of tocopherols and sterols in oils extracted from plant parts. Sitosterol, campestrol and stigmasterols have been observed to be major sterols in oils from most of plants belonging to family *Leguminosae* [10,12,13,21].

FTIR-fingerprints, give a quick check of identification, classification and discrimination of food samples by providing a general outline of pattern and trends indicating presence of various chemical compounds in samples. FTIR spectrum of mash bean seed powder indicated the presence of various types of aliphatic and aromatic compounds, especially carboxylic acids, esters, alkyl halides and nitro compounds. The presence of carboxylic acids is indicated by peaks at 2929.40 (O-H stretching), 1249.94 cm^{-1} (C-O stretching). The peak at 1728.04 corresponds to C = O stretching frequency of aldehydic group. Unsaturated compounds presence is indicated by the peak at 1658.57 cm^{-1} (C = C stretching, alkene) and 1556.41 cm^{-1} (C-C stretching, aromatic compounds). Saturated compounds presences is shown by the peak at 1450.17 cm^{-1} (C-H bending, alkane). Nitro compounds and aromatic amine presence is indicated by the peaks at 1343.93 and 1319.41 cm-1. The peaks at 1160.05 and 1074.23 cm-1 showed the presence of aliphatic amines. Alkyl halides presence is pointed out by the peak at 849.49 cm^{-1}.

For mash bean extract, the highly intensified OH region with intensified shoulder peak of amine group was present. A new peak in the region of 1700–1800 was observed which may be attributed to presence of ester. Saturated compounds presence is indicated by the peaks at 2925.31, 2859.93 (C-H stretching, alkanes) and 1384.79 cm^{-1} (C-H rocking, alkane). Carboxylic acid presences is confirmed by the peaks at 3023.38 (O-H stretching), 1736.21 and 1695.35 cm^{-1} (C = O stretching). Nitro compounds presences are indicated by the peak at 1515.55 cm^{-1} (N-O asymmetric stretching). Primary aliphatic amines presences is indicated by peaks at 1466.52 (N-H bending) and 1221.34 cm^{-1} (C-N stretching).

Table 2 Mineral content (mg/100 g) of seeds of mash bean cultivars

Minerals	NARC-Mash-1	NARC-Mash-2	NARC-Mash-3	NARC-Mash-97
Phosphorus	461.24 ± 0.22^{c}	480.47 ± 3.02^{b}	500.15 ± 2.91^{a}	440.90 ± 0.80^{d}
Potassium	1603.39 + 1.66^{b}	1638.88 ± 2.86^{a}	1646.11 ± 3.17^{a}	1600.03 ± 2.61^{b}
Sodium	227.01 ± 4.55^{c}	244.90 ± 1.41^{b}	284.08 ± 2.01^{a}	261.33 ± 1.79^{a}
Calcium	462.90 ± 2.07^{a}	375.01 ± 3.66^{b}	485.38 ± 1.14^{a}	394.19 ± 2.04^{b}
Magnesium	263.83 ± 3.56^{a}	239.70 ± 1.36^{b}	208.45 ± 1.21^{c}	221.77 ± 1.18^{c}
Iron	5.89 ± 0.25^{b}	6.14 ± 0.21^{b}	6.38 ± 0.18^{a}	6.55 ± 0.33^{a}
Manganese	2.39 ± 2.07^{b}	3.27 ± 0.05^{a}	3.32 ± 0.11^{a}	3.22 ± 0.18^{a}
Zinc	2.40 ± 0.14^{a}	2.28 ± 0.12^{a}	2.50 ± 0.22^{a}	1.94 ± 0.76^{b}
Copper	3.92 ± 0.47^{b}	4.03 ± 0.83^{b}	4.26 ± 0.66^{a}	4.51 ± 0.34^{a}
Na/K	0.14	0.15	0.17	0.16
Ca/P	1.00	0.78	0.97	0.89

Values in the same row having different letters differ significantly with least significant difference (LSD) at probability ($p < 0.05$).

Aromatic compounds presence is shown by peak at 1466.52 cm^{-1} (C-C stretching in ring). The observed bands for amines, amides, amino acids confirmed the presence of proteins, whereas presence of other bio-molecules like carboxylic acids, carbohydrates and oil was indicated by other absorption bands. Bhat et al. [24] and Zia-Ul-Haq et al. [25] have reported previously similar functional groups in *Gnetum gnemon* L. and *Pisum sativum* L. respectively.

Especial attention is being given to the identification of phenolic acids, flavonoids and tannins from extracts of legume seeds. Total phenolic content (TPC, mg GAE/g) of seed extracts from selected mash bean cultivars are presented in Table 7. The TPC was observed in highest amount in NARC-Mash-97 (86 mg GAE/g), whereas the lowest TPC was noted for NARC-Mash-1 (75 mg GAE/g). The total flavonoids contents (TFCs) and condensed tannins (CTC) were expressed in catechin equivalents (CAE/g). The cultivars differed significantly ($P < 0.05$) in TFCs and CTCs. The chromatograms (RP-HPLC) of extracts of seeds of mash bean were recorded at 330 nm and two dominant peaks (1–2) with a retention times of 28 and 28.8 min respectively (Figure 3), were observed.

Table 3 Percentage composition of amino acids in seeds

Amino acid	NARC-Mash-1	NARC-Mash-2	NARC-Mash-3	NARC-Mash-97
Alanine	4.63 ± 0.17^{b}	5.20 ± 0.07^{a}	4.35 ± 0.05^{c}	4.17 ± 0.21^{c}
Arginine	6.03 ± 0.27^{c}	6.30 ± 0.04^{b}	6.53 ± 0.03^{a}	6.64 ± 0.15^{a}
Aspartic acid	13.20 ± 0.27^{a}	12.40 ± 0.08^{b}	11.98 ± 0.07^{bc}	11.53 ± 0.11^{c}
Cystine	0.75 ± 0.29^{b}	0.90 ± 0.04^{a}	0.45 ± 0.03^{c}	0.72 ± 0.23^{b}
Glutamic acid	21.07 ± 0.65^{a}	21.49 ± 0.07^{a}	20.44 ± 0.09^{b}	19.19 ± 0.62^{c}
Glycine	4.39 ± 0.12^{a}	4.61 ± 0.05^{a}	3.73 ± 0.03^{b}	4.34 ± 0.24^{a}
Histidine	2.36 ± 0.31^{b}	2.13 ± 0.02^{b}	3.21 ± 0.01^{a}	3.26 ± 0.26^{a}
Isoleucine	4.48 ± 0.17^{a}	4.37 ± 0.07^{a}	3.79 ± 0.05^{b}	4.25 ± 0.09^{a}
Leucine	8.89 ± 0.12^{a}	7.31 ± 0.03^{b}	7.79 ± 0.04^{b}	7.54 ± 0.45^{b}
Lycine	4.19 ± 0.88^{d}	7.69 ± 0.01^{a}	6.90 ± 0.08^{b}	5.07 ± 0.74^{c}
Methionine	1.92 ± 0.74^{a}	1.12 ± 0.05^{b}	1.42 ± 0.09^{b}	1.29 ± 0.29^{b}
Phenylalanine	5.59 ± 0.18^{a}	4.88 ± 0.06^{b}	5.80 ± 0.07^{a}	5.67 ± 0.12^{a}
Proline	4.30 ± 0.21^{b}	3.69 ± 0.03^{c}	5.01 ± 0.01^{a}	4.20 ± 0.08^{b}
Serine	5.18 ± 0.30^{a}	5.31 ± 0.05^{a}	4.14 ± 0.08^{c}	4.78 ± 0.07^{b}
Threonine	3.95 ± 0.35^{b}	3.80 ± 0.04^{b}	4.50 ± 0.03^{a}	3.99 ± 0.28^{b}
Tryosine	1.01 ± 0.14^{c}	1.70 ± 0.09^{b}	2.80 ± 0.02^{a}	3.15 ± 0.28^{a}
Tryptophan	2.97 ± 0.19^{a}	2.40 ± 0.06^{c}	2.67 ± 0.02^{b}	2.92 ± 0.07^{a}
Valine	5.09 ± 0.11^{a}	4.80 ± 0.08^{b}	4.94 ± 0.04^{a}	5.08 ± 0.04^{a}

Values in the same row having different letters differ significantly with least significant difference (LSD) at probability ($p < 0.05$).

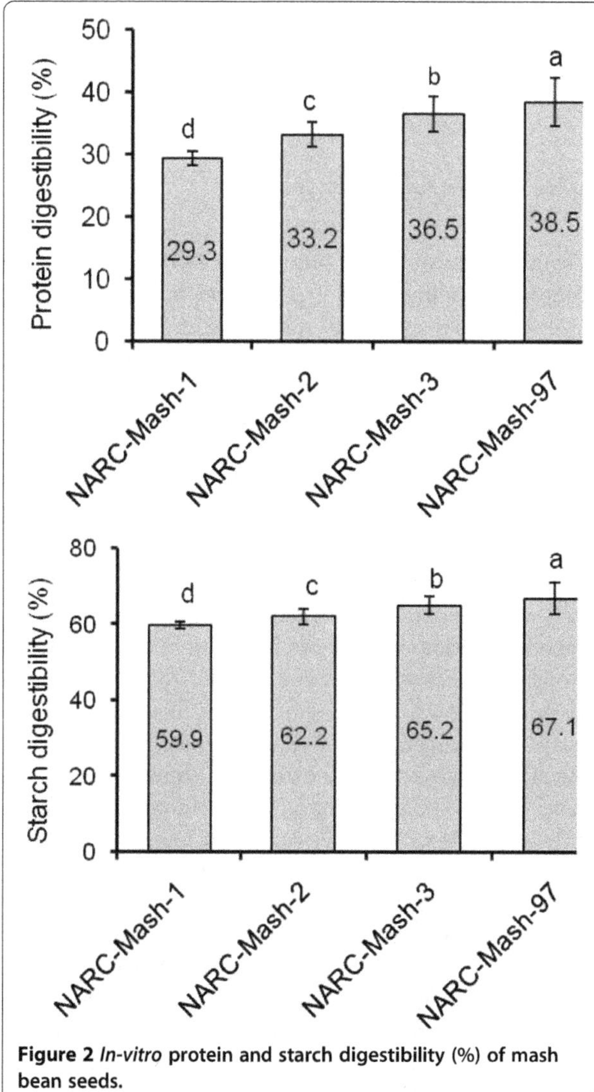

Figure 2 ***In-vitro*** **protein and starch digestibility (%) of mash bean seeds.**

The spectra (UV) of both compounds (peaks 1–2) displayed maxima at 269 and 334 nm. Compounds **1**, and **2** were identified as chlorogenic acid and caffeic acids when compared with standards run simultaneously. The mash bean extracts investigated in this study were characterized by several times higher content of flavonoids and condensed tannins when compared to desi chickpea, kabuli chickpea, lentil, cowpea, *Albizia lebbeck* and *Acacia leucophloea* varieties [11-13,25,26]. Presence of higher contents of various phenolic compounds was noted in extracts (Table 8). Various phenolic acids have been identified earlier in extracts from other legumes like chickpea, cowpea and pea [11,25,26]. Consumption of phenolic-rich foods is associated with low risk of several chronic diseases such as cardiovascular disease, ageing, cancer, neurodegenerative disease and Alzheimer disease as is evident from various epidemiological studies which highlights importance of presence of ample contents of phenolic acids noted in mash bean extract.

The human body has several mechanisms to shield bio-molecules against damage caused by reactive oxygen and nitrogen species. However, the instinctive protection may not be adequate to counter the rigorous or continuous oxidative stress. Hence, certain amounts of exogenous antioxidants are frequently required to maintain sufficient antioxidants level to balance the reactive nitrogen and oxygen species-pressure in the human body. Scientists are exploring antioxidants from natural sources like legume seeds as these are natural, cost effective and without side effects. The scavenging activity of mash bean extracts was expressed by antiradical assays against DPPH$^{\bullet}$ and ABTS$^{\bullet+}$ assay as well as by FRAP and reducing power assays as shown in Table 9. DPPH values of mash bean varieties ranged from 34.72 in NARC-Mash-2 to 39.49 μmol Trolox/g in NARC-Mash-3. Sufficient scavenging of DPPH radical was observed by extracts. It indicates that antioxidants present in extracts quench free radicals by donating them hydrogen atoms thereby converting them to nontoxic species. Although assessment of antiradical activity of an extract by DPPH protocol is fast and trouble-free, it usually has a relatively small linear reaction range therefore antiradical activity against ABTS$^{\bullet+}$ was measured. The ABTS$^{\bullet+}$ scavenging data indicated that the extracts may scavenge free radicals by hydrogen/electron donation mechanism and may protect biomatrices from oxidative degradation resulting from free radicals. Substantial antiradical activity for DPPH and ABTS$^{\bullet+}$ was observed with same order of scavenging in both protocols. It was noted

Table 4 Fatty acid composition (%) of oil of mash bean seeds

Fatty acid	NARC-Mash-1	NARC-Mash-2	NARC-Mash-3	NARC-Mash-97
Palmitic acid	11.31 ± 2.20^{b}	10.99 ± 1.99^{c}	11.23 ± 1.87^{b}	12.09 ± 1.58^{a}
Stearic acid	2.09 ± 0.63^{b}	2.70 ± 0.24^{a}	2.89 ± 0.43^{a}	2.17 ± 0.77^{b}
Behenic acid	0.99 ± 0.14^{a}	1.00 ± 0.29^{a}	0.87 ± 0.22^{a}	0.93 ± 0.30^{a}
Oleic acid	26.62 ± 0.07^{b}	26.74 ± 0.15^{b}	27.34 ± 0.25^{a}	26.65 ± 0.35^{b}
Linoleic acid	07.19 ± 4.47^{b}	08.93 ± 5.11^{a}	07.08 ± 3.74^{b}	08.64 ± 3.87^{a}
α-Linolenic acid	51.80 ± 0.03^{a}	49.64 ± 0.06^{a}	50.59 ± 0.05^{a}	49.52 ± 0.09^{a}

Values in the same row having different letters differ significantly with least significant difference (LSD) at probability ($p < 0.05$).

Table 5 Tocopherol content (mg/100 g) in oil of seeds of mash bean cultivars

Tocopherols	NARC-Mash-1	NARC-Mash-2	NARC-Mash-3	NARC-Mash-97
α-Tocopherol	3.04 ± 0.89^{b}	2.97 ± 0.55^{b}	3.49 ± 0.17^{a}	3.17 ± 0.34^{b}
γ-Tocopherol	722.09 ± 2.17^{a}	724.34 ± 4.13^{a}	722.21 ± 1.16^{a}	720.33 ± 2.01^{a}
δ-Tocopherol	16.69 ± 3.3^{b}	16.18 ± 4.2^{c}	17.14 ± 2.66^{a}	17.12 ± 4.0^{a}

Values in the same row having different letters differ significantly with least significant difference (LSD) at probability ($p < 0.05$).

that reducing potential of extracts increased with increasing amount of extracts. Butylated hydroxanisole was used as standard to compare the reducing power of extracts. Mechanistic studies indicate that antioxidant potential of extracts is closely linked with their reducing power. The results were close to reported earlier [27-29].

The FRAP assay determines antioxidant activity of extracts as their potential to reduce ferric ions to ferrous ions. The FRAP values of the extracts of seeds of selected cultivars are presented in Table 9. Similar to that in $DPPH^{\bullet}$ analyses, high variations of FRAP values were observed and FRAP values of cultivars ranged from 9.65 mmol Fe^{2+}/g in NARC-Mash-3 to 13.76 mmol Fe^{2+}/g in NARC-Mash-97. Our results for FRAP are different from those reported earlier [27-29]. Antioxidant activity of the extracts of seeds of other plant like pea, cowpea, lentil, garden cress, capper and chickpea has been reported in several studies [25,26,30-32] by our research group.

It is generally believed that diabetes can be cured with more consumption of legumes however the mechanism behind this remained unexplored till now. The recent studies indicated that legumes cure diabetes by reducing AGE-formation. Advanced glycation end products (AGE) formation is increased in diabetes mellitus, so search for (AGEs)-inhibitor is a new approach in diabetes treatment. Two models used mostly for quantification of AGE-inhibtion of plant extracts are BSA-MGO and BSA-glucose models. In advanced glycation end (AGE) products inhibition activity, NARC-Mash-97 exhibited the highest inhibition (86.67%), followed by NARC-Mash-3 (74.84%) in BSA-glucose method. BSA-MGO inhibition model showed the same trend like that of BSA-glucose model (Figure 4). It is believed that phenolic compounds present in legume seeds inhibit the AGE-formation by inhibiting production of free radical during glycation process and subsequently inhibiting protein modification. The results (Figure 4) obtained in our study are in agreement with those reported previously for other legume seeds [33-38]. Same trend was observed in tyrosinase inhibition activity as was for AGE inhibition. Tyrosinase inhibition potential of extracts of seeds of mash bean may be ascribed to the presence of phenolic contents since hydroxyl groups present in various phenolic acids make a hydrogen bond at active site of the tyrosinase and as a result tyrosinase activity is decreased or stopped. Tyrosinase inhibitors have potential applications in food and cosmetic industry because they are used to stop or slow-down browning of various food commodities like fruits, vegetable and fisheries products and impart whitening effects to skin by stopping human skin hyper-pigmentation. The browning of food commodities leads to decrease in attractive appearance and loss of nutritional quality. It rationalized traditional use of mashbean in facial massages by indigenous communities and proves its anti-freckles, anti-wrinkling, anti-ageing and skin-whitening activity. For the first time tyrosinase inhibition activities of extracts of seeds of mash beans are being reported.

Statistically non-significant and very low correlations were found between different parameters viz. FRAP with BSA-G, BSA-MGO and TI; FRAP with BSA-MGO and TI at $P < 0.05$ (Table 10). TPC was having statistically no correlation with FRAP. Similarly, there was very low correlation between ABST and BSA-G. Medium correlations ranged between 0.28-0.55. Most of the correlations in this range were statistically non-significant except for DPPH and BSA-MGO ($p < 0.05$) and DPPH and TI ($p < 0.05$).

Table 6 Sterol content (mg/100 g) in oil of seeds of mash bean cultivars

Sterols	NARC-Mash-1	NARC-Mash-2	NARC-Mash-3	NARC-Mash-97
β-Sitosterol	56.5 ± 0.2^{a}	55.1 ± 0.6^{a}	55.1 ± 0.61^{a}	56.1 ± 0.51^{a}
Stigmasterol	34.0 ± 0.4^{a}	34.4 ± 0.8^{a}	34.4 ± 0.8^{a}	33.4 ± 0.1^{b}
Δ^{5}- Venasterol	$4.00 + 0.22^{b}$	$3.64 + 0.38^{b}$	$3.64 + 0.38^{b}$	$4.51 + 0.12^{a}$
Stigmastanol	$2.66 + 0.05^{c}$	$3.79 + 0.12^{a}$	$3.79 + 0.12^{a}$	$3.21 + 0.38^{b}$
Δ^{7}- avenasterol	1.01 ± 0.17^{a}	1.05 ± 0.69^{a}	1.05 ± 0.69^{a}	1.09 ± 0.43^{a}
Campesterol	0.87 ± 0.40^{b}	0.98 ± 0.80^{a}	0.98 ± 0.80^{a}	0.66 ± 0.18^{c}
Unidentified	1.00 ± 0.03^{a}	1.00 ± 0.27^{a}	1.00 ± 0.27^{a}	1.00 ± 0.13^{a}

Values in the same row having different letters differ significantly with least significant difference (LSD) at probability ($p < 0.05$).

Table 7 Total phenolic contents, total flavonoid contents and condensed tannin contents in extracts of seeds of mash bean cultivars

Cultivar	Total phenolic contents	Total flavonoid contents	Condensed tannin contents
NARC-Mash-1	75.91 + 2.72[c]	51.78 + 1.85[b]	86.79 + 1.56[b]
NARC-Mash-2	79.33 + 1.52[b]	47.11 + 2.47[c]	89.14 + 1.11[b]
NARC-Mash-3	82.22 + 1.36[b]	42.66 + 1.81[d]	93.68 + 1.65[a]
NARC-Mash-97	86.99 + 1.19[a]	55.73 + 1.92[a]	79.20 + 1.77[c]

Values in the same row having different letters differ significantly with least significant difference (LSD) at probability ($p < 0.05$).

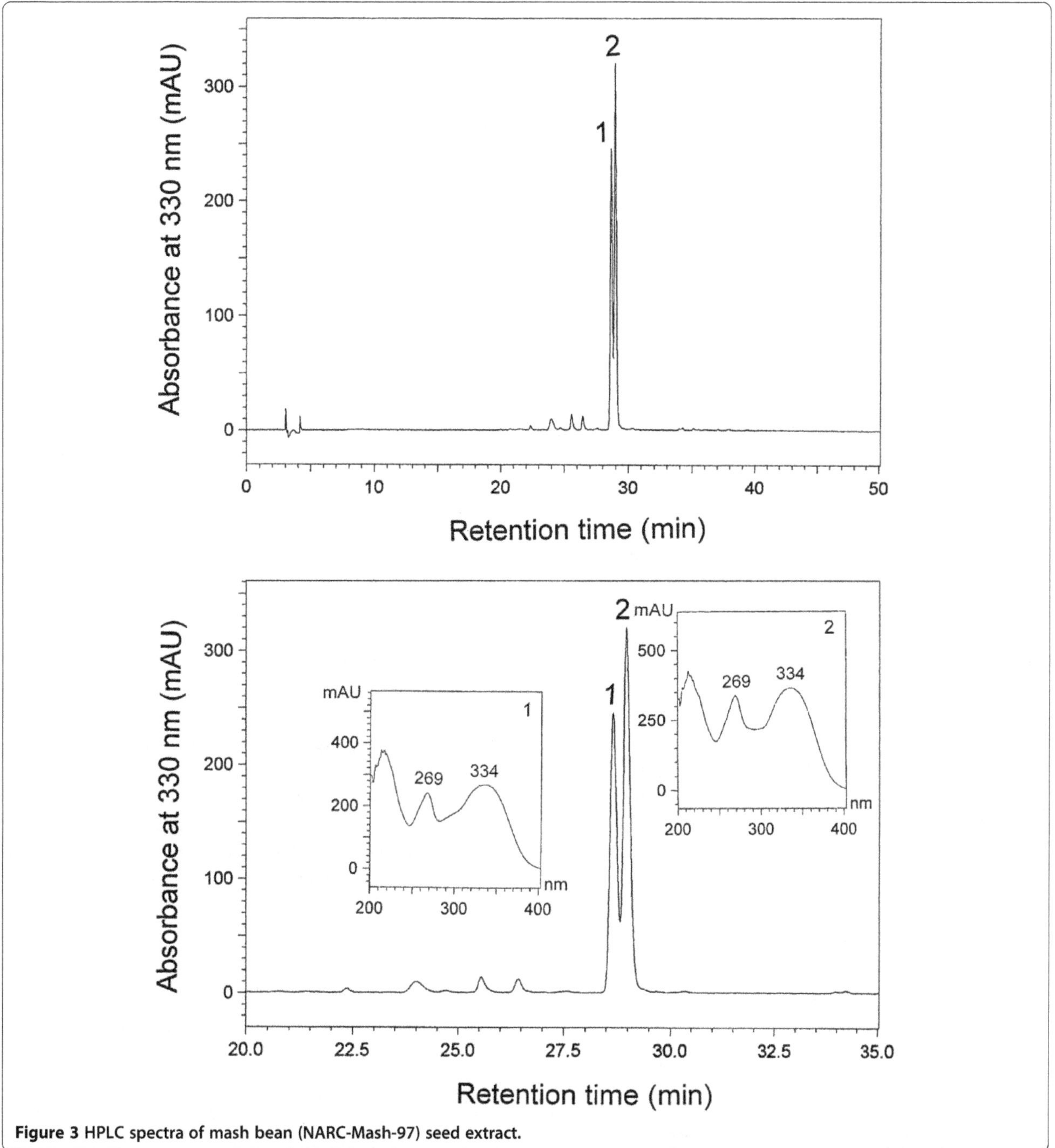

Figure 3 HPLC spectra of mash bean (NARC-Mash-97) seed extract.

Table 8 Content of two main phenolic compounds in the extracts and seeds of mash bean cultivars

Cultivar	Compound 1 (mg/g extract)	Compound 2 (mg/g extract)	Compound 1 (mg/g fresh seeds)	Compound 2 (mg/g fresh seeds)
NARC-Mash-1	4.09 ± 0.20^d	5.22 ± 0.26^d	0.39 ± 0.02^c	0.50 ± 0.02^c
NARC-Mash-2	5.94 ± 0.30^a	8.09 ± 0.40^a	0.55 ± 0.03^a	0.74 ± 0.04^a
NARC-Mash-3	5.48 ± 0.27^b	7.35 ± 0.37^b	0.48 ± 0.02^b	0.65 ± 0.03^b
NARC-Mash-97	5.07 ± 0.25^c	6.94 ± 0.35^c	0.44 ± 0.02^b	0.44 ± 0.02^c

Values in the same row having different letters differ significantly with least significant difference (LSD) at probability ($p < 0.05$).

Very high and positive correlations were found among certain variables like BSA-G and BSA-MGO, BSA-G and TI, BSA-MGO and TI, TPC and BSA-G, TPC and BSA-MGO and TPC and TI. The correlation between DPPH and FRAP was also statistically significant at $p < 0.05$, DPPH and ABST at $p < 0.01$, FRAP and ABST were also highly correlated ($p < 0.01$).

Conclusion

The results suggested mash bean seeds as a rich source of nutrients and extracts of seeds exhibited good antioxidant and biological activities. Seeds are rich source of protein and carbohydrate and good source of dietry fibre. These also contain ample amount of essential minerals like Ca, K, Na, Mg, Cu and Zn and various essential and non-essential amino acids. Seeds also have acceptable fatty acids, tocopherol and sterol profile. Various functional groups were detected in FTIR of seeds and extracts. Antioxidant results suggested them as rich source of phenolic acids, flavonoids and condensed tannin contents. The extracts indicated good tyrosinase and AGE-inhibition activity. These results suggest that mash bean seed may be used in food industry as functional food and nutraceutical as well as in cosmetic and pharmaceutical industry as ingredient of skin-whitening creams and as cure for diabetes respectively. The data obtained will be helpful for labeling of nutrients as well as for monitoring the quality and authenticity of foods containing mash bean in indigenous markets. Further investigations are necessary to evaluate the toxic effects (if any), to determine the ant-nutrients factors present and to understand mechanism of action of tyrosinase-inhibitory and AGE-inhibitory potential of extracts.

Methods

Material

Analytical grade solvents were used. All chemicals were were purchased from Sigma except where indicated. The seeds of four mash bean cultivars namely, NARC-Mash-1, NARC-Mash-2, NARC-Mash-3 and NARC-Mash-97 were procured from National Agricultural Research Centre, Islamabad (Pakistan). Seeds of four cultivars were stored in stainless-steel containers at 4°C prior to analysis.

Proximate analysis

Proximate chemical analysis of seeds was carried out according to AOAC International methods as per our previous studies [39]. Results are shown in Table 1.

Vitamin contents

Powdered sample (5 g) was steamed with concentrated H_2SO_4 (30 ml) for half an hour. After cooling, distilled H_2O was added to this suspension to make its volume up to 50 ml and filtered. Basic lead acetate (60%, 5 ml) was added to this filtrate (25 ml). The pH was adjusted (9.5) and supernatant was collected after centrifugation. To this supernatant, concentrated H_2SO_4 (2 ml) was added. After 1 hr, this mixture was centrifuged and then $ZnSO_4$ (5 ml, 40%) was added. The pH was adjusted (8.4) and supernant was collected after centrifugation. The pH of resulting supernatant was adjusted (7) and this was utilized as niacin extract. One ml of this extract was made 6 ml by distilled H_2O; after addition of cyanogen bromide (3 ml) and shaking, aniline (4%, 1 ml) was added. After 5 min, yellow color formed was spectrophotometrically measured at 420 nm against blank and niacin contents were calculated by a standard graph [16]. Thiochrome method and fluorescence method were

Table 9 Antioxidant capacity of extracts of seeds of mash bean cultivars

Cultivar	Reducing power (mg/g)	DPPH˙ scavenging capacity (μmol Trolox/g)	FRAP (mmol Fe^{2+}/g)	ABTS scavenging capacity (μmol trolox/g)
NARC-Mash-1	1.09 ± 0.18^a	41.64 ± 0.18^b	12.81 ± 0.03^a	33.81 ± 0.45^a
NARC-Mash-2	0.87 ± 0.02^a	34.72 ± 0.29^c	11.70 ± 0.19^a	27.09 ± 0.58^b
NARC-Mash-3	1.02 ± 0.09^a	39.49 ± 0.11^b	9.65 ± 0.37^b	29.74 ± 0.83^b
NARC-Mash-97	0.95 ± 0.06^a	46.56 ± 0.05^a	13.76 ± 0.57^a	35.93 ± 0.22^a

Values in the same row having different letters differ significantly with least significant difference (LSD) at probability ($p < 0.05$).

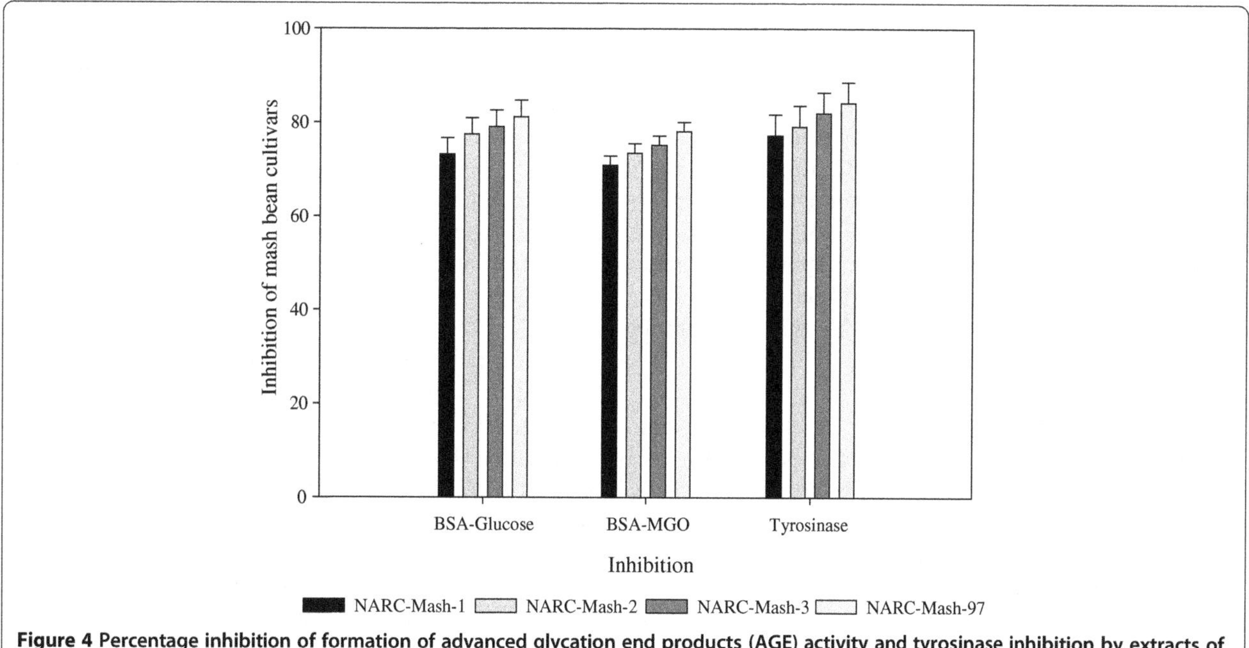

Figure 4 Percentage inhibition of formation of advanced glycation end products (AGE) activity and tyrosinase inhibition by extracts of seeds of mash bean cultivars.

used for determination of thiamine and riboflavin contents respectively [40,41] Figure 1.

Minerals contents

A muffle furnace was used to incinerate seeds (450°C; 12 h) and the resulting samples were digested by acid mixture (nitric/perchloric;2:1). Na and K were estimated by taking aliquots from this digested material by flame photometer. Other minerals like Mn, Mg, Ca, Fe, Cu and Zn were estimated spectrophotometrically (AAS; Perkin-Elmer 5000) while phosphovanado-molybdate method was used to measure phosphorus contents. Standard solutions of known concentration were run concurrently to quantify the samples [31,32] (Table 2).

Table 10 Correlation coefficient of total phenolics contents, DPPH, FRAP, ABTS BSA-MGO, BSA-Glucose and tyrosinase inhibition assay

	TPC	DPPH	FRAP	ABST	BSA-G	BSA-MGO	TI
TPC	-	0.499	0.087	0.282	0.981[b]	0.999[b]	0.992[b]
DPPH		-	0.603[a]	0.967[b]	0.321	0.513[a]	0.551[a]
FRAP			-	0.718[b]	−0.034	0.132	0.046
ABTS				-	0.092	0.303	0.330
BSA-G					-	0.977[b]	0.960[b]
BSA-MGO						-	0.987[b]
TI							-

[a]Correlation is significant at $p < 0.05$ level (2-tailed).
[b]Correlation is significant at $p < 0.01$ level (2-tailed).

Amino acid analysis

HCl (6 M) was used to hydrolyze samples (300 mg) in an evacuated test tube (105°C; 24 h). Citrate buffer (pH 2.2) was used to dissolve the dried residue resulting from flash evaporation. Hitachi Perkin-Elmer (KLA 3B) amino acid analyzer was utilized to quantify amino acids by taking aliquots from above solution. After treatment with performic acid followed by hydrolysis (HCl), cystine and methionine were analyzed separately from same solution. Alkali hydrolysis (NaOH) method was used to measure tryptophan [22,23] (Table 3).

Protein and starch digestibility (*In-vitro*)

In-vitro digestibility of protein was evaluated enzymetically while starch digestibility was evaluated as starch hydrolyzed (%) out of total starch present in sample [42-44] (Figure 2).

Fatty acid (FA) composition

Petroleum ether as solvent was used to extract oil from seeds by Soxhlet apparatus (6 hr) as per official AOCS method [39]. The fatty acid profile of oils obtained was evaluated by a method reported earlier [45]. Briefly, *n*-heptane (1 mL) was used to dissolve oil (1 drop), sodium methanolate (50 μL; 2 M) was added, and shaken in a closed tube (1 min). Water (100 μL) was added and the tube was centrifuged (4500 g; 10 min) and resulting aqueous phase was separated. To remaining heptane phase, HCl (50 μL; 1 M) was added, both phases were mixed for short period of time and resulting aqueous phase was discarded. After addition of sodium hydrogen sulphate (20 mg) and centrifugation (4500 g; 10 min), n-heptane phase was stored in a vial and inserted in a gas chromatograph (Varian 5890) having CP-Sil88 capillary column (ID: 0.25 mm, 100 m, film: 0.2 μm). The temperature setup was as follows: heated (155°C- 220°C; 1.5°C/min), isotherm (10 min); detector and injector (250°C), carrier gas (H_2: 1.07 mL/min), split ratio of 1:50; detector gas (hydrogen: 30 mL/min). Peaks were computed with help of integration software and fatty acid methyl esters (%) were obtained as weight percent by direct internal normalization (Table 4).

Tocopherol contents

Twenty five ml of n-heptane was mixed with oil (250 mg) and tocopherol contents were was analyzed by HPLC system (Merck-Hitachi), containing a pump (L-6000), a fluorescence spectrophotometer (Merck-Hitachi F-1000), excitation wavelength (295); emission wavelength (330 nm) and a D-2500 integration system; 20 μl of samples were inserted by a Merck 655-A40 autosampler in a dual phase HPC (Merck) having column column (25 cm × 4.6 mm) while flow rate was adjusted at 1.3 mL/min. Mobile phase used was n-heptane: tert-butyl methyl ether (99:1) [46] (Table 5).

Sterol composition

The sterols were quantified by a gas chromatograph (Perkin Elmer model 8700), having flame-ionization detector (FID) and OV-17 capillary column (methyl phenyl polysiloxane coated; ID: 30 m × 2.25 mm, film: 20 μm). The column was operated isothermally (255°C) while temperature for injector and detector were 275 and 290°C, respectively. Carrier gas selected was extra pure nitrogen with 3 mL/min as flow rate. Sterols were recognized and quantified by comparing with a sterol standard mixture [21,31] (Table 6).

Extraction

The mash bean seeds were ground to flour by a mill (IKA Works Inc.) and were sieved (60-mesh). After maceration with 5 L solvent mixture of aqueous: methanol (80:20) for 15 days at room temperature and extracts were collected. The process was carried out three times. The resulting extracts were collected and filtered by filter paper. The extra solvent present was evaporated under reduced pressure by using a rotary evaporator. A thick gummy mass was obtained which was then dried in a dessicator and utilized for assessment of biological activities.

FTIR of Mash bean powder and crude extract

Functional groups present in flour and extracts of seeds of mash bean cultivars were identified by FTIR spectroscopy (Perkin Elmer; UK) [24,25] Figure 5 and 6.

Total phenolic, flavonoid and condensed tannin contents (TPC, TFC, CTC)

Total phenolics were were estimated using the Folin and Ciocalteau's phenol reagent [47] and results were reported as gallic acid equivalents [48,49]. A previously reported method was used for estimation of flavonoids contents [50]. Condensed tannins were quantified by acidified vanillin reagent [51] and results were expressed as mg of CAE/g (Table 7).

RP-HPLC

Phenolic acids were finger printed by using HPLC (Shimadzu Corp., Kyoto, Japan) fitted with a pre-packed LUNA C-18 column (4 × 259 mm, 5 μm) equipped with two LC-10 AD pumps, photodiode array detector (SPD-M 10), and a SCTL 10A system. Flow rate was adjusted at 1 mL/min and gradient elution of acetonitrile:water acetic acid (5:93:0) as solvent A and and acetonitrile:water acetic acid (40:58:2) as solvent B was used [52]. Samples were dissolved in methanol (10 mg/mL) while injection volume used was 20 μL. Separated compounds were measured at 330 nm Figure 3, Table 8.

DPPH radical scavenging assay

Scavenging potential of extracts of mash bean seeds against $DPPH^{\bullet}$ was estimated by a previously reported method [53]. The absorbance of extracts (A_{sample}) was measured spectrophotometrically (Shimadzu, Kyoto, Japan) at 517 nm and ethanol was used as blank. The extraction solvent (0.2 mL) after addition of $DPPH^{\bullet}$ was used as negative control ($A_{control}$). Following equation was used to assess antiradical activity:

$$Antiradical\ activity\ \% = \left(1 - \frac{Absorbance_{sample}}{Absorbance_{control}}\right) \times 100$$

Calibration curve of Trolox was used to calculate results and indicated as micromoles of Trolox equivalent (μmol Trolox/g) Table 9.

Ferric reducing antioxidant power (FRAP) activity

FRAP assay was carried out to assess antioxidant activity [54]. Deionized water was used to dilute properly the sample solution to fit within the linearity range of Fe^{2+}. The calibration curve of Fe^{2+} was used to calculate FRAP value as mmoles of Fe^{2+} equivalent (mmol Fe^{2+}/g) Table 9.

Reducing power

Reducing potential of investigated extracts was determined by a reported method [55]. Aliquotes (2.5 ml) of extracts dissolved in phosphate buffer (pH 6.6, 0.2 M) were mixed with $C_6N_6FeK_3$ (10 mg/ml; 2.5 ml) and resulting solution was incubated (20 min; 50°C). To this reaction mixture, trichloroacetic acid (100 mg/ml solution;

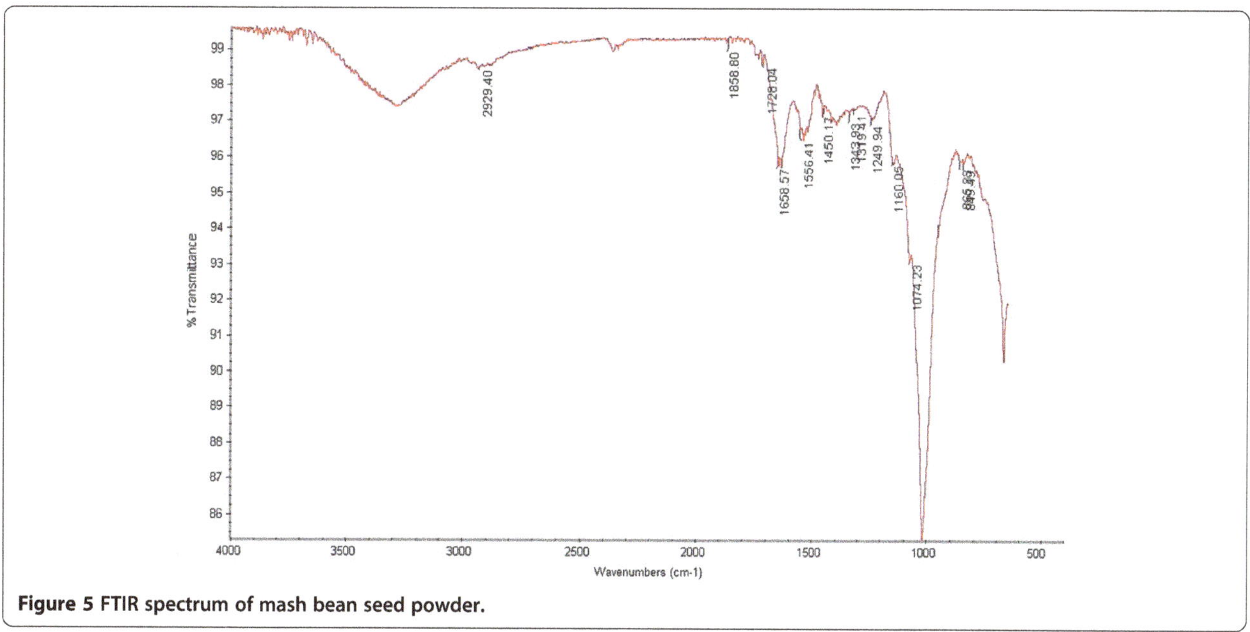

Figure 5 FTIR spectrum of mash bean seed powder.

2.5 ml) was added and centrifuged (1000 rpm; 10 min). The resulting supernant (2.5 ml) was mixed with an equal volume of H_2O (distilled) and $FeCl_3$ (1 mg/ml solution; 0.5 ml) was added. Spectrophotometer was used to measure absorbance at 700 nm against ascorbic acid Table 9.

ABTS$^{•+}$ scavenging assay

Scavenging activity of extracts of seeds was also evaluated against ABTS$^{•+}$ [56]. ABTS aqueous solution (5 mM) was passed from the oxidizing reagent (MnO_2), on filter paper (Fisher Brand P8) to prepare ABTS$^{•+}$. The solution was filtered from fisher membrane (0.2 mm) to remove extra MnO_2. Phosphate buffered saline (5 mM; pH 7.4) was used to dilute extracts to an absorbance of approximately 0.700 (±0.020) at 734 nm. The extracts (1.0 mL) were added to ABTS$^{•+}$ solution (5 mL), and the absorbance was measured after 10 min. The blank used was PBS Table 9.

Evaluation of AGE inhibition activity

Inhibitory potential of mash bean extracts on the formation of advanced glycation end (AGE) products was determined by BSA-MGO and BSA-glucose models (Table 10). Briefly, BSA (5 g) and D-glucose (14.4 g) were dissolved in phosphate buffer (1.5 M; pH 7.4) to get a control solution

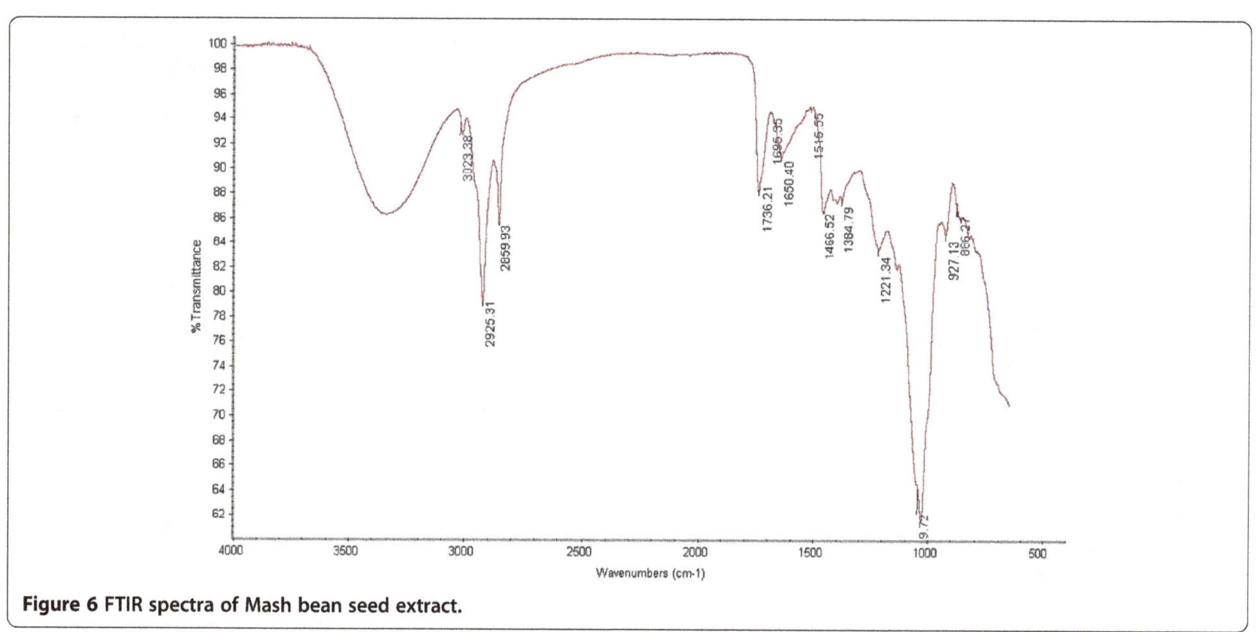

Figure 6 FTIR spectra of Mash bean seed extract.

containing D-glucose (0.8 M) and BSA (50 mg/mL). Two mL of this solution was incubated at 37°C (1 week) in the absence or presence of bean extracts (1 mL) in phosphate buffer. After one week, fluorescent intensity (excitation: 330 nm; emission: 410 nm) was measured. The BSA-MGO assay was performed as reported elsewhere.Briefly, MGO (31 μL) was mixed with BSA (40 mg) in phosphate buffer (pH 7.4; 0.1 M) to make a control solution of MGO (5 mM) and BSA (1 mg/mL). Two mL of control solution was incubated (6 days) with or without bean extracts (1 mL) in phosphate buffer [48-52]. Inhibition (%) of formation of AGE by extract for both models was calculated using the following equation:

$$\text{Percent Inhibition} = 1-\left[\frac{\text{fluorescence with extract}}{\text{fluorescence without extract}}\right] \times 100$$

Measurement of tyrosinase inhibition activity

Microtiter plates (96-well) were used to perform assays and absorbance was measured (475 nm) by a plate reader. Each well contained sample (40 μL) and phosphate buffer (pH 6.8, 80 μL 0.1 M), tyrosinase (31 units/mL, 40 μL) and L-DOPA (2.5 mM; 40 μL), the samples were incubated (37°C) for half an hour and results are shown in Table 10. A control was prepared having all ingredients except tyrosinase [48-52]. The tyrosinase inhibition percentage was calculated as follows:

$$\text{Percent Inhibition} = \frac{(A_{\text{control}} - A_{\text{sample}})}{A_{\text{control}}} \times 100$$

Statistical analysis

All experiments were performed in triplicate and values marked by same letter in same column are not significantly different $(P < 0.05)$. Data are expressed as the mean ± standard deviation. Data were analyzed by using the "MSTATC" statistical computer package [57].

Competing interests

The authors state that there are no competing interests.

Authors' contributions

MZUH and RA contributed to the experiment. SA and SAB contributed in statistical interpretation of the data while SE and HZJ helped in preparation of the manuscript. All authors approved the final form of the manuscript.

Author details

[1]The Patent Office, Karachi, Pakistan. [2]Department of Agronomy, Bahauddin Zakariya University, Multan 60800, Pakistan. [3]Department of Applied Chemistry and Biochemistry, Government College University, Faisalabad, Pakistan. [4]Institute of Animal Reproduction and Food Research of the Polish Academy of Sciences, Tuwima Str. 10, 10-747 Olsztyn, Poland. [5]Agricultural Faculty, Department of Horticulture, Ataturk University, Erzurum, Turkey. [6]Department of Crop Science, Faculty of Agriculture, 43400 UPM Serdang, Selangor, Malaysia.

References

1. Bhattacharya S, Latha RB, Bhat KK: **Controlled stress rheological measurement of black gram flour dispersions.** *J Food Eng* 2006, **63:**135–139.
2. Anitha K, Ranjith K, Vakula K, Thirupathi G, Balaji B: **Protective effect of *Vigna mungo* (L.) against carbon tetrachloride induced hepatotoxicity.** *Int J Pharm Res* 2012, **2:**29–34.
3. Battu G, Anjana CKVLSN, Priya TH, Malleswari VN, Reeshm S: **A phytopharmacological review on *Vigna* species.** *Pharmanest* 2011, **2:**62–69.
4. Nitin M, Ifthekar S, Mumtaz M: **Hepatoprotective activity of methanolic extract of *Vigna mungo* (Linn.) Hepper in ethanol-induced hepatotoxicity in rats.** *Rgush J Pharm Sci* 2012, **2:**62–67.
5. Zia-Ul-Haq M, Ahmed S, Rizwani GH, Qayum M, Ahmad S, Hanif M: **Platelet aggregation inhibition activity of selected flora of Pakistan.** *Pak J Pharm Sci* 2012, **25:**863–865.
6. Zia-Ul-Haq M, Landa P, Kutil Z, Qayum M, Ahmad S: **Evaluation of anti-inflammatory activity of selected legumes from Pakistan: in vitro inhibition of cyclooxygenase-2.** *Pak J Pharm Sci* 2013, **26:**185–187.
7. Ahmed ZI, Ansar M, Saleem A, Arif ZU, Javed HI, Saleem R: **Improvement of mash bean production under rainfed conditions by rhizobium inoculation and low rates of starter nitrogen.** *Pak J Agric Res* 2012, **25:**154–160.
8. Achakzai AKK, Taran SA: **Effect of seed rate on growth, yield components and yield of mash bean grown under irrigated conditions of arid uplands of Balochistan, Pakistan.** *Pak J Bot* 2011, **43:**961–969.
9. Zia-Ul-Haq M, Iqbal S, Ahmad M: **Characteristics of oil from seeds of 4 mungbean (*Vigna radiata* (L.) wilczek] cultivars grown in Pakistan.** *J Am Oil Chem Soc* 2008, **85:**851–856.
10. Zia-Ul-Haq M, Ahmad S, Chiavaro E, Ahmed S: **Studies of oil from cowpea (*Vigna unguiculata* (L) Walp.) cultivars commonly grown in Pakistan.** *Pak J Bot* 2010, **43:**1333–1341.
11. Zia-Ul-Haq M, Ahmad S, Amarowicz R, DeFeo V: **Antioxidant activity of the extracts of some cowpea (Vigna unguiculata (L) Walp.) cultivars commonly consumed in Pakistan.** *Molecules* 2013, **18:**2005–2017.
12. Zia-Ul-Haq M, Cavar S, Qayum M, Khan I, Ahmad S: **Compositional studies and antioxidant potential of *Acacia leucophloea* Roxb.** *Acta Bot Croat* 2013, **72:**27–31.
13. Zia-Ul-Haq M, Ahmad S, Qayum M, Erçişli S: **Compositional studies and antioxidant potential of *Albizia lebbeck* (L.) Benth.** *Turk J Bio* 2013, **37:**25–32.
14. Khalid H, Intikhab J: **Characterization and biochemical studies of the oils extracted from four cultivars of *Vigna mungo* grown in Pakistan.** *J Rashid Latif Med College* 2013, **1:**19–24.
15. Singh N, Kaur M, Sandhu KS, Sodhi NS: **Physicochemical, cooking and textural characteristics of some Indian black gram (*Phaseolus mungo* L.) varieties.** *J Sci Food Agri* 2006, **84:**977–982.
16. Zia-Ul-Haq M, Ahmad S, Amarowicz R, Ercisli S: **Compositional studies of some pea (*Pisum sativum* L.) seed cultivars commonly consumed in Pakistan.** *Ital J Food Sci* 2013, **25:**295–302.
17. Soris TP, Kala KB, Mohan VR, Vadivel V: **The biochemical composition and nutritional potential of three varieties of *Vigna mungo* (L.) Hepper.** *Adv Biores* 2010, **1:**6–16.
18. Miyamoto Y, Kajikawa A, Zaidi JH, Nakanishi T, Sakamoto K: **Minor and trace element determination of food spices and pulses of different origins by NAA and PAA.** *J Radioanal Nucl Chem* 2000, **243:**747–765.
19. NRC/NAS B: *Recommended Dietary Allowances.* 10th edition. Washington DC, USA: National Academy Press; 1989.
20. Jamil A, Lubna B, Hamid Y: **Studies on *Vigna mungo* L., effect of processing on carbohydrate fractionation and influence of grain starch on protein utilization in albino rats.** *Pak J Biol Sci* 1999, **2:**1258–1262.
21. Zia-Ul-Haq M, Ahmad S, Ahmad M, Iqbal S, Khawar KM: **Effects of cultivar and row spacing on tocopherol and sterol composition of chickpea (*Cicer arietinum* L) seed oil.** *Tarim Bilimleri Dergisi* 2009, **15:**25–30.
22. Gopala KAG, Prabhakar JV, Aitzetmuller K: **Tocopherol and fatty acid composition of some Indian pulses.** *J Am Oil Chem Soc* 1997, **74:**1603–1606.
23. Akihisa T, Nishimsura Y, Nakamura N, Roy K, Gosh P, Thakur S, Tamura T: **Sterols of *Cajanus cajan* and three other *Leguminosae* seeds.** *Phytochem* 1992, **31:**1765–1768.
24. Bhat R, Yahya NB: **Evaluating belinjau (*Gnetum gnemon* L.) seed flour quality as a base for development of novel food products and food formulations.** *Food Chem* 2014, **156:**42–49.

25. Zia-Ul-Haq M, Amarowicz R, Ahmad S, Riaz M: **Antioxidant potential of some pea (*Pisum sativum* L.) cultivars commonly consumed in Pakistan.** *Oxid Commun* 2013, **36:**1046–1057.
26. Imran I, Zia-Ul-Haq M, Calani L, Mazzeo T, Pellegrini N: **Phenolic profile and antioxidant potential of selected plants of Pakistan.** *J Appl Bot Food Qual* 2014, **87:**30–35.
27. Girish TK, Pratape VM, Rao UJSP: **Nutrient distribution, phenolic acid composition, antioxidant and alpha-glucosidase inhibitory potentials of black gram (*Vigna mungo* L.) and its milled by-products.** *Food Res Int* 2012, **46:**370–377.
28. Girish TK, Pratape VM, Rao UJSP: **Protection of DNA and erythrocytes from free radical induced oxidative damage by black gram (*Vigna mungo* L.) husk extract.** *Food Chem Toxic* 2012, **50:**1690–1696.
29. Marathe SA, Rajalakshmi V, Jamdar SN, Sharma A: **Comparative study on antioxidant activity of different varieties of commonly consumed legumes in India.** *Food Chem Toxic* 2011, **49:**2005–2012.
30. Zia-Ul-Haq M, Amarowicz R, Ahmad S, Qayum M, Ercişli S: **Antioxidant potential of mungbean cultivars commonly consumed in Pakistan.** *Oxid Commun* 2013, **36:**15–25.
31. Zia-Ul-Haq M, Ahmad S, Calani L, Mazzeo T, Rio DD, Pellegrini N, DeFeo V: **Compositional study and antioxidant potential of *Ipomoea hederacea* Jacq. and *Lepidium sativum* L. seeds.** *Molecules* 2012, **17:**10306–10321.
32. Zia-Ul-Haq M, Ćavar S, Qayum M, Imran I, DeFeo V: **Compositional studies, antioxidant and antidiabetic activities of *Capparis decidua* (Forsk.) Edgew.** *Int J Mol Sci* 2011, **12:**8846–8861.
33. Tiwari AK, Swapna M, Ayesha SB, Zehra A, Agawane SB, Madhusudana K: **Identification of proglycemic and antihyperglycemic activity in antioxidant rich fraction of some common food grains.** *Int Food Res J* 2011, **18:**915–923.
34. Peng XF, Zheng ZP, Cheng KW, Shan F, Ren GX, Chen F, Wang MF: **Inhibitory effect of mung bean extract and its constituents vitexin and isovitexin on the formation of advanced glycation endproducts.** *Food Chem* 2008, **106:**475–481.
35. Yao Y, Cheng X, Wang S, Wang L, Ren G: **Influence of altitudinal variation on the antioxidant and antidiabetic potential of azuki bean (*Vigna angularis*).** *Int J Food Sci Nutr* 2011, **63:**117–124.
36. Yao Y, Cheng X, Wang L, Wang S, Ren G: **Biological potential of sixteen legumes in China.** *Int J Mol Sci* 2011, **12:**7048–7058.
37. Yao Y, Cheng X, Wang L, Wang S, Ren G: **Major phenolic compounds, antioxidant capacity and antidiabetic potential of rice bean (*Vigna umbellata* L.) in china.** *Int J Mol Sci* 2012, **13:**2707–2716.
38. Lim TY, Lim YY, Yule CM: **Evaluation of antioxidant, antibacterial and anti-tyrosinase activities of four Macaranga species.** *Food Chem* 2009, **114:**594–599.
39. Association of Official Analytical Chemists (AOAC): *Official Methods of Analysis of the Association of Official Analytical Chemists.* 14th edition. Washington, DC, USA: AOAC; 1990.
40. Gstirner F: *Chemisch-Phisikalische Vitamin Estimmungs-Methoden.* Stuttgart, Germany: Ferdinand Enke Verlag; 1965.
41. Arinathan V, Mohan VR, Britto D, John A: **Chemical composition of certain tribal pulses in South India.** *Int J Food Sci Nutr* 2003, **54:**209–217.
42. Ekpenyong TE, Borchers RL: **Digestibility of proteins of winged bean seed.** *J Food Sci Tech* 1979, **16:**92–95.
43. Hsu HW, Vavak DL, Satterlee LD: **A multienzyme technique for estimating protein digestibility.** *J Food Sci* 1977, **42:**1269–1271.
44. Goni I, Garcia-Alonso A, Saura-Calixto FA: **A starch hydrolysis procedure to estimate glycemic index.** *Nutr Res* 1997, **17:**427–437.
45. ISO/FIDS 5509: *International Standards.* 1st edition. Genève, Switzerland: International Organization for Standardization; 1997.
46. Balz M, Shulte E, Their HP: **Trennung von Tocopherol und Tocotrienolendurch HPLC.** *Fat Sci Tech* 1992, **94:**209–213.
47. Singleton VL, Rossi JA: **Colorimetry of total phenolic with phosphomolybdicphosphotungstic acid reagents.** *Am J Eno Viticul* 1965, **16:**144–158.
48. Heimler D, Vignolini P, Dini MG, Romani A: **Rapid tests to assess the antioxidant activity of *Phaseolus vulgaris* L. dry bean.** *J Agric Food Chem* 2005, **53:**3053–3056.
49. Xu BJ, Chang SKC: **A comparative study on phenolic profiles and antioxidant activities of legumes as affected by extraction solvents.** *J Food Sci* 2007, **72:**S159–S166.
50. Jia Z, Tang M, Wu J: **The determination of flavonoid contents in mulberry and their scavenging effects on superoxide radicals.** *Food Chem* 1999, **64:**555–559.
51. Broadhurst RB, Jones WT: **Analysis of condensed tannins using acified vanillin.** *J Sci Food Agri* 1978, **29:**788–794.
52. Crozier A, Jensen E, Lean MEI, Mcdonald MS: **Quantitative analysis of flavonoids by reverse-phase high performance liquid chromatography.** *J Chromatogr A* 1997, **761:**315–321.
53. Chen CW, Ho CT: **Antioxidant properties of polyphenols extracted from green and black teas.** *J Food Lipids* 1995, **2:**35–46.
54. Benzie IFF, Strain JJ: **The ferric reducing ability of plasma (FRAP) as a measure of "antioxidant power": the FRAP assay.** *Anal Biochem* 1996, **239:**70–76.
55. Pin-Der D, Gow-Chin Y, Wen-Jye Y, Lee-Wen C: **Antioxidant effects of water extracts from barley (*Hordeum vulgare* L.) prepared under different roasting temperatures.** *J Agri Food Chem* 2001, **49:**1455–1463.
56. Reo DD, Pellegrini N, Proteggente A, Pannala A, Yang M, Rice-Evans C: **Antioxidant activity applying an improved ABTS$^{\cdot+}$ radical cation decolourisation assay.** *Free Radic Biol Med* 1999, **26:**1231–1237.
57. Freed R, Eisensmith SP, Goetz S, Reicosky D, Smail VW, Welberg P: *User's Guide to MSTAT-C.* East Langing, MI, USA: Michigan State University; 1991.

4

The effects of cadmium chloride on secondary metabolite production in *Vitis vinifera* cv. cell suspension cultures

Emine Sema Cetin[1*], Zehra Babalik[2], Filiz Hallac-Turk[3] and Nilgun Gokturk-Baydar[4]

Abstract

Background: Plant secondary metabolites are possess several biological activities such as anti-mutagenic, anti-carcinogenic, anti-aging, etc. Cell suspension culture is one of the most effective systems to produce secondary metabolites. It is possible to increase the phenolic compounds and tocopherols by using cell suspensions. Studies on tocopherols production by cell suspension cultures are seldom and generally focused on seed oil plants. Although fresh grape, grape seed, pomace and grape seed oil had tocopherols, with our best knowledge, there is no research on tocopherol accumulation in the grape cell suspension cultures. In this study, it was aimed to determine the effects of cadmium chloride treatments on secondary metabolite production in cell suspension cultures of grapevine. Cell suspensions initiated from callus belonging to petiole tissue was used as a plant material. Cadmium chloride was applied to cell suspension cultures in different concentration (1.0 mM and 1.5 mM) to enhance secondary metabolite (total phenolics, total flavanols, total flavonols, *trans*-resveratrol, and α-, β-, γ- δ-tocopherols) production. Cells were harvested at two days intervals until the 6th day of cultures. Amounts of total phenolics, total flavanols and total flavonols; *trans*-resveratrol and tocopherols (α-, β-, γ- and δ-tocopherols) and dry cell weights were determined in the harvested cells.

Results: Phenolic contents were significantly affected by the sampling time and cadmium concentrations. The highest values of total phenolic (168.82 mg/100 g), total flavanol (15.94 mg/100 g), total flavonol (14.73 mg/100 g) and *trans*-resveratrol (490.76 μg/100 g) were found in cells treated with 1.0 mM $CdCl_2$ and harvested at day 2. Contents of tocopherols in the cells cultured in the presence of 1.0 mM $CdCl_2$ gradually increased during the culture period and the highest values of α, β and γ tocopherols (145.61, 25.52 and 18.56 μg/100 g) were detected in the cell cultures collected at day 6.

Conclusions: As a conclusion, secondary metabolite contents were increased by cadmium chloride application and sampling time, while dry cell weights was reduced by cadmium chloride treatments.

Keywords: Cadmium chloride, Cell, Secondary metabolite

Background

Plant cell suspension culture is one of the most effective systems to produce secondary metabolites with high amount and purity. Using this system could ensure a continuous supply of uniform quality, specialized, natural components [1,2] compared to traditional extraction methods. Unfortunately, the yield of the desired end-product is often too low to make this a commercially viable alternative to extraction from field grown plants. It is possible to increase the secondary metabolite accumulation in the cell culture by using some treatments such as light irradiation, UV, jasmonic acid, ozon, heavy metal, ethylene and sucrose [3,4].

Plant cell culture is considered to be a potential means of producing valuable plant products in a factory setting. Among these products, food additives such as anthocyanins, shikonin compounds, safflower yellow, saffron and colorants are of high interest [5]. With the help of this technique, the presence of anthocyanins and *trans*-resveratrol were shown in *Vitis vinifera* suspension culture [6].

* Correspondence: esemacetin@gmail.com
[1]Department of Horticulture, Faculty of Agriculture and Natural Science, Bozok University, 66200 Yozgat, Turkey
Full list of author information is available at the end of the article

The phenolic compound family is huge and comprises a complex group of compounds varying from simple phenols to highly polymerised compounds. Polyphenols have been extensively studied and are reported to possess several biological activities. Numerous studies have focused on their anti-mutagenic chemopreventive and anti-carcinogenic activities [7-9].

Resveratrol (3, 4, 5-trihydroxystilbene), a natural polyphenol, is found in some plants that are used in human nutrition. Grape is the major source of resveratrol, and a significant amount can also be found in red wine. Several experimental studies have demonstrated biological properties of resveratrol, especially its anti-inflammatory, antioxidant, anti-platelet and antitumor effects [10]. Many reports have shown that resveratrol can prevent or slow the progression of a wide variety of illnesses, including cancer, cardiovascular disease and ischaemic injuries as well as enhance stress resistance and extend the lifespans of various organisms [11,12]. Interest in this compound has been renewed in recent years, first from its identification as a chemopreventive agent for skin cancer, and subsequently from reports that it activates sirtuin deacetylases and extends the lifespans of lower organisms [13].

Tocopherols are another group of secondary metabolites produced in plant cell suspension cultures. Tocopherols as liposoluble, naturally occurring non-polar antioxidants comprising α (5, 7, 8-trimethyltocol), β (5, 8-dimethyltocol), γ (7, 8-dimethyltocol) and δ (8-methyltocol) tocopherols. Tocopherols appear to possess high antioxidant activity by donating the hydrogen of the hydroxyl group to the lipid peroxyl radical [14] and α-tocopherol is considered to be extremely important because of its vitamin E activity [15]. Tocopherols are ubiquitous, even if at different concentrations, in oil seeds, leaves and other green parts of higher plants [16].

Studies conducted on the production of tocopherols by cell suspension cultures are seldom and generally focused on seed oil plants [2,17]. Although fresh grape [18], grape seed, pomace [19] and grape seed oil [20-22] had tocopherols in different amounts, with our best knowledge, there is no research on tocopherol accumulation in the grape cell suspension cultures.

Within this study, it was aimed to determine the effects of cadmium chloride ($CdCl_2$) treatments on secondary metabolite production in the grape cell suspension cultures to be able to provide a preliminary reference for researchers who study or willing to study on this topic.

Results and discussion

In the $CdCl_2$ treatments, cell suspension cultures were exposed to increasing concentrations of $CdCl_2$ for up to 6 days. Changes in the dry cell weights were determined as g/L and data are showed in Figure 1. Dry cell weights varied depending on the $CdCl_2$ concentrations and sampling time. It was determined that when the $CdCl_2$ concentration increased and the sampling time extended, regular decreases were observed in the dry cell weights. 1.0 and 1.5 mM $CdCl_2$ inhibited the growth of cells after 4 days and more dramatically after 6 days.

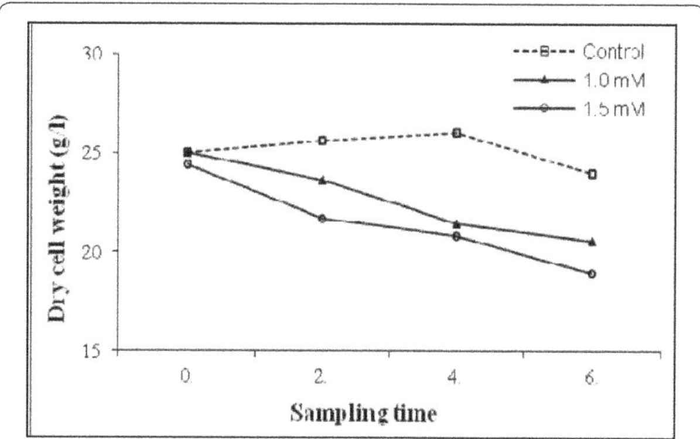

Figure 1 Effects of cadmium chloride on the dry cell weight on cell suspension culture.

The data about the effects of cadmium chloride treatments on the phenolic contents of cell suspensions are presented in Table 1. Phenolic contents were significantly affected by the sampling time and cadmium concentrations ($p \leq 0.05$). The highest values of total phenolic (168.80 mg/100 g), total flavanol (15.90 mg/100 g) and total flavonol (14.70 mg/100 g) were found in cells treated with 1.0 mM $CdCl_2$ and harvested at day 2. Cell suspension cultures treated with 1.5 mM $CdCl_2$ and sampled at days 4 and 6 gave the lowest amounts of total phenolics, total flavanols and total flavonols.

In this study, the production of the *trans*-resveratrol in the cell cultures was significantly changed depending on the $CdCl_2$ concentrations and sampling time ($p \leq 0.05$). In the control group, *trans*-resveratrol showed a gradual increase during the sampling times and the highest value was detected in cell cultures harvested at day 6. In the cell cultures treated with 1.5 mM $CdCl_2$, *trans*--resveratrol significantly increased from day 0 to day 4, but it showed a sharp decrease at day 6. On the other hand the greatest *trans*-resveratrol content (490.76 μg/100 g) was detected in cell cultures treated with 1.0 mM $CdCl_2$ and collected at the 2nd day of the application followed by the cells cultured in the media containing 1.0 mM $CdCl_2$ and harvested at day 4. But it subsequently exhibited a strong decrease in the cells collected at the 6th day of the culture.

Tocopherol composition of the callus samples was affected significantly depending on the $CdCl_2$ concentrations and the sampling times ($P \leq 0.05$) as shown in Table 2. The effect was dose-dependent both in terms of time and $CdCl_2$ concentration applied. In terms of tocopherols, α, β and γ-tocopherols were found at different concentrations depending on the treatments, while

Table 1 The effects of $CdCl_2$ treatments on phenolic contents in Öküzgözü cell suspension cultures

$CdCl_2$ concentrations	Sampling time (day)	Total phenolics (mg/100 g)	Total flavanols (mg/100 g)	Total flavonols (mg/100 g)	*trans*-resveratrol (µg/100 g)
Control	0.	102.84 e*	7.14 d	10.16 bc	138.12 e
	2.	128.32c	5.48 e	7.78 d	139.24 e
	4.	133.00 c	7.96 d	6.56 d	210.65 d
	6.	146.8 1b	9.42 c	6.65 d	216.42 d
1.0 mM	0.	100.64 ef	5.45 e	10.87 b	251.74 c
	2.	168.82 a	15.94 a	14.73 a	490.76 a
	4.	146.24 b	12.51 b	7.82 d	344.62 b
	6.	105.66 de	9.82 c	4.54 e	215.14 d
1.5 mM	0.	112.00 d	5.43 e	9.17 c	119.46 f
	2.	108.67 de	5.85 e	4.00 e	137.16 e
	4.	93.11 f	3.92 f	1.86 f	250.82 c
	6.	77.27 g	4.00 f	2.23 f	100.62 f

*Differences between means indicated by the same letters are not statistically significant ($p \leq 0.05$).

δ-tocopherol was not detected in the cell cultures of Öküzgözü. Contents of tocopherols in the cells cultured in the presence of 1.0 mM $CdCl_2$ gradually increased during the culture period and the highest values of tocopherols were detected in the cell cultures collected at day 6. The analyses conducted in the present work revealed that these conditions are the most convenient conditions to stimulate the biochemical pathways leading to tocopherols. However, when the $CdCl_2$ concentrations and the culture period increased, tocopherol contents of the cells decreased. Our data indicate that the lowest tocopherol concentrations were found in the cells treated 1.5 mM $CdCl_2$ and sampled after 6 day.

$CdCl_2$ application inhibited the growth of cells. Because it is well known that when cadmium is excess in plants, it inhibits and disturbs various biochemical and physiological processes such as respiration, photosynthesis, cell elongation, plant-water relationships, nitrogen metabolism and mineral nutrition, resulting in poor growth, low biomass, cell death and inhibition of growth [23-25]. Similarly, treatment with Cd^{2+} at 0.05 mM appeared to inhibit cell division and induce either mitotic or total cell death in the sensitive tobacco cell subpopulation as reported by Kuthanova et al. [26].

Cell suspension cultures treated with 1.5 mM $CdCl_2$ and sampled at days 4 and 6 gave the lowest amounts of total phenolics, total flavanols and total flavonols. These decreases can be resulted from the low cell weight in the presence of high $CdCl_2$ concentrations and long exposure time. Our results confirmed that cytotoxic effects of cadmium in cells were concentration dependent and followed a distinct time course [27]. Similarly, Kuthanova et al. [26] found that treatment with Cd^{2+} in 1 mM concentration caused total and rapid cell death after 6 h, while

Table 2 The effects of $CdCl_2$ treatments on tocopherol contents in Öküzgözü cell suspension cultures

$CdCl_2$ concentrations	Sampling time (day)	α- tocopherol (µg/100 g)	β- tocopherol (µg/100 g)	γ- tocopherol (µg/100 g)
Control	0.	118.52 c*	13.53 d	6.00 g
	2.	83.00 f	12.50 de	6.54 fg
	4.	82.41 f	17.00 c	12.46 b
	6.	86.12 f	11.55 e	11.34 c
1.0 mM	0.	97.18 e	11.00 e	4.00 h
	2.	113.00 cd	17.12 c	9.55 e
	4.	125.54 b	17.24 c	10.53 cd
	6.	145.61 a	25.52 a	18.56 a
1.5 mM	0.	102.58 e	17.00 c	7.14 f
	2.	109.55 d	21.24b	11.00 c
	4.	38.52 g	13.35 d	10.64 de
	6.	37.58 g	5.57f	2.20 i

*Differences between means indicated by the same letters are not statistically significant ($p \leq 0.05$).

application of 0.05 mM Cd^{2+} induced a marked decline of cell viability during the first 24 h of the cultivation in tobacco cells.

The accumulation of phenolic compounds represents a major key factor in the inducible defense mechanisms of plants through the phenylpropanoid pathway [28,29]. The induction of the phenylpropanoid metabolism could also be achieved experimentally by treatments with elicitors or exposure to specific stress conditions [30,31]. With respect to the high abundance of phenolic metabolites in plant tissues, regulation of phenylalanine ammonia lyase activity represents an important step in tolerance to stress factors [32]. Phenolic synthesis is recognized as a result of signalling processes initiated very quickly after injury, an attack of pathogens or elicitation [33,34]. Kuthanova et al. [26] reported that 0.05 mM Cd^{2+} treated cells correlated with the stimulation of the activity of PAL, key enzyme in phenylpropanoid biosynthesis. They also found that Cd^{2+} treatment significantly stimulated PAL activity during the whole culture period which was 25 times higher on day 3 when compared to the control cells.

Cadmium is not an essential nutrient for plants and it is normally toxic [35,36]. Its toxicity can promote altered metabolism [37] which can include the formation of reactive oxygene species (ROS) in plants under stress situations. Evidence confirmed that Cd stress induced the production of ROS such as superoxide, hydroxyl radicals (OH·) and hydrogen peroxide (H_2O_2) in plants [36]. However the interaction of Cd and antioxidative systems such as catalase, superoxide dismutase and glutathion reductase only recently have been studied in plant species [38,39]. The degree of plant antioxidant enzyme activities under Cd stress was found in several distinct patterns, which varied according to Cd concentration, duration, the species and tissues [36]. On the other hand, very little is known about the the responses of grapevine cell cultures in terms of stress defence mechanisms under Cd [40].

Conclusions

As a conclusion the results of our study showed that $CdCl_2$ treatment can be used for enhancing phenolic compounds and tocopherols in grape cell cultures depending on the $CdCl_2$ concentrations and exposure times. These increases might be explained by hypothesizing that Cd act as a stres factor on grape cell cultures which stimulate and alter the patways responsible for phenolics and tocopherol biyosynthesis. But high Cd concentrations and long exposure time had negative effects on cell viability and cell weight. When the treatment is used in high concentrations for long exposure durations, not only the cell division and cell viability but also the secondary metabolite accumulation decreases. To the best of our knowledge, this study reports the use of $CdCl_2$ treatment to enhance phenolics and tocopherols in grape cell suspension culture for the first time. However, further investigations with various strategies for the phenolic and tocopherol contents should also be carried out in grape cell lines.

Methods

In this research, callus tissues obtained from leaf petioles of Öküzgözü grape cultivar were used. Petioles were surface sterilised with commercial bleach (15%) for 15 min and rinsed three times with sterile distilled water. Petioles were then cut into 1 cm pieces and placed onto a solid B5 [41] culture medium with 30 g/L sucrose and 8 g/L bacto agar supplemented with 0.5 mg/L benzylaminopurine (BA), 0.5 mg/L indole acetic acid (IAA) and 2,4 dichlorophenoxyacetic acid (2,4-D). The pH was adjusted to 5.75. Explants were maintained at 25°C under dark conditions. Induced calli were subcultured on the same media in order to maintain sufficient stock cultures.

Cell suspension cultures

Cell suspensions were initiated by inoculating fresh friable fragments of calli (2.5 g each) into 50 mL of liquid media in 250 mL Erlenmeyer flasks. Media were supplemented with macro elements (B5 medium), micro elements [42], vitamins [43], 0.1 mg/L naphtalen acetic acid, 0.2 mg/L kinetin, 250 mg/L casein hydrolizate and 20 g sucrose. Then, they were placed in a rotary shaker (100 rpm). The analyses were replicated three times.

Cadmium chloride ($CdCl_2$) treatment

At day 7, cell cultures were supplied with 1.0 and 1.5 mM $CdCl_2$ dissolved in water. Control treatment did not contain $CdCl_2$. Cells were harvested every 2d by filtration, rapidly washed, weighed and stored until day 6.

Determination of dry cell weight

Growth kinetics were detrmined by obtaining the dry weight of the cell suspensions as g/L. Dry cell weights were determined after drying the biomass for 48 hours at 75°C.

Determination of phenolic compounds

Phenolic compound extraction were carried out as previously described by Caponio et al. [44]. Total phenolic, total flavanols and total flavonol contents of the samples were determined spectrophotometrically using a PG Instruments T70 Plus Dual Beam Spectrophotometer (Arlington, MA, USA). Total phenolic contents were determined according to the Folin-Ciocalteu colorimetric method [45], calibrating against gallic acid standards and expressing the results as mg gallic acid equivalents (GAE) (mg/100 g). Total flavanol contents were determined according to the Arnous et al. [46], calibrating against catechin standards and expressing the results as mg catechin equivalents (CE) (mg/100 g). Total flavonol contents were determined according to Dai et al. [47], calibrating against rutin standards and expressing the results as mg rutin equivalents (RE) (mg/100 g). Data presented are average of three measurements.

HPLC analyses were performed on the HPLC system, Shimadzu Corp., Kyoto, Japan. The HPLC system was equipped with a pump (LC 10 AD), auto-sampler (SIL 10 AD), column oven (CTO 10A) and diode-array UV/VIS detector (DAD-λmax = 278). The separation was executed on a Agilent Eclipse XB C-18 (5 μm, 4.6 × 250 mm, Wallborn, Germany). The mobile phase was composed of acetic acid (2%) and methanol with the gradient elution system at a flow rate of 0.8 mL/min. For gradient elution, mobile phase A contained 3% acetic acid in water; solvent B contained methanol. The following gradient was used: 0–3 min, from 100% A to 95% A, 5% B; 3–20 min, from 95% A, 5% B to 80% A, 20% B; 20–30 min, from 80% A, 20% B to 75% A, 25% B; 30–40 min, from 75% A, 25% B to 70% A, 30% B; 40–50 min 70% A, 30% B to 60% A, 40% B; 50–55 min, 60% A, 40% B to 50% A, 50% B; 55–65 min, 50% A, 50% B to 100% B. The injection volume was 20 μL. Samples, standard solution of *trans*-resveratrol and mobile phases were filtered by a 0.45 μm pore size membrane filter (Millipore Co. Bedford, MA). The detection UV wavelength was set at 280 nm. The column temperature was set at 30°C. The compounds were quantified using Shimadzu CLASS-VP software. The contents of *trans*-resveratrol were determined on HPLC and expressing the results as μg/100 g. Data presented are average of three measurements.

Determination of tocopherols

The extraction of tocopherols (α, β, γ and δ-tocopherol) were carried out as previously described by Caretto et al. [2]. Briefly, the method consisted of an alkaline hydrolysis (potassium hydroxide 60%) followed by extraction with *n*-hexane-ethyl acetate (9:1). Chromatography separation was performed by using a Beckman HPLC Analytical System. Luna Silica (250 × 4.6 mm) 5 μ column was used with heptane/tetrahydrofuran (95:5) as the mobile phase. RF-10AXL Floresan dedector was used to determine tocopherols. The tocopherol content was calculated as μg/100 g Fresh Cell Weight (FCW). Each experiment was carried out with at least three replicates.

Statistical analysis

Data were subjected to analysis of variance with mean separation by Duncan's multiple range test. Differences were considered statistically significant at the $p \leq 0.05$ levels. Statistical analysis was performed using SPSS 16.0 for Windows.

Competing interests

The authors declare that they have no competing interests.

Authors' contributions

All authors contributed the manuscript at all stages-eg, literature search, laboratory analysis and interpretation of data, writing etc. All authors read and approved the final manuscript.

Author details

[1]Department of Horticulture, Faculty of Agriculture and Natural Science, Bozok University, 66200 Yozgat, Turkey. [2]Fruit Research Station, Republic of Turkey Ministry of Food, Agriculture and Livestock, Egirdir, Isparta, Turkey. [3]Department of Horticulture, Faculty of Agriculture, Suleyman Demirel University, Isparta, Turkey. [4]Department of Agricultural Biotechnology, Faculty of Agriculture, Suleyman Demirel University, Isparta, Turkey.

References

1. Dörnenburg H, Knorr D: **Challenges and opportunities for metabolite production from plant cell and tissue cultures.** *Food Technol* 1997, **51**:47–53.
2. Caretto S, Speth B, Fachechi C, Gala R, Zacheo G, Giovinazzo G: **Enhancement of vitamin E production in sunflower cell cultures.** *Plant Cell Rep* 2004, **23**:174–179.
3. Rakwal R, Tamogami S, Kodama O: **Role of jasmonic acid as a signaling molecule in copper chloride elicited rice phytoalexin production.** *Biosci Biotech Bioch* 1996, **60**:1046–1048.
4. Qu JG, Zhang W, Jin MF, Yu XJ: **Effect of homogeneity on cell growth and anthocyanin biosynthesis in suspension cultures of *Vitis vinifera*.** *Chinese J Biotechnol* 2006, **22**:805–810.
5. Misawa M: **Plant tissue culture: an alternative for production of useful metabolites.** *FAO Agric Serv Bull* 1994, **108**:89.
6. Waffo Teguo P, Hawthorne ME, Cuendet M, Merillon JM, Kinghorn AD, Pezzuto JM, Mehta RG: **Potential cancer-chemopreventive activities of wine stilbenoid and flavans extracted from grape (*Vitis vinifera*) cell cultures.** *Nutr Cancer* 2001, **40**:173–179.
7. Brown JP: **A review of the genetic effects of naturally occurring flavonoids, anthraquinones and related compounds.** *Mutat Res* 1980, **75**:243–277.
8. Stich HF, Ohshima H, Pignatelli B, Michelon J, Bartsch H: **Inhibitory effect of betel nut extracts on endogenous nitrosation in humans.** *J Natl Cancer I* 1983, **70**:1047–1050.
9. Steel VE, Kelloff GJ, Balentine D, Boone CW, Mehta R, Bagheri D, Sigman CC, Zhu S, Sharma SVE: **Comparative chemopreventive mechanisms of green tea, black tea and selected polyphenol extracts measured by in vitro bioassays.** *Carcinogenesis* 2000, **21**:63–67.
10. Olas B, Wachowicz B, Saluk Juszczak J, Zielinski T: **Effect of resveratrol, a natural polyphenolic compound, on platelet activation induced by endotoxin or thrombin.** *Thromb Res* 2002, **107**:141–145.
11. Hsieh TC, Wu JM: **Resveratrol: Biological and pharmaceutical properties as anticancer molecule.** *Biofactors* 2010, **36**:360–369.
12. Wu JM: **Resveratrol alleviates some cardiac dysfunction indexes in an SHR model of essential hypertension.** *Am J Hypertens* 2010, **23**:115.
13. Baur JA, Sinclair DA: **Therapeutic potential of resveratrol: The *in vivo* evidence.** *Nat Rev Drug Discov* 2006, **5**(6):493–506.
14. Tomeo AC, Geller M, Watkins TR: **Antioxidant effects of tocotrienols in patients with hyperlipidemia and carotid stress.** *Lipids* 1995, **30**:1179–1183.
15. Diplock AT: **The role of antioxidant nutritions in disease: Healt and nutrition.** *Inform* 1992, **3**:1214–1217.
16. Kamal Eldin A, Appelqvist LA: **The chemistry and antioxidant properties of tocopherols and tocotrienols.** *Lipids* 1996, **31**:671–701.
17. Fachechi C, Nisi R, Gala R, Leone A, Caretto S: **Tocopherol biosynthesis is enhanced in photomixotrophic sunflower cell cultures.** *Plant Cell Rep* 2007, **26**:525–530.
18. Göktürk Baydar N: **Organic acids, tocopherols and phenolic compositions of some Turkish grape cultivars.** *Chem Nat Compd* 2006, **42**:156–159.
19. Göktürk Baydar N, Özkan G: **Tocopherol contents of some Turkish wine by-products.** *Eur Food Res Technol* 2001, **223**:290–293.
20. Oomah DB, Liang J, Godfrey D, Mazza G: **Microwave heating of grape seed: Effect on oil quality.** *J Agric Food Chem* 1998, **46**:4017–4021.

21. Göktürk Baydar N, Akkurt M: **Oil content and oil quality properties of some grape seeds.** *Turk J Agric For* 2001, **25**:163–168.
22. Gliszcynska Swiglo A, Sikorska E: **Simple reversed phase liquid chromatography method for determination of tocopherols in edible plant oils.** *J Chromatogr A* 2004, **1048**:195–198.
23. Sanitá-diToppi L, Gabbrielli R: **Response to cadmium in higher plants.** *Environ Exp Bot* 1999, **41**:105–130.
24. Popova LP, Maslenkova LT, Yordanova RY, Ivanova AP, Krantev AP, Szalai G, Janda T: **Exogenous treatment with salicylic acid attenuates cadmium toxicity in pea seedlings.** *Plant Physiol Bioch* 2009, **47**:224–231.
25. Xu J, Yin HX, Li X: **Protective effects of proline against cadmium toxicity in micropropagated hyperaccumulator, *Solanum nigrum* L.** *Plant Cell Rep* 2009, **28**:325–333.
26. Kuthanova A, Gemperlova L, Zelenkova S, Eder J, Machackovai Opatrny Z, Cvikrova M: **Cytological changes and alterations in polyamine contents induced by cadmium in tobacco BY-2 cells.** *Plant Physiol Biochem* 2004, **42**:149–156.
27. Fojtova M, Kovarik A: **Genotoxic effect of cadmium is associated with apoptotic changes in tobacco cells.** *Plant Cell Environ* 2000, **23**(5):531–537.
28. Matern U, Grimmig B: **Natural Phenols as Stress Metabolites. Natural Phenols in Plant Resistance.** In *Acta Hort*, Volume 381. Edited by Geibel M, Treutter D, Feucht W. ; 1994:448–462.
29. Dangl JL, Dietrich RA, Thomas H: **Senescence and Programmed Cell Death.** In *Biochemistry and Molecular Biology of Plants.* Edited by Buchanan B, Gruissem W, Jones R. Rockville, MD: American Society of Plant Physiologists; 2000:1044–1100.
30. Saltveit ME: **Wound induced changes in phenolic metabolism and tissue browning are altered by heat shock.** *Postharvest Biol Technol* 2000, **21**:61–69.
31. Cisneros Zevallos L: **The use of controlled postharvest abiotic stresses as a tool for enhancing the nutraceutical content and adding-value of fresh fruits and vegetables.** *J Food Sci* 2003, **68**:1560–1565.
32. Kacperska A: **Water Potential Alteration: A Prerequisite or a Triggering Stimulus for the Development of Freezing Tolerance in Overwintering Herbaceous Plants.** In *Advances in Plant Cold Hardiness.* Edited by Li PH, Christerson L. Boca Raton: CRC Press; 1993:73–91.
33. Bais HP, Walker TS, Schweizer HP, Vivanco JM: **Root specific elicitation and antimicrobial activity of rosmarinic acid in hairy root cultures of *Ocimum basilicum*.** *Plant Physiol Biochem* 2002, **40**:983–995.
34. Li W, Koike K, Asada Y, Yoshikawa T, Nikaido T: **Rosmarinic acid production by *Coleus forskohlii* hairy root cultures.** *Plant Cell Tiss Org* 2005, **80**:151–155.
35. Leita L, Contin M, Maggioni A: **Distribution of Cd and induced Cd-binding proteins in roots, stems and leaves of *Phaseolus vulgaris*.** *Plant Sci* 1991, **77**:139–147.
36. Wang H, Zhao SC, Liu RC, Zhou W, Jin JY: **Changes of photosynthetic activities of maize (Zea mays L.) seedlings in response to cadmium stress.** *Photosynthetica* 2009, **47**:277–283.
37. Bergmann H, Machelett B, Lippmann B, Friedrich Y: **Influnce of heavy metals on the accumulation of trimethylglycine, putrescine and spermine in food plants.** *Amino Acids* 2001, **20**:325–329.
38. Ferreira RR, Fornaizer RF, Vitöria AP, Lea PJ, Azevedo RA: **Changes in antioxidant enzyme activities in soybean under cadmium stress.** *J Plant Nutr* 2002, **25**:327–342.
39. Pereira GJG, Molina SMG, Lea PJ, Azevedo RA: **Activity of antioxidant enzymes in response to cadmium in *Crotalaria juncea*.** *Plant Soil* 2002, **239**:123–132.
40. Fornazier RF, Ferreira RR, Pereira GJG, Molina SMG, John Smith R, Lea PJ, Azevedo RA: **Cadmium stres in sugar cane callus cultures: Effect on antioxidant ezymes.** *Plant Cell Tiss Org* 2002, **71**:25–131.
41. Gamborg OL, Miller RA, Okajima K: **Nutrient requirements of suspension cultures of soybean root cells.** *Exp Cell Res* 1968, **50**:151–156.
42. Murashige T, Skoog F: **A revised medium for rapid growth and bioassays with tobacco tissue cultures.** *Physiol Plantarum* 1962, **15**:472–497.
43. Morel G: **Le probleme de la transformation tumorale chez les végétaux.** *Physiol Veg* 1970, **8**:189–191.
44. Caponio F, Alloggio V, Gomes T: **Phenolic compounds of virgin olive oil: Influence of paste preparation techniques.** *Food Chem* 1999, **64**:203–209.
45. Singleton VL, Rossi JR: **Colorimetry of total phenolics with phospho molybdic phosphotungstic acid.** *Am J Enol Viticult* 1965, **16**:144–158.
46. Arnous A, Makris DP, Kefalas P: **Effect of principal polyphenolic components in relation to antioxidant characteristics of aged red wines.** *J Agr Food Chem* 2001, **49**(12):5736–5742.
47. Dai GH, Andary C, Mondolot L, Boubals D: **Involment of phenolic compounds in the resistance of grapevine callus to downy mildew (*Plasmopara viticola*).** *Eur J Plant Pathol* 1995, **101**:541–547.

5

Effects of different water management options and fertilizer supply on photosynthesis, fluorescence parameters and water use efficiency of *Prunella vulgaris* seedlings

Yuhang Chen[1,2], Li Liu[1], Qiaosheng Guo[1*], Zaibiao Zhu[1] and Lixia Zhang[1]

Abstract

Background: *Prunella vulgaris* L. is a medical plant cultivated in sloping, sun-shaded areas in China. Recently, owing to air-environmental stress, especially drought stress strongly inhibits plant growth and development, the appropriate fertilizer supply can alleviate these effects. However, these is little information about their effects on *P. vulgaris* growing in arid and semi-arid areas with limited water and fertilizer supply.

Results: In this study, water stress decreased the photosynthetic pigment contents, inhibited photosynthetic efficiency, induced photodamage in photosystem 2 (PS2), and decreased leaf instantaneous WUE (WUEi). The decreased net photosynthetic rate (Pn) under medium drought stress compared with the control might result from stomatal limitations. However, fertilizer supply improved photosynthetic capacity by increasing the photosynthetic pigment contents and enhancing photosynthetic efficiency under water deficit. Moreover, medium fertilization also increased WUEi under the two water conditions, but fertilizer supply did little to alleviate the PS2 photodamage caused by drought stress. Hence, drought stress was the primary limitation in the photosynthetic process of *P. vulgaris* seedlings, while the photosynthetic characteristics of the seedlings exhibited positive responses to fertilizer supply.

Conclusions: Appropriate fertilizer supply is recommended to improve photosynthetic efficiency, enhance WUEi and alleviate photodamage under drought stress.

Keywords: *Prunella vulgaris* L, N, P and K fertilizer, Drought, Photosynthesis, Water use efficiency

Background

Prunella vulgaris L. (Labiatae), also known as "self-heal," is a popular ingredient in the preparation of traditional Chinese medicine [1]. The dried spica of *P. vulgaris*, i.e., Prunellae Spica, is traditionally used as herbal medicine to alleviate fever, reduce sore throats, and accelerate wound healing [2, 3]. The spicae have been shown to possess antioxidant, anticancer, anti-lipid peroxidation, anti-inflammatory, anti-hyperglycemia and hepatoprotective activities [2]. In traditional Chinese medicine, the air-dried plants are widely used to prepare functional tea, and the leaves are also consumed as medicinal vegetables [4, 5].

Due to its medicinal and industrial importance, the demand for *P. vulgaris* has increased steadily in recent years [6]. The wild population of *P. vulgaris* cannot meet this growing need, and therefore it was proposed in the 1990s that *P. vulgaris* be cultivated to allow for more efficient resource utilization in China [1]. *P. vulgaris* was originally classified as a moderate shade species, especially during the seedling stage. Strong irradiation at midday usually induces severe photoinhibition and photo-oxidative damage of the photosynthetic apparatus of *P. vulgaris* leaves [7]. A number of environmental stresses, including drought and malnutrition, may increase *P. vulgaris* plant sensitivity to photoinhibition and photodamage, inducing

*Correspondence: gqs@njau.edu.cn
[1] Institute of Chinese Medicinal Materials, Nanjing Agricultural University, Nanjing 210095, People's Republic of China
Full list of author information is available at the end of the article

cellular damage and thus decreasing their productivity [1, 4, 5].

Prunella vulgaris plants require moderate levels of nutrients and are sensitive to drought [1, 6]. The growth and yield of *P. vulgaris* are restricted by water and nutrient deficiencies because most *P. vulgaris* plantations are located in mountainous areas in China. Soil water deficit in the dry season is one of the most important limitations to photosynthesis and consequently, *P. vulgaris* productivity [1, 6]. However, there are numerous well-documented photosynthetic responses of plants to N, P, and K fertilization, which include significant and positive correlations between photosynthetic capacity and leaf N, P, and K content, suggesting that a large proportion of these elements is used for the synthesis of various components in the photosynthetic apparatus [6, 8]. Furthermore, fertilization (e.g., N, P, K) frequently increases cell wall rigidity and osmotic adjustment [6, 8]. Increased fertilization might improve the photosynthetic capacity or stomatal control under water and nutrient deficit conditions.

Under drought stress, disturbances in photosynthesis at the molecular level are connected with low electron transport through photosystem 2 (PS2) and/or with structural injuries of PS2 and the light-harvesting complexes [9, 10]. Restricted CO_2 may lead to increased susceptibility to photodamage (due to stomatal closure) and, subsequently, to photoinhibition [11]. Photoinhibition is characterized by parallel decreases in Pn and quantum yield of PS2 (Φ PS2) and is accompanied by a decline in the maximum quantum yield of photosynthesis (Fv/Fm) associated with loss of PS2 activity [11, 12] and an increase in minimal Chl fluorescence (F0) [13]. Chl fluorescence is a useful tool for quantifying the effect of abiotic stress on photosynthesis [12]. Fertilizer could increase Fv/Fm and the effective quantum yield of photochemical energy conservation in PS2 (Fv′/Fm′) [14]. The Chl content and photosynthetic rates might also be enhanced through the supply of N, P, and K [6, 8]. These authors suggested that fertilization might alleviate photoinhibition and photodamage caused by drought stress.

The interactive effects of nutrition and water availability on *P. vulgaris* growth and the production of secondary metabolites have been well documented [1, 6, 7]. However, the physiological and biochemical characteristics of *P. vulgaris* plants under drought and nutrient-limited conditions have been less thoroughly studied. The objectives of this study were to (1) determine the photosynthetic adaptation of *P. vulgaris* seedlings to various water supply and fertilization conditions, and (2) determine whether fertilization could improve the photosynthetic capacity of seedlings under dry conditions.

Results

Diurnal variation of environmental factors

On the measurement day, the PAR increased steeply from 09:00 to 11:00, remained at high levels until 15:00, and then decreased sharply (Fig. 1a). Under the impacts of PAR diurnal variation, Ca (Fig. 1a) and RH (Fig. 1b) were at high levels in the early morning, followed by a sharp decrease, remaining at relatively low levels during the midday period, and then began to increase from 15:00. In contrast, Ta exhibited a diurnal trend similar to PAR (Fig. 1b).

The contents of the photosynthetic pigments exhibited significant ($P < 0.05$) responses to water stress and fertilization supply, and the interaction between water and fertilization affected Chl*a* and *b*, Chl*a* + *b* and Car (Table 1). Water stress decreased the contents of Chl*a* and *b*, Chl*a* + *b* and Car. On the other hand, significantly higher values were measured in Fl compared with the other fertilization treatments (F0 and F2) under well-watered and drought stress conditions.

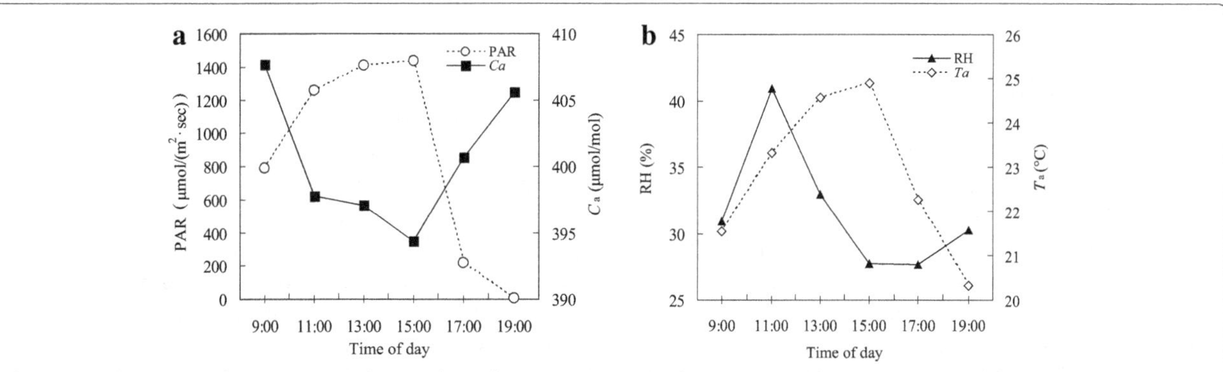

Fig. 1 Diurnal variations of environmental photosynthetically active radiation (PAR) and ambient CO_2 concentration (Ca) (**a**); air temperature (Ta) and air relative humidity (RH) (**b**) from 09:00 to 19:00 on the measuring day

Table 1 Chlorophyll (Chl) a and b contents, Chl *a* + *b* content, carotenoid (Car) content of *P. vulgaris* seedlings under different water (W) and fertilization (F) supply regimes

Treatment	Chl*a* [g kg^{-1} (FM)]	Chl*b* [g kg^{-1} (FM)]	Chl*a* + *b* [g kg^{-1} (FM)]	Car [g kg^{-1} (FM)]
W1/F0	0.40 ± 0.06 d	0.16 ± 0.03 d	0.57 ± 0.10 d	0.10 ± 0.01 d
W1/F1	0.86 ± 0.09 b	0.34 ± 0.04 b	1.19 ± 0.13 b	0.20 ± 0.02 b
W1/F2	0.61 ± 0.10 c	0.25 ± 0.04 c	0.86 ± 0.14 c	0.15 ± 0.02 c
W2/F0	0.42 ± 0.01 d	0.17 ± 0.00 d	0.58 ± 0.02 d	0.12 ± 0.00 d
W2/F1	1.03 ± 0.10 a	0.41 ± 0.03 a	1.45 ± 0.13 a	0.24 ± 0.02 a
W2/F2	1.00 ± 0.04 a	0.40 ± 0.02 a	1.40 ± 0.06 a	0.21 ± 0.01 b
Water (W)	43.68**	43.83**	44.24**	28.78**
Fertilization (F)	120.30**	116.13**	120.52**	89.46**
W × F	14.24**	12.91**	14.01**	3.51

W1 and W2 correspond to soil water contents between 45–50 and 70–75 % of the field water capacity, respectively; F0: no fertilization, F1: 0.12 g N + 0.2 g P_2O_5 + 0.1 g K_2O kg^{-1} soil, F2: 0.24 g N + 0.4 g P_2O_5 + 0.2 g K_2O kg^{-1} soil. Different letters indicate significant differences between treatments at $P < 0.05$ (ANOVA)

Mean ± SD, fresh mass (FM). n = 6, *$P < 0.05$, **$P < 0.01$

Water supply and fertilization effects on the diurnal variation of photosynthetic parameters

Diurnal variation of Tr

Transpirational water loss was compensated for at dusk every day. Hence, based on the diurnal variation of the main environmental factors affecting transpiration (Fig. 1), under drought stress, leaf Tr was maintained at a relatively high level from 09:00 to 15:00, followed by a significant and continuous decrease, reaching the lowest value of the day at 19:00 (Fig. 2a1). Under well-watered conditions, leaf Tr was maintained at a relatively high level from 09:00 to 15:00, followed by a significant and continuous decrease, reaching the lowest value of the day at 19:00 (Fig. 2a2). Irrespective of fertilizer supply, Tr decreased with decreasing soil water availability at each measurement point during the day. For a particular water content, fertilizer supply significantly affected Tr (Table 2).

Diurnal variation of Gs

Stomata are the main portals for carbon dioxide (CO_2) and vapor water exchange between plant leaves and the atmosphere; thus, Gs directly controls photosynthesis and transpiration. The diurnal variation of Gs under all water and fertilization treatments showed similar trends (Fig. 2b). At 09:00, Gs was maintained at relatively high levels due to compensation for transpirational water loss at dusk the day before. Then, Gs decreased significantly until dusk. Irrespective of the fertilization level, Gs decreased with decreasing soil water content. At comparable soil water contents, the Gs of plants receiving the moderate level of fertilization was consistently higher than that of the high fertilization plants and those plants that did not receive fertilizer. In the afternoon, the effect of fertilizer supply on Gs under non-limiting water conditions and medium drought was statistically significant (Table 2).

Diurnal variation of Ci

In general, Ci is dependent on Gs and the ability of the mesophyll cells to assimilate intracellular CO_2. At 09:00, Ci was high (Fig. 2c), consistent with high Ca, and associated with low PAR and high Gs. During the period from mid-morning to early evening (09:00–17:00), Ci reached a constant low value due to high PAR, depletion of Ca in the plant canopy, and high Gs, which facilitated Ca depletion (Fig. 1a), followed by a return to the early morning levels as the light and Gs decreased. Increasing soil water content did not change the trend in diurnal Ci variation, leading to only a weak increase in Ci under both moderate and high fertilizer supply in the late afternoon (Fig. 2c1, c2). Compared with the drought stress treatments (Fig. 2c1), the Ci of well-watered treatments (Fig. 2c2) was higher during the middle of the day but significantly lower in the late afternoon under moderate fertilizer supply (Table 2).

Diurnal variation of Ls

The diurnal changes in Ls displayed a similar pattern under the well-watered and drought conditions (Fig. 3a). Ls increased with an increase in water stress, and moderate fertilization also increased Ls under drought stress and well-watered conditions. Compared with the well-watered treatment (Fig. 3a2), the high fertilization treatment and the control (Fig. 3a1) reduced stomatal limitations during the middle of the day and in the early morning in the drought stress treatments. This suggests that moderate fertilizer supply modified the stomatal limitation of CO_2 diffusion for most of the day, especially in the seedlings grown under drought stress conditions (Table 2).

Diurnal variation of Pn

Diurnal variations of Pn are shown in Fig. 3b. Overall, the pattern of Pn mirrored that of Gs. Pn and Gs maxima occurred at 09:00 as PAR and Gs increased, even though Ci was reduced. Subsequently, Pn steadily declined in all water and fertilization treatments until a minimum value was reached in the early evening (19:00), similar to Gs. Hence, Gs was the main factor limiting the photosynthesis of mesophyll cells at this time of day. At 19:00, Pn

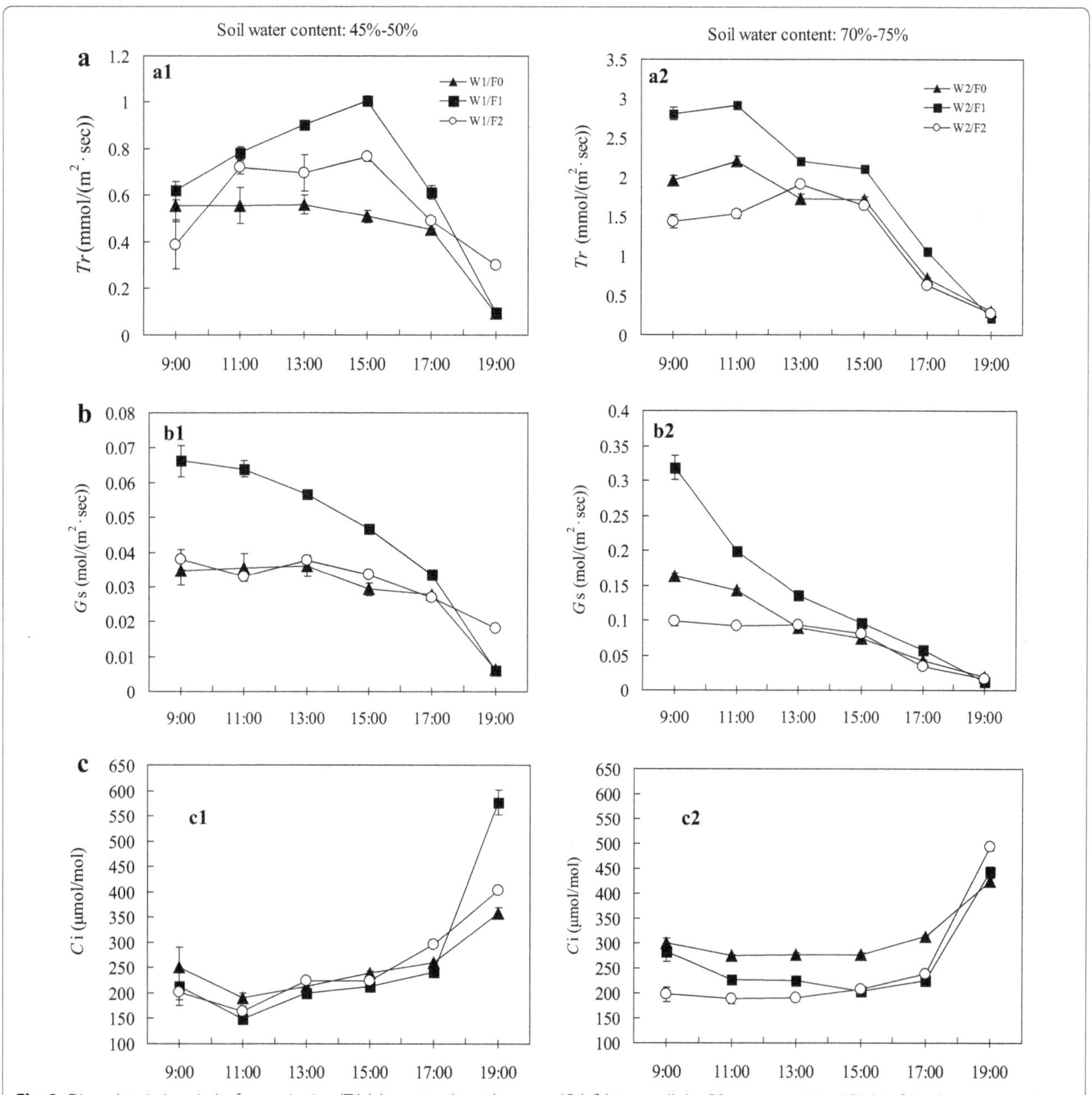

Fig. 2 Diurnal variations in leaf transpiration (Tr) (**a**), stomatal conductance (Gs) (**b**), intercellular CO_2 concentration (Ci) (**c**) of *P. vulgaris* exposed to two soil water (W) conditions and three levels of fertilizer (F), n = 6–8. F0: no fertilization, F1: 0.12 g N + 0.2 g P_2O_5 + 0.1 g K_2O kg^{-1} soil, F2: 0.24 g N + 0.4 g P_2O_5 + 0.2 g K_2O kg^{-1} soil

was at the lowest level of the day due to low PAR and Gs, despite the high Ci.

Water and fertilization did not affect the trend in the diurnal variation of Pn (Fig. 3b). Independent of soil N + P + K, increasing soil water availability increased Pn. Under all soil water contents, the Pn of the moderate fertilization treatment was higher than that of the high fertilization treatment and the control. The fertilization effects under non-limiting water conditions and medium drought were significantly different at each measurement point (Table 2).

Diurnal variation of WUEi

Similar to the trend in the diurnal variation of Pn, WUEi (Fig. 3c) reached a maximum level at 09:00 due to high Pn and PAR. Throughout the remainder of the day, WUEi declined as Ta increased and RH (Fig. 1b) and Pn decreased, while Tr remained at relatively high levels

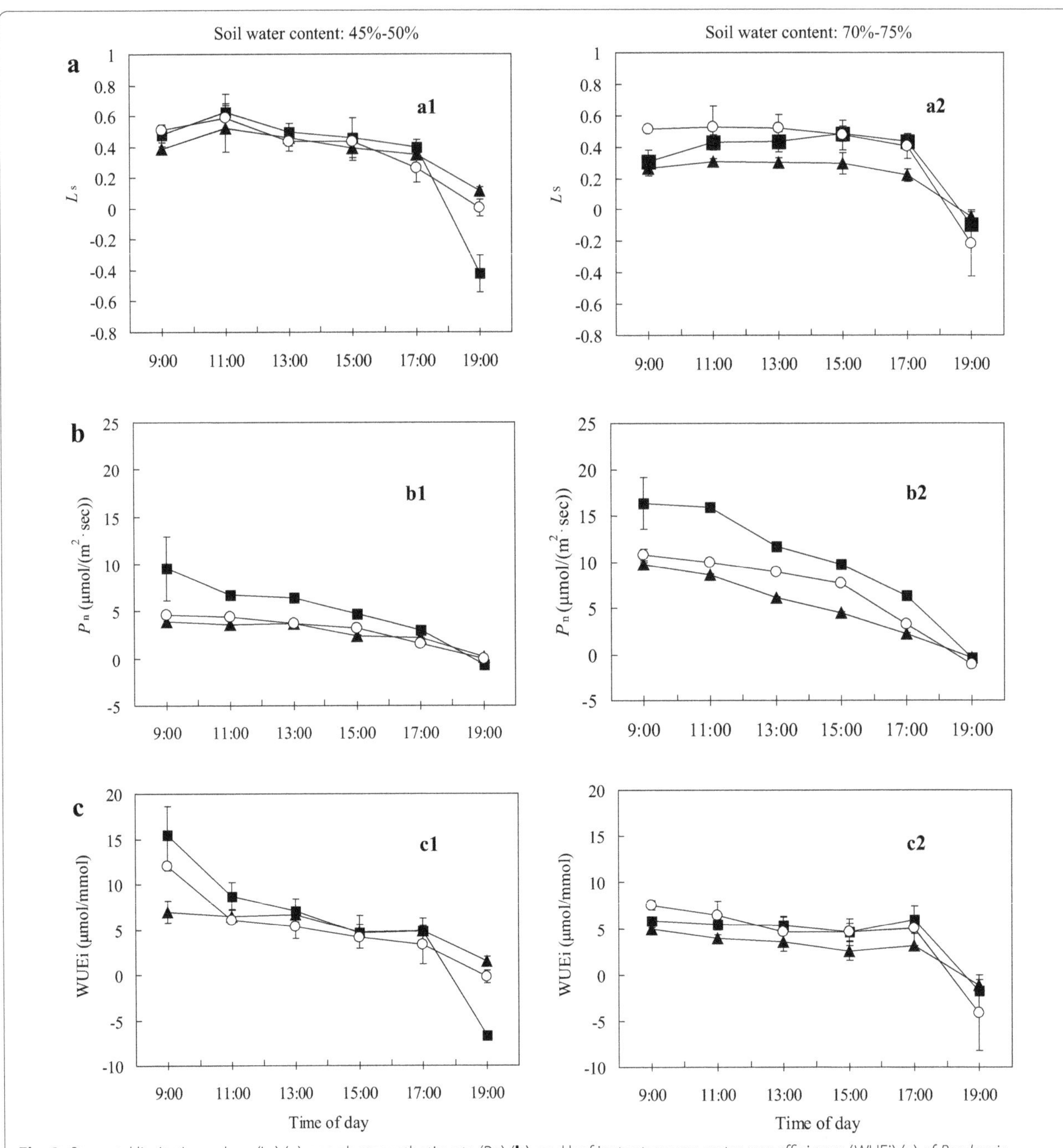

Fig. 3 Stomatal limitation values (Ls) (**a**), net photosynthetic rate (Pn) (**b**), and leaf instantaneous water use efficiency (WUEi) (**c**) of *P. vulgaris* exposed to two soil water (W) conditions and three levels of fertilizer (F), n = 6–8. F0: no fertilization, F1: 0.12 g N + 0.2 g P_2O_5 + 0.1 g K_2O kg^{-1} soil, F2: 0.24 g N + 0.4 g P_2O_5 + 0.2 g K_2O kg^{-1} soil

(Fig. 2a). A decrease in the soil water content led to an increase in the WUEi when the plants were grown under moderate fertilizer levels, compared with a decrease in the WUEi when the plants were grown under high fertilization (Fig. 3c1). A comparison of the control seedlings and those exposed to high fertilization under drought conditions indicated that an increase in fertilization enhanced the WUEi of *P. vulgaris* seedlings (Table 2).

Chl fluorescence

Fertilization significantly affected the Chl fluorescence parameters (Table 3). In addition, water stress

Table 2 Statistical tests of the effects of three fertilization levels on Tr, Gs, Ci, Ls, Pn, and WUEi of *P. vulgaris* seedlings at various measuring times and soil water availabilities

Treatment %	09:00	11:00	13:00	15:00	17:00	19:00
Effects of fertilization supply on Tr						
45–50	–	–	*	*	*	*
70–75	*	*	–	*	–	–
Effects of fertilization supply on Gs						
45–50 %	*	*	*	*	–	*
70–75 %	*	*	–	–	*	*
Effects of fertilization supply on Ci						
45–50	*	–	–	*	*	*
70–75	*	*	*	*	*	–
Effects of fertilization supply on Ls						
45–50	*	–	–	*	*	*
70–75	*	*	*	*	*	–
Effects of fertilization supply on Pn						
45–50	*	*	*	*	*	*
70–75	*	*	*	*	*	–
Effects of fertilization supply on WUEi						
45–50	*	*	*	–	–	*
70–75	*	*	–	*	*	–

The seedlings were exposed to two soil water conditions (non-limiting soil water content and medium drought, corresponding to soil water contents between 70–75 % and 45–50 % of the field water capacity, respectively) and three levels of fertilization (F0: no fertilizer, F1: 0.12 g N + 0.2 g P_2O_5 + 0.1 g K_2O kg^{-1} soil, F2: 0.24 g N + 0.4 g P_2O_5 + 0.2 g K_2O kg^{-1} soil; n = 6 for each treatment)

* Significant differences between fertilization treatments at a specific time of day for a particular soil water content (ANOVA, $P < 0.05$), – indicates no significant difference

significantly affected the following Chl fluorescence parameters: ΦPS2, qP, qN, Fv′/Fm′ and ETR, although qP and ETR showed significant responses to the interaction of water and fertilization. Water stress decreased Fv/Fm, ΦPS2 and ETR, while Fl increased these parameters under drought and well-watered conditions. qP was slightly decreased with an increase in drought stress, whereas qN was increased. Fv′/Fm′ decreased under drought stress but responded somewhat positively to Fl.

Discussion

As expected, the photosynthetic characteristics exhibited strong responses to drought stress and fertilization, which is in agreement with many previous studies [10, 15, 16]. Nevertheless, we found that drought stress seemed to play the primary limiting role in photosynthetic capacity, which was improved by fertilization, but the decreasing tendency could not be altered.

Lower photosynthetic performance of *P. vulgaris* seedlings may be associated with decreasing Chl and Car contents under water stress. However, the highest photosynthetic pigment contents were measured in the moderate fertilization treatment under water stress, which implied that moderate fertilization could alleviate the damages caused by water stress and improve photosynthetic performance under water deficit. Similar results have been observed in *Sophora davidii* seedlings [10] and *Doritaenopsis* seedlings [17].

Both drought stress and fertilization slightly affected the diurnal fluctuation patterns of *P. vulgaris* seedlings, which are strongly related to the biological rhythm of the plant [10, 16]. Most studies have reported positive effects of fertilizer supply on plant photosynthesis and WUE [16, 18], although negative effects on WUE in response to fertilizer supply have been observed in some experiments [18, 19]. In this study, an increase in soil nutrient availability did not alter the trend in the diurnal variations of the photosynthetic parameters measured in response to soil water availability. However, compared with the plants grown under drought stress conditions, fertilization generally increased Gs and Ci, diminished Ls, and enhanced Pn. Fertilization enhanced WUEi under non-limiting water and medium drought conditions, which was in agreement with the results of previous study [20] and the general theory that the supply of a limited resource can enhance the use efficiency of other resources. Generally, an increase in soil water availability is more effective than an increase in nutrient availability in improving the growth of *P. vulgaris* seedlings. Although differences between pot studies and field experimental conditions

Table 3 The maximum quantum yield of photosystem 2 (PS2) photochemistry (Fv/Fm), effective quantum yield of PS2 (ΦPS2), photochemical quenching (qP), non-photochemical quenching (qN), effective quantum yield of photochemical energy conservation in PS2 (Fv′/Fm′) and electron transport rate (ETR) of *P. vulgaris* seedlings under different water (W) and fertilization (F) regimes

Treatment	Fv/Fm	ΦPS2	QP	QN	Fv′/Fm′	ETR
W1/F0	0.80 ± 0.03 b	0.18 ± 0.02 c	0.46 ± 0.06 c	1.82 ± 0.06 ab	0.39 ± 0.01 e	90.45 ± 4.26 c
W1/F1	0.84 ± 0.01 a	0.26 ± 0.01 b	0.62 ± 0.02 ab	1.91 ± 0.04 a	0.43 ± 0.01 cd	131.12 ± 3.74 b
W1/F2	0.82 ± 0.01 a	0.24 ± 0.04 b	0.58 ± 0.05 b	1.91 ± 0.14 a	0.41 ± 0.03 de	114.19 ± 20.03 b
W2/F0	0.80 ± 0.01 b	0.26 ± 0.01 b	0.58 ± 0.01 b	1.64 ± 0.03 c	0.45 ± 0.02 bc	92.56 ± 9.95 c
W2/F1	0.84 ± 0.01 a	0.32 ± 0.01 a	0.68 ± 0.00 a	1.75 ± 0.03 bc	0.48 ± 0.01 a	168.02 ± 7.14 a
W2/F2	0.83 ± 0.01 a	0.30 ± 0.05 a	0.56 ± 0.17 b	1.69 ± 0.07 c	0.45 ± 0.02 ab	150.70 ± 22.53 a
Water (W)	0.69	56.76**	19.77**	69.68**	110.67**	38.14**
Fertilization (F)	29.97**	25.42**	27.43**	6.87**	14.45**	72.94**
W x F	1.33	0.79	4.56*	0.41	0.58	8.00**

W1 and W2 correspond to soil water contents between 45–50 and 70–75 % of the field water capacity, respectively; F0: no fertilization, F1: 0.12 g N + 0.2 g P_2O_5 + 0.1 g K_2O kg^{-1} soil, F2: 0.24 g N + 0.4 g P_2O_5 + 0.2 g K_2O kg^{-1} soil. Different letters indicate significant differences between treatments at $P < 0.05$ (ANOVA)

Mean ± SD, n = 5, *$P < 0.05$, **$P < 0.01$

often limit the practical application of pot experimental results, this study provides useful data for the management of the early phase of *P. vulgaris* seedlings.

The efficiency and stability of PS2 have been widely monitored through the measurement of Fv/Fm in dark-adapted leaves [14]. In our study, water stress increased Fv/Fm, which implies that drought stress had a greater effect on the energy cycling (Fm) between the reaction center (RC) and the Chl pool compared with the energy absorption rate of the leaves (F0) [21]. Moderate fertilization led to apparent modifications of F0, Fm and Fv/Fm, which might alleviate photoinhibition or other types of PS2 injuries caused by drought stress. This is consistent with the findings in *Sophora davidii* seedlings [10].

For monitoring the efficiency of photochemical processes in PS2 in a light-adapted state, Φ PS2 and Fv′/Fm′ are usually used [22]. In our measurements, both parameters exhibited negative responses to drought stress and positive responses to moderate fertilization under well-watered and drought stress conditions, suggesting that water stress decreased the efficiency of excitation energy capture of open PS2 RCs, whereas fertilization supply might improve this under medium drought stress. Similar results were also observed in *Sophora davidii* seedlings responding to drought stress [10]. However, excess fertilizer supply might strongly aggravate the damage caused by drought stress.

Two basic parameters describe the quenching of maximum variable Chl fluorescence yield during the irradiation induction period: qP and qN. In our study, decreased qP under drought stress suggested that drought stress might damage PS2 RCs, resulting in their closure. Higher qN under drought stress indicated that plants efficiently dissipated the energy trapped in PS2 in the form of heat. This is the photoprotective mechanism under stress [23].

In the present study, the ETR of PS2 decreased under the drought condition, which indicated that the proportion of open reaction centers of PS2 and CO_2 fixation were reduced. The results indicated that photosynthetic electron transport ability was reduced and the dark reaction was blocked, which decreased the photosynthetic rate [24].

Conclusion

In conclusion, drought stress not only decreases the contents of the photosynthetic pigments, the photosynthetic capacity, and the WUEi but also affects the efficiency of PS2 of *P. vulgaris* seedlings. However, the photosynthetic pigments and gas exchange responded positively to fertilization. In addition, fertilization alleviated the degree of photo-inhibition and the injury caused by drought stress by slightly improving Fv/Fm and increasing Fv′/Fm′. Thus, appropriate fertilization is recommended for *P. vulgaris* seedlings to improve photosynthesis inhibited by drought stress and to facilitate seedling establishment under water deficit.

Methods

Plants and their growth

Seeds of *P. vulgaris* were collected in July 2009 in Queshan County, Henan Province, P.R. China. Apparently healthy seeds were air-dried and then stored at ambient laboratory temperature until the experimental pretreatment was initiated in October 2009. Surface soil from an experimental field at Nanjing Agricultural University was used as the growth substrate. The collected soil was combined and thoroughly mixed. Soil (4.0 kg) was placed in each 4.5 L plastic pot. The organic matter content of the soil was 21.32 g kg^{-1}; available N was 34.65 g kg^{-1}; available P was 12.07 g kg^{-1}; and available K was 16.34 g kg^{-1}. The field capacity of the soil was 25 %.

Before sowing, *P. vulgaris* seeds were soaked in 2.5 % sodium hypochlorite solution for 1 h. Twenty seeds of approximately the same size were sown in each pot on 10 October 2009. All pots were moved into a rain shelter located at the Institute of Chinese Medicinal Materials, Nanjing Agricultural University, Nanjing, Jiangsu Province, P.R. China. All pots were well watered to ensure germination. After one month, the seedlings were thinned to four uniform plants per pot.

Experimental design

The experiment was arranged using a randomized design consisting of two water regimens [70–75 and 45–50 % of field water capacity (FWC)] and three fertilizer treatments (N0P0K0 = no fertilization (control), N1P1K1 = 0.12 g N + 0.2 g P_2O_5 + 0.1 g K_2O kg^{-1} soil, N2P2K2 = 0.24 g N + 0.4 g P_2O_5 + 0.2 g K_2O kg^{-1} soil). The N, P_2O_5, and K_2O were applied as urea, superphosphate and potassium sulfate, respectively. One third of the N and all of the P and K were applied basally. The remaining N fertilizer was applied on 5 March 2010, before rapid growth of the plants. Each treatment group had ten replicates. A total of 60 pots were established. All pots were measured gravimetrically by weighing and watered with distilled water every other day at 18:00 pm. On 25 March 2010, the drought treatments were initiated in half of the pots by withholding irrigation; the remaining pots continued to be well watered. The experimental treatments were conducted from 25 March to 15 June (when the plants were harvested), 2010.

Photosynthetic parameters

On 24 April 2010, a cloudless day, the diurnal variation in the leaf net photosynthetic rate (Pn), stomatal conductance (Gs), inter-cellular CO_2 concentration (Ci), transpiration rate (Tr), as well as ambient CO_2 concentration (Ca), air temperature (Ta), air relative humidity (RH) and photosynthetically active radiation (PAR), were measured every two hours from 09:00 to 19:00 using a portable photosynthesis system (LI-6400, Li-Cor, Lincoln, NE, USA). The measurements were conducted on the second fully expanded leaves from 6 individual plants per treatment. The stomatal limitation value (Ls) was calculated using the formula: Ls = 1 – Ci/Ca [20]

Chl fluorescence

Chl fluorescence was determined on fully expanded and exposed leaves (one leaf per plant) using a modulated fluorometer (PAM 2100, Walz, Effeltrich, Germany) on April 25 2014, according to [15]. Initial fluorescence (F0) and maximal fluorescence (Fm) were measured after a 30 min dark adaptation. The intensity of the saturation pulses used to determine the maximal fluorescence emission in the presence (Fm′) and absence (Fm) of quenching was 8000 μmol (photon) m^{-2} s^{-1}, 0.8 s, whereas the "actinic light" was 336 μmol (photon) m^{-2} s^{-1}. Steady-state fluorescence (Fs), basic fluorescence after light induction (F0′), maximal PS2 photochemical efficiency (Fv/Fm), effective quantum yield of PS2 (ΦPS2), and photochemical (qP) and non-photochemical (qN) fluorescence quenching coefficients were also recorded. The effective quantum yield of photochemical energy conservation in PS2 (Fv′/Fm′) was calculated as (Fm′ – Fs)/Fm′ according to previous research [25].

Determination of WUE and photosynthetic pigments

In this experiment under controlled conditions, WUE was studied at the level of the leaf instantaneous ratio of Pn to Tr (WUEi) [16]. After the determination of photosynthetic activity, all leaves were harvested. Fresh leaves (0.1 g) were collected for the determination of Chl content. The leaves were ground in 80 % acetone for the extraction of Chl and carotenoids (Car). The absorbance of the extract was measured at 645 and 663 nm using a UV/visible spectrophotometer (Lambda 25, Perkin Elmer, CT, USA).

Statistical analysis

Significant differences between the water and fertilization treatments (n = 6) at a particular measurement point were analyzed with one-way ANOVA using SPSS 16.0 for Windows (Chicago, USA). The main effects of water and fertilizer availability and their interactions were determined using two-way analysis of variance (ANOVA). The differences were considered significant at $P < 0.05$.

Authors' contributions

QSG planned the study. YHC and LXZ perform the experiments. YHC experimental design and write the manuscript. LL and ZBZ supervised the study. All authors read and approved the final manuscript.

Author details

[1] Institute of Chinese Medicinal Materials, Nanjing Agricultural University, Nanjing 210095, People's Republic of China. [2] College of Pharmaceutical Sciences, Chengdu Medical College, Chengdu 610083, People's Republic of China.

Acknowledgements

This study was funded by the programs of the National Nature Science Foundation of China (Nos. 30772730, 81072986, 31500263 and 81202867), the project funded by the China Postdoctoral Science Foundation (2014M560726), and the project supported by the Scientific Research Fund of Sichuan Provincial Education Department (13ZB0219).

Competing interests

The authors declare that they have no competing interests.

References

1. Chen YH, Yu MM, Zhu ZB, Zhang LX, Guo QS. Optimisation of potassium chloride nutrition for proper growth, physiological development and bioactive component production in *Prunella vulgaris* L. PLoS ONE. 2013;8(7):e66259. doi:10.1371/journal.pone.0066259.
2. Psotová J, Kolář M, Soušek J, Švagera Z, Vičar J. Biological activities of *Prunella vulgaris* extract. Phytother Res. 2003;17(9):1082–7. doi:10.1002/ptr.1324.
3. Psotova J, Svobodova A, Kolarova H, Walterova D. Photoprotective properties of *Prunella vulgaris* and rosmarinic acid on human keratinocytes. J Photochem Photobiol B. 2006;84(3):167–74. doi:10.1016/j.jphotobiol.2006.02.012.
4. Chen YH, Guo QS, Zhu ZB, Zhang LX. Changes in bioactive components related to the harvest time from the spicas of *Prunella vulgaris*. Pharm Biol. 2012;50(9):1118–22. doi:10.3109/13880209.2012.658477.
5. Chen YH, Guo QS, Zhu ZB, Zhang LX, Zhang XM. Variation in concentrations of major bioactive compounds in *Prunella vulgaris* L. related to plant parts and phenological stages. Biol Res. 2012;45(2):181–5. doi:10.4067/S0716-97602012000200009.
6. Chen YH, Guo QS, Liu L, Liao L, Zhu ZB. Influence of fertilization and drought stress on the growth and production of secondary metabolites in *Prunella vulgaris* L. J Med Plants Res. 2011;5(9):1749–55.
7. Zhou LJ, Shi HZ, Guo QS, Han BQ, Xian WY. Effects of light intensity on photosynthetic characteristics and seedling growth of *Prunella vulgaris*. China J Chin Mater Med 3. 2011;36(13):1693–6.
8. Zhu ZB, Yu MM, Chen YH, Guo QS, Zhang LX, Shi HZ, et al. Effects of ammonium to nitrate ratio on growth, nitrogen metabolism, photosynthetic efficiency and bioactive phytochemical production of *Prunella vulgaris*. Pharm Biol. 2014;52(12):1518–25. doi:10.3109/13880209.2014.902081.
9. Hura T, Hura K, Grzesiak M, Rzepka A. Effect of long-term drought stress on leaf gas exchange and fluorescence parameters in C3 and C4 plants. Acta Physiol Plant. 2007;29(2):103–13. doi:10.1007/s11738-006-0013-2.
10. Wu FZ, Bao WK, Li FL, Wu N. Effects of water stress and nitrogen supply on leaf gas exchange and fluorescence parameters of *Sophora davidii* seedlings. Photosynthetica. 2008;46(1):40–8. doi:10.1007/s11099-008-0008-x.
11. Powles SB. Photoinhibition of photosynthesis induced by visible light. Ann Rev Plant Physiol. 1984;35:15–44.
12. Tezara W, Marín O, Rengifo E, Martínez D, Herrera A. Photosynthesis and photoinhibition in two xerophytic shrubs during drought. Photosynthetica. 2005;43(1):37–45. doi:10.1007/s11099-005-7045-5.
13. Osmond CB, Grace SC. Perspectives on photoinhibition and photorespiration in the field: quintessential inefficiencies of the light and dark reactions of photosynthesis? J Exp Bot. 1995;46:1351–62. doi:10.1093/jxb/46.
14. Wang KY, Kellomäki S. Effects of elevated CO_2 and soil-nitrogen supply on chlorophyll fluorescence and gas exchange in scots pine, based on a branch-in-bag experiment. New Phytol. 1997;136(2):277–86. doi:10.1046/j.1469-8137.1997.00744.x.
15. Flexas J, Ribas-Carbó M, Bota J, Galmés J, Henkle M, Martínez-Cañellas S, et al. Decreased rubisco activity during water stress is not induced by decreased relative water content but related to conditions of low stomatal conductance and chloroplast CO_2 concentration. New Phytol. 2006;172(1):73–82. doi:10.1111/j.1469-8137.2006.01794.x.
16. Liu XP, Fan YY, Long JX, Wei RF, Kjelgren R, Gong CM, et al. Effects of soil water and nitrogen availability on photosynthesis and water use efficiency of *Robinia pseudoacacia* seedlings. J Environ Sci (China). 2013;25(3):585–95. doi:10.1016/S1001-0742(12)60081-3.
17. Jeon MW, Ali MB, Hahn EJ, Paek KY. Photosynthetic pigments, morphology and leaf gas exchange during ex vitro acclimatization of micropropagated CAM doritaenopsis plantlets under relative humidity and air temperature. Environ Exp Bot. 2006;55(1–2):183–94. doi:10.1016/j.envexpbot.2004.10.014.
18. Brueck H. Effects of nitrogen supply on water-use efficiency of higher plants. J Plant Nutr Soil Sci. 2008;171(2):210–9. doi:10.1002/jpln.200700080.
19. Górny AG. Garczyński AGS. Genotypic and nutrition-dependent variation in water use efficiency and photosynthetic activity of leaves in winter wheat (*Triticum aestivum* L.). J Appl Genet. 2002;43(2):145–60.
20. Yin CY, Berninger F, Li CY. Photosynthetic responses of *Populus przewalski* subjected to drought stress. Photosynthetica. 2006;44(1):62–8. doi:10.1007/s11099-005-0159-y.
21. Havaux M, Strasser RJ, Greppin H. A theoretical and experimental analysis of the qP and qN coefficients of chlorophyll fluorescence quenching and their relation to photo-chemical and nonphotochemical events. Photosynth Res. 1991;27(1):41–55. doi:10.1007/BF00029975.
22. Roháček K, Barták M. Technique of the modulated chlorophyll fluorescence: basic concepts, useful parameters, and some applications. Photosynthetica. 1999;37(3):339–63. doi:10.1023/A:1007172424619.
23. Bigras FJ. Photosynthetic response of white spruce families to drought stress. New Forest. 2005;29(2):135–48. doi:10.1007/s11056-005-0245-9.
24. Dannehl H, Wietoska H, Heckmann H, Godde D. Changes in D1 protein turnover and recovery of photosystem II activity precede accumulation of chlorophyll in plants after release from mineral stress. Planta. 1996;199(1):34–42. doi:10.1007/BF00196878.
25. Zhang LX, Guo QS, Chang QS, Zhu ZB, Liu L, Chen YH. Chloroplast ultrastructure, photosynthesis and accumulation of secondary metabolites in *Glechoma longituba* in response to irradiance. Photosynthetica. 2015;53(1):144–53. doi:10.1007/s11099-015-0092-7.

6

Compositional studies and Biological activities of *Perovskia abrotanoides* Kar. oils

Sadaf Naz Ashraf[1], Muhammad Zubair[1], Komal Rizwan[1], Rasool Bakhsh Tareen[2], Nasir Rasool[1], Muhammad Zia-Ul-Haq[3] and Sezai Ercisli[4*]

Abstract

Background: Current study has been designed to evaluate the chemical composition of essential and fixed oils from stem and leaves of *Perovskia abrotanoides* and antioxidant and antimicrobial activities of these oils.

Results: GC-MS analysis of essential oil identified 19 compounds with (E)-9-dodecenal being the major component in stem and hexadecanoic acid in leaves. In contrast, GC-MS analysis of fixed oil showed 40 constituents with α-amyrin the major component in stem and α-copaene in leaves. The antioxidant activity showed the highest value of 76.7% in essential oil from leaves in comparison with fixed oil from stem (45.9%) through inhibition of peroxidation in linoleic acid system. The antimicrobial assay tested on different microorganisms (e.g. *E. coli, S. aureus, B. cereus, Nitrospira, S. epidermis, A. niger, A. flavus and C. albicans*) showed the higher inhibition zone at essential oil from leaves (15.2 mm on *B. cereus*) as compared to fixed oil from stem (8.34 mm on *S. aureus*) and leaves (11.2 mm on *S. aureus*).

Conclusions: The present study revealed the fact that essential oil analyzed from Perovskia abrotanoides stem and leaves could be a promising source of natural products with potential antioxidant and antimicrobial activities, as compared to fixed oil.

Keywords: *Perovskia abrotanoides*, Essential oil, Fixed oil, Antioxidant capacity, Antimicrobial activity

Introduction

Perovskia abrotanoides (Lamiaceae) locally known as *hoosh, visk, brazambal, domou,* and *gevereh* [1-4] is a medicinally important plant found in Baluchistan province and northern areas of Pakistan. The plant is used by local communities for treatment of typhoid, headache, gonorrhea, vomiting, motion, toothache, atherosclerosis, cardiovascular diseases, liver fibrosis, and cough [5-7]. It has sedative, analgesic, antiseptic and cooling effect [5,8,9]. Herbal tea of this plant is used in curing infection problems and painful urination [10].

Some of the pharmacological effects of plant such as antiplasmodial, antiinflammatory and cytotoxic effects have also been reported [11-13]. Its antioxidant performance including heart enhancing and optimized performance as cell toxicity in pathogens, viruses and cancer cells were also reported [5,14]. Essential and fixed oils of *Perovskia abrotanoides* plays an important role in protection of stored grains and showed to be effective in washing wounds, anti-ring worm, dermal parasites, anti-fungus and anti-hypoxia [14-17]. Despite its multipurpose usage, little data exists on chemical composition as well as biological activities of this plant. Therefore this study was designed to investigate composition of essential and fixed oil and antioxidant and antimicrobial activities of stem and leaves of *Perovskia abrotanoides.*

Results and discussion

The diversity of agrogeoclimatic conditions of Pakistan offers the broadest array of flora. This rich floral biodiversity of Pakistan constitutes an impressive pool for 'natural food and healing' from which the indigenous communities select ingredients for food and prepare herbal recipes for the treatment, management and control of their various ailments. It is believed that antimicrobials and antioxidants of plant origin are without side effects as compared to synthetic drugs and have an enormous therapeutic potential to heal many infections and diseases. Much of the potential of these botanicals is however still unearthed. In current research, a less explored

* Correspondence: sercisli@gmail.com
[4]Ataturk University Agricultural Facultu Department of Horticulture, 25240 Erzurum, Turkey
Full list of author information is available at the end of the article

medicinal plant found in Pakistan has been investigated for compositional studies and biological activities.

GC-MS analysis of essential and fixed oils

Qualitative and quantitative GC-MS analysis of the essential and fixed oils were performed in order to identify different compounds in the oils. The GC-MS analysis identified 13 and 15 compounds in the essential oils of *P. abrotanoides* stem and leaves respectively. Essential oils of plant (stem, leaves) consisted of a mixture of different classes of compounds. The major components found in essential oil of stem were (E)-9-dodecenal (66.5%), octadecanoic acid, methyl ester (8.37%), 2,2,5,5-tetramethylhexane (3.96%), while in leaves: hexadecanoic acid, methyl ester (27.79%), lupeol (21.5%), octadecenoic acid, methyl ester (18.45%), eicosane (6.22%) and tetradecane (5.19%) were present in higher concentrations. Considerable amount of some other constituents was also present in the plant essential oils. (Table 1, Figure 1). The major component found in essential oil was (E)-9-dodecenal followed by 5,6-dimethylheptadecane, octadecanoic acid and tetradecane in stem and hexadecanoic acid followed by 5-β-cholestan-3α-ol, and hexatricontane in leaves (around 51% of the total compounds). The fixed oils of stem and leaves were analyzed by GC-MS to monitor their compositions (Table 1). A total of fifty constituents were identified in the fixed oils of *P. abrotanoides* which represented the 86% (stem) and 86.35% (leaves) composition of the total oil. In stem oil the major constituents (> 5%) were α-amyrin (47.01%), α-amyrenone (11.8%) and isopropyl-hexadecanoate (6.56%) while in leaves α-copaene (10.99%), trans-phytol (7.33%), isopropyl-hexadecanoate (6.67%), unidentified (5.63%) and α-amyrenone (5.19%) were present in major concentration. The composition of essential and fixed oils content showed variations in different parts (stem and leaves) of the plant. There are some reports in the literature on the chemical composition of the different chemotypes of *P. abrotanoides* essential oil from different countries. Camphor, α-pinene, o-cimene, 1,8-cineol, camphene, borneol, β-pinene, α-humulene, caryophyllene and some other components have already been reported from aerial parts of *P. abrotanoides*. Camphor, borneol, α-terpineol, bornyl acetate, α-humulene and α-cadinol were present in high concentration in the oil of stem, leaves, flowers and roots of *P. abrotanoides* [18-22]. Our results are in partial agreement with the earlier reports of this plant. It is well known that the essential and fixed oil contents and composition depend on several factors such as different genotype, agronomic practices employed, climatological factors, development stages of plant parts analysed, growing condition, season, post-harvest storage and processing conditions and solvent used for extraction. These factors may explain the differences found among our samples and those analyzed in previous studies. Thereof we have identified some of the similar compounds which reported before in oils of this plant and some new compounds have also reported in our study which are not reported earlier.

Antioxidant activity

Antioxidants are an important part of the defense system of the human body and help to cope with oxidative stress caused by reactive oxygen species. Plants are important sources of antioxidants and there is increasing interest in antioxidant analysis of plants [23]. DPPH• is increasingly used for quickly assessing the ability of antioxidants to transfer the labile H atoms to radicals [24]. This hydrogen donation ability leads towards formation of stable complex of free radicals, resulting in termination of damages caused by these radicals. The essential and fixed oils were screened for their possible antioxidant activity by DPPH radical scavenging (IC_{50}) (Table 2). The stem essential oil showed potential DPPH radical scavenging activity (IC_{50} = 17.9 μg/mL) then leaves essential oil (IC_{50} = 45.4 μg/mL) and the fixed oil of leaves exhibited high DPPH radical scavenging activity (IC_{50} = 62.5 μg/mL) than fixed oil of stem (IC_{50} = 73.1 μg/mL). Standard Antioxidant compound BHT showed highest DPPH radical scavenging activity (IC_{50} = 8.78 μg/mL). Antioxidant effect of a plant extract and its fixed or essential oil is mainly due to various bioactive compounds like flavonoids, phenolic acids, tannins and diterpenes.

The percent inhibition of linoleic acid peroxidation was observed for plant essential and fixed oils whereas synthetic BHT provided inhibition at the level of 90.4%. Essential oil of leaves exhibited highest % inhibition of peroxidation (76.4%) followed by stem essential oil (61.1%), leaves fixed oil (47.4%) and stem fixed oil (45.9%). When the results of DPPH scavenging activity and the percent inhibition of linoleic acid oxidation were compared with standard BHT, all the oils showed significantly ($p < 0.05$) minor activity. Essential oils have the efficacy to reduce the peroxide formation during incubation in linoleic acid system as per our previous studies [25-27].

The reducing potential of essential and fixed oils of *P. abrotanoides* was also investigated at various concentrations (2.5-10 mg/mL) and absorbance recorded at 700 nm (Figure 2). The order of reducing potential of *P. abrotanoides* was found as: leaves essential oil > leaves fixed oil > stem essential oil > stem fixed oil. Reducing power of different plants and essential oils has already been reported in literature. Plants have reducing power due to the presence of phenolic compounds [28-31]. The results showed that antioxidant activities of essential oil was much higher in respect with fixed oil, which could be due to the higher concentration of (E)-9-dodecenel from stem and hexadecanoic acid from leaves determined by GC-MS analysis.

Table 1 GC-MS product distribution and relative proportion (wt% of total compounds) of dry weight essential oil (stem and leaves) from *Perovskia abrotanoides* (in mg/g oil)

RI	Chemical constituents	% Composition			
		Essential oil		Fixed oil	
		Stem	Leaves	Stem	Leaves
820	2,2,3,4-tetramethylpentane	0.46	–	–	–
850	t*	–	0.77	–	–
922	2,4,6-trimethyloctane	0.76	–	–	–
973	Nonane, 5-(2-methylpropyl)	–	1.07	–	–
1000	t*	–	–	0.33	–
1033	Eucalyptol	–	–	–	1.26
1052	2,2,5,5- tetramethylhexane	3.96	–	–	–
1125	Ethylbenzene	–	–	–	3.08
1143	Camphor	–	–	–	1.54
1165	Borneol	–	–	–	2.12
1175	S-propyl methanesulfonothioate	3.0	–		
1189	α terpineol	–	–	–	1.07
1200	1-bromo dodecane	–	–	0.11	–
1236	2-pentylfuran	–	–	0.52	3.05
1291	Bornyl acetate	–	–	–	0.7
1352	Terpinyl acetate	–	–	–	1.26
1362	Geranyl acetate	–	–	–	0.83
1399	Tetradecane	2.41	5.19	–	–
1403	(E)-9-dodecenal	66.5	–	–	–
1404	δ-caryophyllene	–	–	–	0.7
1416	α-copaene	–	–	–	10.99
1424	Caryophyllene	t*	–	–	0.95
1454	1,4,7-cycloundecatriene,1,5,9,9-tetramethyl, Z,Z,Z-	–	–	–	0.82
1500	Pentadecane	–	3.97	t*	–
1532	Epiglobulol	–	–	0.65	4.95
1642	Cubenol	–	–	–	1.64
1654	α-cadinol	–	–	0.48	–
1659	β-eudesmol	–	–	0.42	–
1719	8-hexylpentadecane	–	–	–	1.16
1720	Tetradecanoic acid	–	–	0.17	–
1729	4-(E)-3-hydoxyprop-1-enyl)-2-methoxyphenol	–	–	1.26	1.93
1750	t*	–	–	0.33	–
1800	t*	–	–	–	0.81
1800	Cis-9-hexadecenal	–	–	2.23	–
1822	5,6-dimethylheptadecane	2.94	–	–	–
1834	Unidentified	–	–	2.61	2.06
1847	2-pentadecanone, 6,10,14-trimethyl	–	–	–	0.79
1854	5-methyloctadecane	–	–	0.48	3.21
1914	Anthranilic acid	–	–	–	1.04
1949	Heptadec-1-ol	–	–	–	0.99
1962	Hexadecanoic acid, methyl ester	–	27.79	–	–

Table 1 GC-MS product distribution and relative proportion (wt% of total compounds) of dry weight essential oil (stem and leaves) from *Perovskia abrotanoides* (in mg/g oil) *(Continued)*

2000	Eicosane	–	6.22	–	–
2000	Unidentified	–	–	–	5.63
2010	Unidentified	–	–	–	1.28
2012	Isopropyl hexadecanoate	–	–	6.56	6.67
2080	Octadecanoic acid, 17- methyl ester	–	2.19	–	–
2088	Octadecanoic acid, methyl ester	8.37	18.45	–	–
2100	Unidentified	1.93	–	–	–
2110	t*	–	–	–	0.69
2111	Trans phytol	–	–	2.03	7.33
2138	7-methoxy-8-(3-methyl-2-butenyl)-2H-chromen-2-one	–	–	2.29	–
2161	Oleic acid	–	–	–	3.31
2200	Stearic acid	–	–	0.33	–
2221	Nonadecanoic acid	–	–	0.6	–
2240	Androst-2-en-17-one	–	–	2.18	—
2330	Trans-ferruginol	–	–	0.48	3.06
2400	Tetracosane	–	–	0.17	–
2430	3-hydroxyandrostan-17-one	–	–	–	2.06
2435	t*	–	–	–	0.88
2526	Docosanoic acid	–	2.47	–	–
2700	Unidentified	–	1.1	–	–
2730	Tetracosanoic acid	–	2.62	–	–
3011	t*	0.53	–	–	
3098	5.β-cholestan-3α-ol	–	1.46	–	–
3400	Tetratriacontane	–	–	0.14	–
3408	β sitosterol	–	–	2.82	–
3654	1-(+)-Ascorbic acid 2,6-dihexadecanoate	3.0	–	–	–
3654	Unidentified	–	–	–	3.3
3600	Hexatriacontane	–	0.69	–	–
4000	Unidentified	2.51	1.99	–	–
4050	α-amyrenone	–	–	11.8	5.19
4160	α-amyrin	–	–	47.01	–
4175	Lupeol	–	21.5	–	–

t* = traces.
Compounds identification was carried on the basis of Retention indices and EI/MS.

Antimicrobial activity

The antimicrobial activity of the *P. abrotanoides* oils was assessed (Table 3). The results from the disc diffusion method, followed by measurement of minimum inhibitory concentration (MIC), indicated that the Stem essential oil showed good inhibitory activity against *Nitrospira sp.* and *A. flavus* (IZ = 9.76, 9.94 mm; MIC = 19.6, 14.5 mg/mL) and it was inactive against *E. coli, S. aureus, B. cereus, S. epidermidis* and *C. albicans*. Stem fixed oil showed potent activity against *C. albicans* (IZ = 25.2 mm; MIC = 1.26 mg/mL) and moderate activity against *E. coli, S. aureus* (IZ = 12.9, 8.34 mm; MIC = 7.95, 14.2 mg/mL). Stem fixed oil exhibited no inhibitory activity against *B. cereus, Nitrospira sp., S. epidermidis, A. niger, A. flavus*. Leaves essential oil was inactive against *E.coli, S. aureus, Nitrospira, S. epidermidis, A. flavus* and it moderately inhibited the growth of *B. cerus. A. niger, C. albicans*. Leaves fixed oil was inactive against *E. coli, B. cerus, Nitrospira*. And it showed potent activity against *C. albicans* (IZ = 24.2 mm, MIC = 1.93 mg/mL) and moderately inhibited the growth of other microbes. For the comparison of results Novidate and Fungone were used as positive control for bacterial and fungal strains respectively. The standard drugs showed higher activity on the microbes than the plant oils (Table 3). The standard antibiotics were highly purified chemical compounds so there activity was more as compared to the oils of leaves and stem. For the comparison of results, Novidate and Fungone were used as positive control for bacterial and fungal strains respectively. The standard drugs showed higher activity on the microbes than the plant oils. Our previous studies have found that essential oil obtained from different parts (flowers, leaves, stem, roots) of *P. abrotanoides* possessed potential antimicrobial activity against *S. aureus, B. cereus, S. typhi* and *C. albicans* and a decreased activity against *A. niger* [25-27], which could be a therapeutically approach for acute kidney injury.

Conclusions

We have investigated two different types of *Perovskia abrotanoides* oils. The presence of different compounds founded by GC-MS analysis, rendered the essential oil very efficient in antioxidant and antimicrobial capacity in respect with fixed oil which could be due to the presence of unsaturated fatty acids and anthocyanins compounds. We believe that the compounds, in particular (E)-9-dodecenal and hexadecanoic acid from essential oil are directly involved in antioxidant and antimicrobial processes. Finally, our study revealed the fact that essential oil analysed from *Perovskia abrotanoides* stem and leaves could be a promising source of natural products with potential antioxidant and antimicrobial activities in respect with fixed oil.

Methods

Collection of plant material

The *Perovskia abrotanoides* whole plant was collected and deposited from Quetta and Ziarat Valley and further

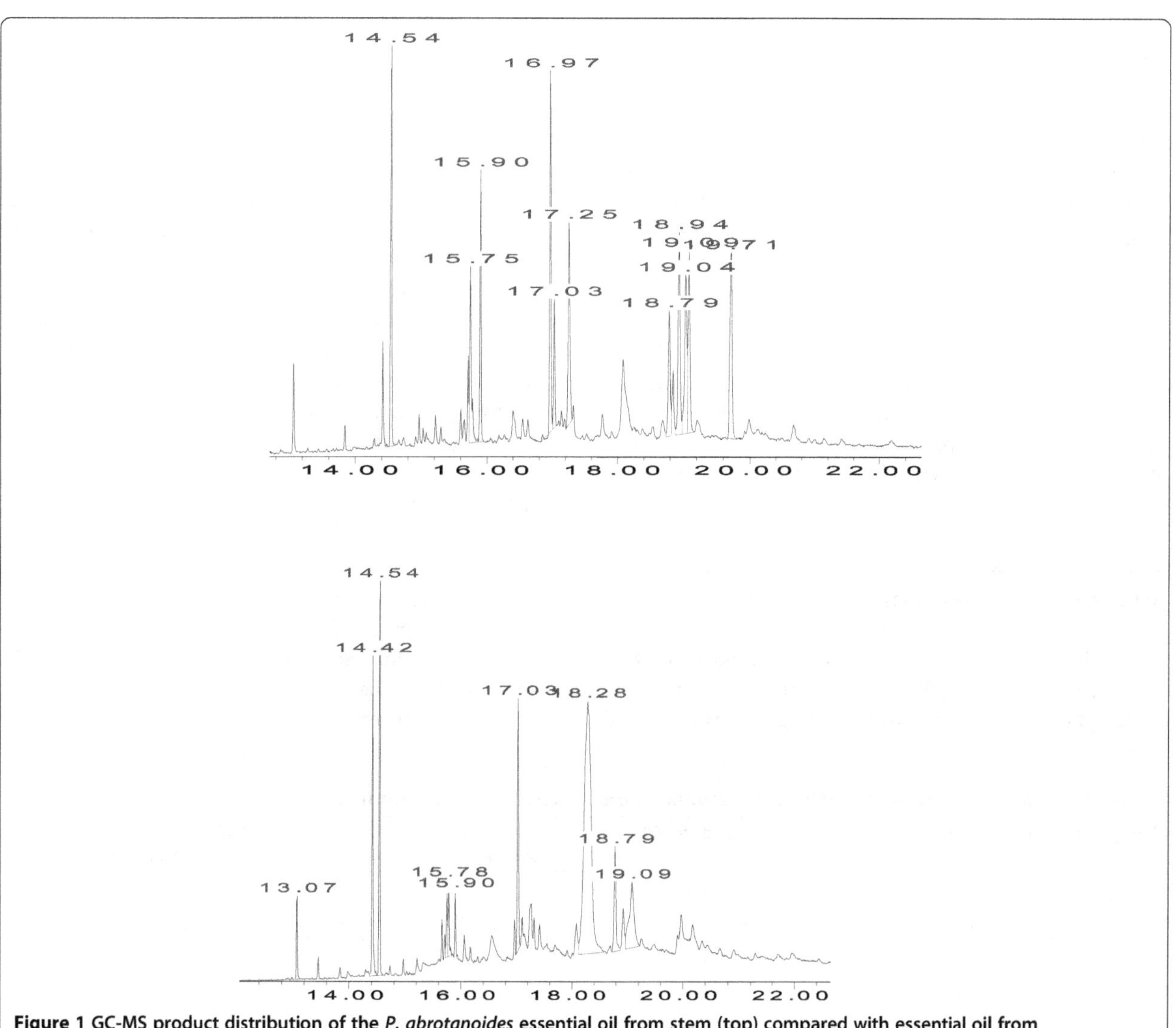

Figure 1 GC-MS product distribution of the *P. abrotanoides* essential oil from stem (top) compared with essential oil from leaves (bottom).

identified by Dr. Rasool Bukhsh Tareen, Department of Botany, University of Baluchistan, Quetta, Pakistan.

Essential oil extraction

Dried powdered of plant material (100 g) was subjected to hydro-distillation for 5 h using a Clevenger type apparatus for extraction of stem and leaves from essential oils. The extracted essential oil was dried over anhydrous Na_2SO_4, filtered and stored in a vial at 4°C until further analysis.

Fixed Oil extraction

The fixed oil was extracted following the method of Ajayi and contributors [32]. Briefly, 100 g of the shade dried powdered stem and leaves of plant were separately extracted in 250 mL *n*-hexane (99.9% purity) solvent by using a Soxhlet extractor for 6 h and then the extra-solvent was removed by distillation under reduced pressure in a rotary evaporator at 35°C and the pure oil was kept at 4°C in the dark.

GC-MS analysis of essential and fixed oils

Essential and fixed oils were analysed by GC-MS (QP2010, SHIMADZU, Japan) using an Agilent GC 6890 N model (1 μL sample injected, split 1:50 column flow 1.0 mL/min., program temp. 200°C, rate 10°C/min) coupled with a quadrupolar MS 5973. GC was equipped with capillary column (30 m × 0.25 mm; film thickness 0.25 μm). Oven temperature was kept at 45°C first for 5 min, and then raised at 325°C at a 15°C/min for another 5 min. Helium gas was then employed at a flow rate of 1.1 mL/min (60 KPa pressure; 38.2 cm/sec linear velocity). The identification of components was based on comparison of their mass spectra with those of NIST mass spectral library [33].

Antioxidant activity

The 2-diphenyl-1-picrylhydrazyl (DPPH) radical scavenging assay

The antioxidant activity of the oils was assessed by their ability to scavenging DPPH stable radicals as reported earlier [34]. The samples (10 to 500 μg/mL) were mixed with DPPH solution (1 mL; 90 μM) and then with methanol (95%) to a final volume of 4 mL. Synthetic antioxidant, butylated hydroxytoluene (BHT) was used as control. After 1 h incubation period at room temperature, the absorbance was recorded at 515 nm. Percent radical scavenging concentration was calculated using the following formula:

$$\text{Radical scavenging } (\%) = 100 \times \left(\frac{A_{blank} - A_{sample}}{A_{blank}}\right)$$

Where:

A_{blank} = absorbance of the control
A_{sample} = absorbance of the test samples

Inhibition of linoleic acid peroxidation

The antioxidant activity of *Perovskia abrotanoides* oils was evaluated in terms of percent inhibition of peroxidation in linoleic acid system [35]. Extract (5 mg) was mixed with linoleic acid solution (0.13 mL), ethanol (10 mL; 99.8%), sodium phosphate buffer (10 mL; 0.2 M; pH = 7) and diluted to 25 mL with distilled water. The solution was incubated at 40°C for 360 h and extent of oxidation was investigated using the colorimetric method [36]. Then at 0.2 mL sample solution, ethanol (10 mL; 75%), ammonium thiocyanate (0.2 mL; 30%) and 0.2 mL of ferrous chloride solution (20 mM in 3.5% HCl w/v) were added consecutively. After stirring for 3 min, the absorbance of mixture was calculated at 500 nm. A control was also performed only with linoleic acid (without any extract). The synthetic antioxidant such as BHT was used as positive control. Inhibition (%) of linoleic acid oxidation was investigated with the following equation:

$$\%\text{ Inhibition} = 100 - \left[\frac{(Abs.\,increase\ of\ sample\ at\ 360\ h)}{(Abs.increase\ of\ control\,at\ 360\ h)} \times 100\right]$$

Analysis of reducing power

The reducing power of the oils was determined according to the procedure described by Yen et al. [36] with little modification. The plant oils at various concentrations (2.5-10 mg) were mixed with sodium phosphate buffer

Table 2 % Yield and Antioxidant activity of *Perovskia abrotanoides* essential and fixed oils*

Plant parts and tested samples		% yield (g/100 g)	DPPH radical scavenging (IC_{50})(μg/mL)	Inhibition of peroxidataion in linoleic acid system (%)
Stem	Essential oil	2.13 ± 0.06	17.9 ± 0.004	61.1 ± 0.87
	Fixed oil	2.96 ± 0.04	73.1 ± 0.007	45.9 ± 1.74
Leaves	Essential oil	2.86 ± 0.05	45.4 ± 0.01	76.7 ± 0.87
	Fixed oil	3.61 ± 0.10	62.5 ± 0.05	47.4 ± 0.87
Standard	BHT	–	8.78 ± 0.10	90.4 ± 0.87

*Values are mean ± S.D. of three separate experiments ($P < 0.05$).

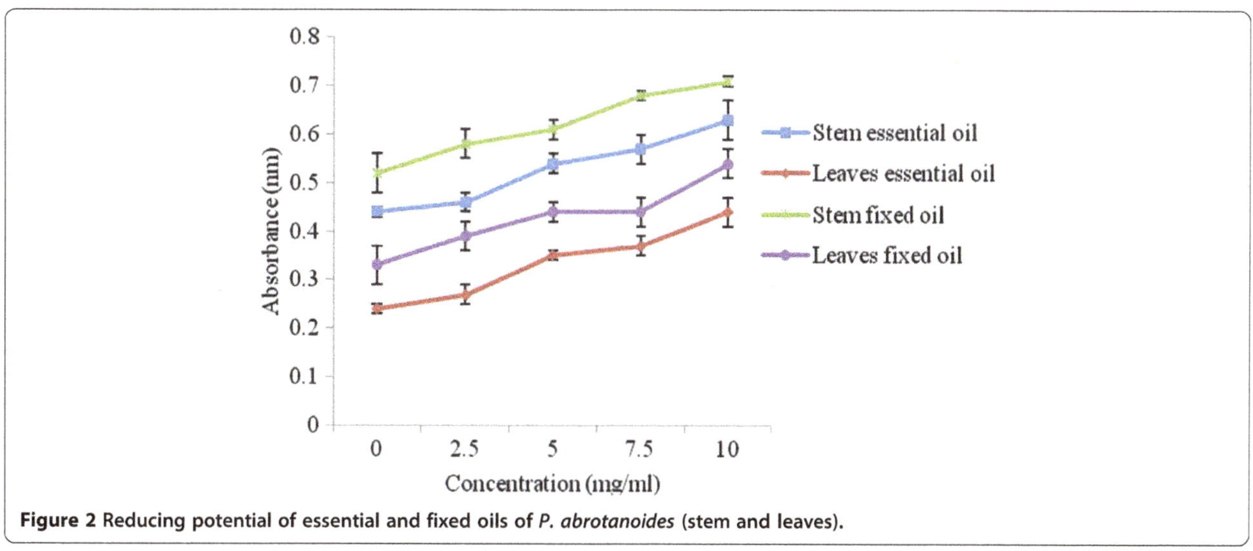

Figure 2 **Reducing potential of essential and fixed oils of *P. abrotanoides* (stem and leaves).**

(5 mL; 0.2 M) of pH 6.6 and potassium ferricyanide (5 mL; 1.0%). The mixture was heated for 20 min at 50°C. Then trichloroacetic acid (5 mL; 10%) was added and centrifuged at 980 rpm at 5°C for 10 min. Further, its upper layer (5 mL) was dissolved in 5 mL distilled water and finally freshly prepared ferric chloride (1 mL; 0.1%) was added. The absorbance was calculated at 700 nm and a result for each sample was recorded in triplicate.

Antimicrobial assay

Microbial strains

Bacillus cereus ATCC 14579, *Escherichia coli* ATCC 25922, *Nitrospira sp.* locally isolated, *Staphylococcus epidermidis* ATCC 12229, *Staphylococcus aureus* API Staph tac 6736153 were used as bacterial strains and *Aspergillus niger* ATCC 10595, *Aspergillus flavus* ATCC *32612*, *Candida albicans* ATCC 10231 as fungal strains. The

Table 3 Antimicrobial activity of *Perovskia abrotanoides* essential and fixed oils*

	Diameter of Inhibition Zone (IZ, mm)					
Tested microrganisms	Stem		Leaves		Standard drugs	
	Essential oil	Fixed oil	Essential oil	Fixed oil	Novidate	Fungone
E. coli	-	12.9 ± 0.03	-	-	27.7 ± 0.02	-
S. aureus	-	8.34 ± 0.03	-	11.2 ± 0.04	21.2 ± 0.03	-
B. cereus	-	-	15.2 ± 0.05	-	24.5 ± 0.02	-
Nitrospira sp.	9.76 ± 0.01	-	-	-	31.7 ± 0.04	-
S. epidermidis	-	-	-	14.2 ± 0.07	23.4 ± 0.01	-
A. niger	-	-	10.6 ± 0.005	8.46 ± 0.01	-	30.2 ± 0.06
A. flavus	9.94 ± 0.03	-	-	9.64 ± 0.02	-	28.4 ± 0.01
C. albicans	-	8.07 ± 0.02	9.78 ± 0.01	8.32 ± 0.03	-	30.2 ± 0.01
Minimum Inhibitory Concentration (MIC, mg/mL)						
E. coli	-	7.95 ± 0.02	-	-	0.77 ± 0.01	-
S. aureus	-	14.2 ± 0.03	-	9.40 ± 0.05	1.06 ± 0.02	-
B. cereus	-	-	5.75 ± 0.04	-	0.95 ± 0.03	-
Nitrospira sp.	19.6 ± 0.05	-	-	-	0.28 ± 0.01	-
S. epidermidis	-	-	-	5.25 ± 0.02	0.88 ± 0.03	-
A. niger	-	-	11.4 ± 0.02	18.4 ± 0.01	-	0.47 ± 0.02
A. flavus	14.5 ± 0.01	-	-	17.1 ± 0.03	-	0.24 ± 0.01
C. albicans	-	1.26 ± 0.01	11.7 ± 0.08	1.93 ± 0.02	-	0.17 ± 0.02

*Values are mean ± S.D of three separate experiments ($P < 0.05$).

pure bacterial and fungal strains were obtained from the Department of Veterinary Microbiology, University of Agriculture, Faisalabad, Pakistan. The bacterial strains were cultured overnight at 37°C in nutrient agar (Oxoid, UK) while fungal strains were cultured overnight at 28°C using potato dextrose agar (Oxoid, UK).

Antimicrobial disc susceptibility test

Antimicrobial activity of the plant oils was determined by using the disc susceptibility test [37]. The discs (6 mm diameter) were impregnated plant oils (100 µL/disc) placed aseptically on the inoculated agar. Discs without injected samples served as a negative control, and Novidate (100 µL/disc) and Fungone (100 µL/disc) (both from Oxoid, UK) as positive control. The Petri dishes were incubated at 37 ± 0.1°C for 20–24 h and 28 ± 0.3 for 40–48 h for both bacteria and fungi. At the end of the period, the inhibition zones were measured. The positive antimicrobial activity was read based on growth inhibition zone.

Minimum inhibitory concentration (MIC)

The MIC of the plant oils was estimated following resazurin microtitre-plate assay reported by Sarker and contributors [38].

Statistical analysis

All the aforementioned experiments were conducted in triplicate. Statistical comparisons were performed by one-way analysis of variance (ANOVA) followed by Dunnett's t-test using SPSS version 12.0 (SPSSS Inc., Chicago, IL, USA). Probability values < 0.05 were considered to indicate significant difference.

Competing interests

The authors declare that they have no competing interests.

Authors' contributions

SNA, MZ, KR, RBT, and NR made a significant contribution to experiment design, acquisition of data, analysis and drafting of the manuscript. MZUH and SE have made a substantial contribution to interpretation of data, drafting and carefully revising the manuscript for intellectual content. All authors read and approved the final manuscript.

Author details

[1]Department of Chemistry, Government College University, Faisalabad 38000, Pakistan. [2]Department of Botany, University of Balochistan, Quetta, Pakistan. [3]The Patent Office, Karachi, Pakistan. [4]Ataturk University Agricultural Facultu Department of Horticulture, 25240 Erzurum, Turkey.

References

1. Mahboubi M, Kazempour N: **The antimicrobial activity of essential oil from *Perovskia abrotanoides* karel and its main components.** *Ind J Pharma Sci* 2009, **71**:343–347.
2. Mazandarani M, Ghaemi E: **Ethnopharmacological investigation of different parts of *Perovskia abrotanoides* Karel.** *Planta Med* 2010, **76**:123–125.
3. Amiri MS, Jabbarzadeh P, Akhondi M: **An ethnobotanical survey of medicinal plants used by indigenous people in Zangelanlo district, Northeast Iran.** *J Med Plants Res* 2012, **6**:749–753.
4. Mahboubi M: **Iranian medicinal plants as antimicrobial agents.** *J Microbiology, Biotech Food Sci* 2013, **2**:2388–2405.
5. Moallem SA, Niapour M: **Study of embryotoxicity of *Perovskia abrotanoides*, an adulterant infolk medicine, during organogenesis in mice.** *J Ethnopharma* 2008, **117**:108–114.
6. Kumar P, Gupta S, Murugan P, Singh SB: **Ethnobotanical studies of Nubra Valley - A cold arid zone of Himalaya.** *Ethnobot Leaflet* 2009, **13**:752–765.
7. Tareen RB, Bib T, Khan MA, Ahmad M, Zafar M: **Indigenous knowledge of folk medicine by the women of Kalat and Khuzdar Regions of Balochistan, Pakistan.** *Pak J Bot* 2010, **42**:1465–1485.
8. Hosseinzade H, Amel S: **Antinociceptive effects of the aerial parts of *Perovskia abrotanoides* extracts in mice.** *Med J Iranian Hos* 2001, **4**:15–27.
9. Nassiri AM, Parvardeh S, Niapour M, Hosseinzadeh H: **Antinociceptive and anti-inflammatory effects of *Perovskia abrotanoides* aerial part extracts in mice and rats.** *Medicinal Plant* 2000, **3**:25–33.
10. Ballabh B, Chaurasia OP, Ahmed Z, Singh SB: **Traditional medicinal plants of cold desert Ladakh used against kidney and urinary disorders.** *J Ethnopharm* 2008, **118**:331–339.
11. Sairafianpour M, Christensen J, Staerk D, Budnik BA, Kharazmi A, Bagherzadeh K, Jaroszewski JW: **Leishmanicidal, antiplasmodial, and cytotoxic activity of novel diterpenoid 1,2-quinones from *Perovskia abrotanoides*: new source of tanshinones.** *J Nat Product* 2001, **64**:1398–1403.
12. Esmaeilli S, Naghibi F, Mosaddegh M, Sahranavard S, Ghafari S, Abdullah NR: **Screening of antiplasmodial properties traditionally used among some Iranian plants.** *J Ethnopharm* 2008, **121**:400–404.
13. Beikmohammadi M: **The evaluation of medicinal properties of *Perovskia abrotanoides* Karel.** *Middle-East J Sci Res* 2012, **11**:189–193.
14. Rustaiyan AH, Masoudi S, Ameri N, Samiee K, Monfared A: **Volatile constituents of *Ballota aucheri* Boiss, *stachys benthamiana* Boiss and *Perovskia abrotanoides* Karel, Growing wild in Iran.** *Essential oil Res* 2006, **1**:3–5.
15. Aoyagi Y, Takahashi Y, Satake Y, Takeya K, Aiyama R, Matsuzaki T, Hashimoto S, Kurihara T: **Cytotoxicity of abietane diterpenoids from *Perovskia abrotanoides* and of their semisynthetic analogues.** *Bioorganic Med Chem* 2006, **14**:5285–5291.
16. Jaafari M, Hooshmand S, Samiei A, Hossainzadeh H: **Evaluation of- leishmanicidal effect of *Perovskia abrotanoides* Karel. root extract by *in vitro* leishmanicidal assay using promastigotes of *Leishmania major*.** *Pharmacologyonline* 2007, **1**:299–303.
17. Arabi F, Moharramipour S, Sefidkon F: **Chemical Composition and insecticidal activity of essential oil from *Perovskia abrotanoides* (*Lamiaceae*) against *Sitophilus oryzae* (*Coleoptera: Curculionidae*) and *Tribolium castaneum* (*Coleoptera: Tenebrionidae*).** *Int J Tropical Insect Sci* 2008, **28**:144–150.
18. Saleh MM, Kating H: **Gas-chromatographic analysis of the volatile oil of *Perovskia abrotanoides* K.** *Planta Med* 1978, **33**:85–88.
19. Inouye S, Uchida K, Yamaguchi H, Miyara T, Gomi T, Amano M: **Volatile aroma constituents of three *Labiatae* herbs growing wild in the Karakoram- Himalaya district and their antifungal activity by vapor contact.** *J Essential Oil Res* 2001, **35**:68–72.
20. Semnani MK: **The essential oil composition of *Perovskia abrotanoides* from Iran.** *Pharm Bio* 2004, **42**:214–216.
21. Nezhadali A, Masrorniab M, Solatib A, Akbarpoura M, Moghaddamb NM: **Analysis of the flower essential oil at different stages of plant growth and *in vitro* antibacterial activity of *Perovskia abrotanoides* Karal in Iran.** *J Chem Soc Comm* 2009, **1**:146–150.
22. Ghomi SJ, Batooli H: **Determination of bioactive molecules from flowers, leaves, stems and roots of *Perovskia abrotanoides* karel growing in central Iran by nano scale injection.** *J Nanomaterial Bio* 2010, **5**:551–556.
23. Zia-Ul-Haq M, Ahmad S, Calani L, Mazzeo T, Del Rio D, Pellegrini N, Defeo V: **Compositional study and antioxidant potential of *Ipomoea hederacea* Jacq. and *Lepidium sativum* L. seeds.** *Molecules* 2012, **17**:10306–10321.
24. Zia-Ul-Haq M, Ahmad S, Amarowicz R, Defeo V: **Antioxidant activity of the extracts of some cowpea (*Vigna unguiculata* (L.) Walp.) cultivars commonly consumed in Pakistan.** *Molecules* 2013, **18**:2005–2017.
25. Riaz M, Rasool N, Bukhari IH, Shahid M, Zubair M, Rizwan K, Rashid U: ***In vitro* antimicrobial, antioxidant, cytotoxicity and GC-MS analysis of *Mazus goodenifolius*.** *Molecules* 2012, **17**:14275–14287.

26. Rizwan K, Zubair M, Rasool N, Riaz M, Zia-Ul-Haq M, De Feo V: **Phytochemical and biological studies of *Agave attenuata*.** *Int J Mol Sci* 2012, **13:**6440–6451.
27. Rasool N, Afzal S, Riaz M, Rashid U, Rizwan K, Zubair M, Ali S, Shahid M: **Evaluation of antioxidant activity, cytotoxic studies and GC-MS profiling of *Metthiola incana* (Stock flower).** *Legume Res* 2013, **36:**21–32.
28. Yildirim A, Oktay M, Bilaloglu V: **The antioxidant activity of the leaves of *Cydonia vulgaris*.** *Turk J Med Sci* 2001, **31:**23–27.
29. Siddhuraju P, Mohan PS, Becker K: **Studies on the antioxidant activity of *Indian laburnum* (*Cassia fistula* L.): a preliminary assessment of crude extracts from stem bark, leaves, flowers and fruit pulp.** *Food Chem* 2002, **79:**61–67.
30. Southwell I, Russell M, Smit SRL, Brophy JJ, Day J: ***Melaleuca teretifolia*, a novel aromatic and medicinal plant from Australia.** *Med Aromatic Plant* 2008, **3:**435–457.
31. Thirugnanasampandan R, Mahendran G, Narmatha BV: **Antioxidant properties of some medicinal *Aristolochiaceae* species.** *African J Biotech* 2008, **7:**357–361.
32. Ajayi GO, Olagunju JA, Ademuyiwa O, Martins OC: **Gas chromatography–mass spectrometry analysis and phytochemical screening of ethanolic root extract of *Plumbago zeylanica*, Linn.** *J Med Plants Res* 2011, **5:**1756–1761.
33. Mass Spectral Library; 2002. Available online: http://www.sisweb.com/software/ms/nist.htm (accessed on 23 May 2010).
34. Mimica-Dukic N, Bozin B, Sokovic M, Mihajlovic B, Matavulj M: **Antimicrobial and antioxidant activities of three *Mentha* species essential oils.** *Planta Med* 2003, **69:**413–419.
35. Iqbal S, Bhanger MI: **Antioxidant properties and components of some commercially available varietes of rice bran in Pakistan.** *Food Chem* 2005, **93:**265–272.
36. Yen GC, Duh PD, Chuang DY: **Antioxidant activity of anthraquinones and anthrone.** *Food Chem* 2000, **70:**307–315.
37. National Committee For Clinical Laboratory Standards: *Performance Standards for Antimicrobial Disc Susceptibility Test.* 5th edition. Wayne, PA, USA: Clinical and Laboratory Standards Institute; 1997.
38. Sarker SD, Nahar L, Kumarasamy Y: **Microtitre plate-based antibacterial assay incorporating resazurin as an indicator of cell growth, and its application in the *in vitro* antibacterial screening of phytochemicals.** *Method* 2007, **42:**321–324.

7

Cadmium toxicity affects chlorophyll a and b content, antioxidant enzyme activities and mineral nutrient accumulation in strawberry

Ferhad Muradoglu[1], Muttalip Gundogdu[1], Sezai Ercisli[2], Tarik Encu[3], Fikri Balta[4], Hawa ZE Jaafar[5*] and Muhammad Zia-Ul-Haq[6*]

Abstract

Background: Cadmium (Cd) is well known as one of the most toxic metals affecting the environment and can severely restrict plant growth and development. In this study, Cd toxicities were studied in strawberry cv. Camarosa using pot experiment. Chlorophyll and malondialdehyde (MDA) contents, catalase (CAT), superoxide dismutase (SOD), ascorbate peroxidase (APX) activities and mineral nutrient concentrations were investigated in both roots and leaves of strawberry plant after exposure Cd.

Results: Cd content in both roots and leaves was increased with the application of increasing concentrations of Cd. We found higher Cd concentration in roots rather than in leaves. Chlorophyll a and b was decreased in leaves but MDA significantly increased under increased Cd concentration treatments in both roots and leaves. SOD and CAT activities was also increased with the increase Cd concentrations. K, Mn and Mg concentrations were found higher in leaves than roots under Cd stress. In general, increased Cd treatments increased K, Mg, Fe, Ca, Cu and Zn concentration in both roots and leaves. Excessive Cd treatments reduced chlorophyll contents, increased antioxidant enzyme activities and changes in plant nutrition concentrations in both roots and leaves.

Conclusion: The results presented in this work suggested that Cd treatments have negative effect on chlorophyll content and nearly decreased 30% of plant growth in strawberry. Strawberry roots accumulated higher Cd than leaves. We found that MDA and antioxidant enzyme (CAT, SOD and APX) contents may have considered a good indicator in determining Cd tolerance in strawberry plant.

Keywords: Antioxidant enzymes, Cadmium, Chlorophyll, Heavy metal stress, Strawberry

Background

Cadmium is believed as one of the most important contaminant in the ecosphere. The main sources of Cd in environment are mining and smelting of Cd-containing ores, municipal wastes, pesticides, trace emissions, burning of fossil fuels and fertilizers [1,2]. In plants, the first organ to contact the toxic metal ions are roots, and therefore roots have greater contents of metal than aerial parts [2]. As compared to other metals like Zn, Cu or Mn, Cd is a non-essential heavy metal that is non-toxic at low concentrations, but it is toxic at higher concentrations [2]. It manifests its toxicity by inhibiting some growth, changing the plant nutrient contents and composition, and by antagonizing the effects on essential elements and several enzymes activities [3,4]. It induces complex changes in plants at genetic, physiological and biochemical levels, leading to phytotoxicity, whose main indications are leaf rolls, chlorosis and reduction of root and stem growth [5,6], limiting transport of metals [7], respiratory and photosynthetic activities, enzyme activities, hormone balance and membrane functions [8], induction of lipid peroxidation and chlorophyll breakdown in plants [9] and generation of oxidative stress [2,10].

* Correspondence: hawazej@gmail.com; ahirzia@gmail.com
[5]Department of Crop Science, Faculty of Agriculture, University Putra Malaysia, 43400 Selangor, Malaysia
[6]The Patent Office, Karachi, Pakistan
Full list of author information is available at the end of the article

Among all side effects induced by Cd, lipid peroxidation is the most harmful as it can lead to bio membrane deterioration. The main indicator of oxidative stress in plants is MDA, which is the decomposition product of polyunsaturated fatty acids of bio membrane [11]. Plants manage the oxidative stress by antioxidant enzymes like CAT, SOD, GPX, APX, GR, and non-enzymatic constituents such as ascorbate and glutathione [12-14]. Among enzymes, SOD is the first line of defense as it converts superoxide radical to hydrogen peroxide (H_2O_2), which is later reduced to water and oxygen either by APX in ascorbate-glutathione cycle or by GPX and CAT in cytoplasm and other cellular compartments [12]. It is well known that the response of plants to Cd-induced depend on several factors such as genotype, root system, growing condition, agronomic practices employed, climatological and geological conditions of soil, and growing season as well as maturity of plants. Root uptake, root-to-shoot translocation and partitioning of Cd between plant organs can vary in both plant species and cultivar belongs to single specie [15].

Strawberry (*Fragaria x ananassa* Duch) has been widely grown worldwide because of adapting to various climate and soil condition. Camarosa cultivar dominates strawberry production in Turkey due to its bigger fruits, high fruit quality and excellent transportation capacity [16,17]. Threats of environmental pollution with heavy metals render stress a general concern for the agricultural crops. Strawberry plants exposed to Cd toxicity may experience severe cellular injury that may lead to cell death within a short period. Cd is easily taken up by strawberry plants and accumulated in organs [18]. Previous studies commonly concerned with the influences of Cd on the upper part of plants. Little is known of Cd toxicity to the root system in strawberry plant. This study was an attempt to understand the effect of Cd treatments on plant growth, antioxidant enzyme activities and mineral nutrients accumulations in both roots and leaves of Camarosa strawberry cultivar.

Results and discussion

Effect of Cd on chlorophyll and Cd accumulation in strawberry

As shown in Figure 1. The chlorophyll content in strawberry plant organs decreased under Cd treatment. There was regularly a reduction attributable to Cd application both chlorophyll a and chlorophyll b in Camarosa (strawberry) cultivar. Chlorophyll a content was found higher than chlorophyll b content. There was nearly 5, 15, 25, and 30% decrease in chlorophyll a and 3, 11, 15 and 18% decrease in chlorophyll b when Cd applications were increased from 0 to 60 mg kg^{-1} respectively. According to Qian et al. [19], cadmium-induced declining effect on chlorophyll and carotenoid contents which could be explained on the basis of inhibitory effect of Cd on enzymes involved in pigment biosynthesis. Furthermore, chlorophyll a and chlorophyll b contents showed significant decline at the applications of Cd and the results were in consist with earlier report where Cd inhibited the biosynthesis of chlorophyll and generated a kind of senescence [19,20]. Our results are in agreement with finding by Yang et al. [21] who reported that leaves of *Potamogeton crispus* under Cd stress showed decreased 35,8% chlorophyll a and 26.7% chlorophyll b and chlorophyll a content was found higher than chlorophyll b content. Several report have shown that under Cd stress decrease chlorophyll content in leaf garden grass [22] and almond seedling [23]. Therefore, chlorophyll pigments seem to be one of the main reasons of heavy-metal injury in plants.

Statistically significant differences among Cd applications for accumulation of Cd in root and leaf of strawberry plants were observed (Figure 2). Increasing Cd concentrations were ensuring significant increase Cd accumulation in both root and leaf. The average Cd concentration in root was approximately four times higher than in leaf. The Cd concentration ranged from 0.74 to 3.77 mg kg^{-1} in root and from 0.27 to 0.79 mg kg^{-1} in leaf. Increasing Cd concentrations were increased

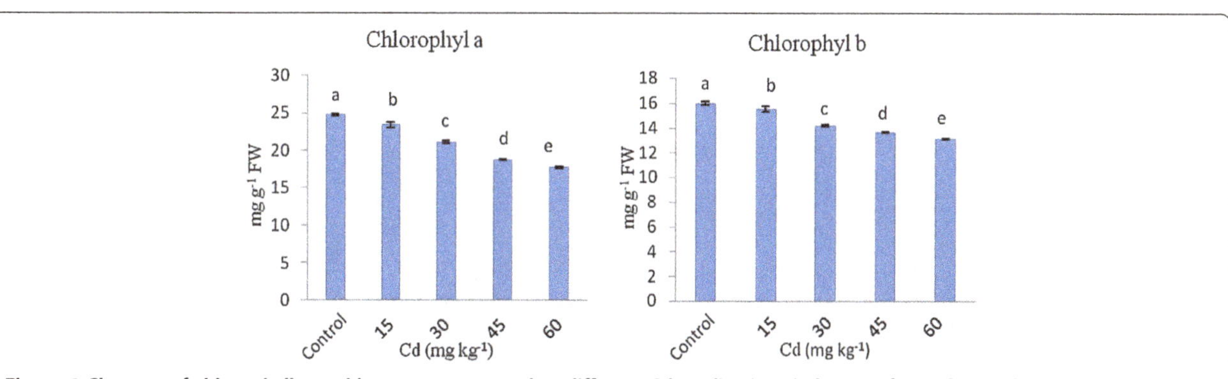

Figure 1 Changes of chlorophyll a and b contents exposed to different Cd applications in leaves of strawberry plants. Same letters are not significantly different according to Duncan test (p≤0.05).

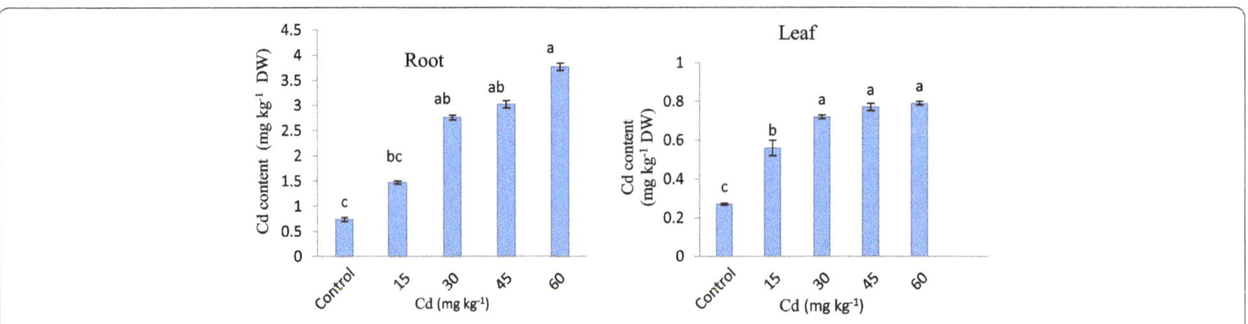

Figure 2 Cadmium accumulation in strawberry plant exposed to different Cd applications. Same letters are not significantly different according to Duncan test ($p \leq 0.05$).

accumulation of Cd approximately 1.98, 3.72, 4.08 and 5.09 times in root and 2.07, 2.26, 2.85 and 2.92 times in leaf as compared with control respectively. Cd uptake and accumulation in plant differences greatly among species and also among different organs and tissues. Cd is usually accumulated in the roots, because this is the first organ exposed to heavy metal and it is also translocated into the shoots. Our results showed that the accumulation of Cd in root was higher than in leaf of strawberry (Figure 2). Similarly, Gill et al. [22] reported that Cd accumulation in root and leaves increased with the increasing Cd concentration in soil and Cd content in root was found higher than leaves in *Lepidium sativum*. Nada et al. [23] observed similar situation in almond seedling.

Effects of Cd on MDA content

The increased contents of lipid peroxides are indication of more production of toxic oxygen species than normal. Strawberry plant showed significant increase in MDA production when treated with Cd applications. Leaf had higher MDA content than root. In root and leaf of strawberry plant, MDA production was increased nearly 30% in root and 33%in leaf compare with control after expose to 60 mg kg^{-1} Cd application (Figure 3). When plants grow in stressed environments, free-radicals generated in excess, accumulate in the cells. It leads to lipid peroxidation of biomembranes, and its end product is MDA. Therefore, the MDA-concentration is an indicator of physiological stresses and the aging process [8]. Our results showed increase in MDA content in both root and leaf depend on Cd concentrations. Nada et al. [23] observed an increase MDA content in both root and leaf of almond seedlings that exposed to Cd treatment. This result is in agreement with our study.

Effect of Cd on antioxidant enzyme activity

The changes in SOD activity were determined in both root and leaf of strawberry with increase in level of Cd concentrations when compared with control (Figure 4). With increase in Cd concentration in strawberry plant, a steadily increase in SOD was determined in both root and leaf. In every Cd concentration increase, SOD activity was higher than control. In root, Cd concentrations caused an increase in SOD activity by 8, 17, 27 and 29% respectively and in leaf 4, 7, 29 and 34% as compared with control.

As shown in Figure 5, significant increases were observed in CAT activity. Increasing Cd concentrations provide to regularly increase CAT activity in root but Cd concentrations provide a severe increasing in leaf. When the increasing CAT activity was compared with to Control, increasing Cd concentrations caused an increase in CAT activity by 1.0, 1.2, 1.7 and 2.0 times in root

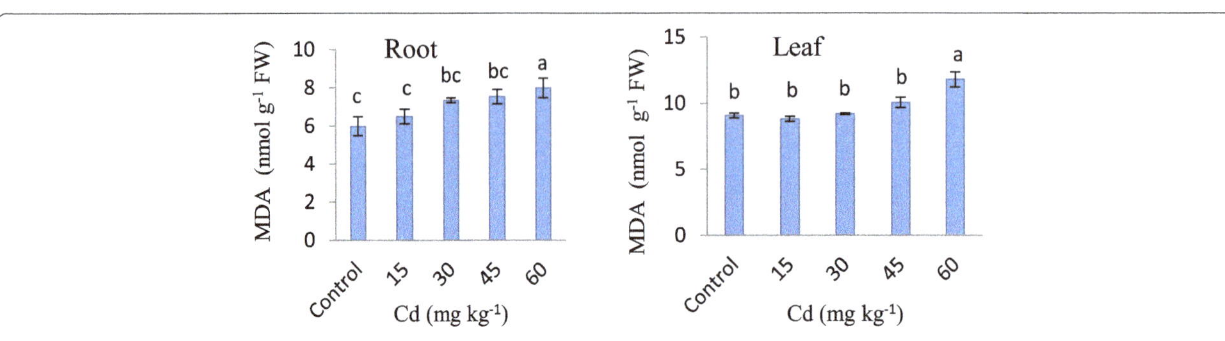

Figure 3 Changes in malondialdehyde content in strawberry plant exposed to different Cd applications. Same letters are not significantly different according to Duncan test ($p \leq 0.05$).

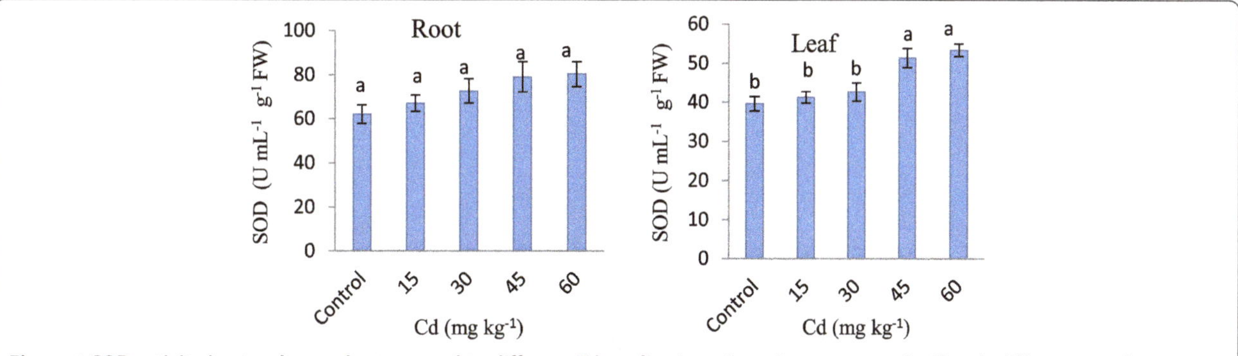

Figure 4 SOD activity in strawberry plant exposed to different Cd applications. Same letters are not significantly different according to Duncan test ($p \leq 0.05$).

respectively but this increase was followed very sharp by 3, 4, 9 and 19 times in leaves.

APX results are shown in Figure 6. Increasing in APX activity belongs to Cd concentrations was found statistically significant. APX activity was monitored regularly increased by 124% in root and 237% in leaf up to 45 mg kg^{-1} Cd concentration as compared control. On the other hand, a decline in APX activity was monitored after 45 mg kg^{-1} Cd concentration that this decline was higher than the control by 28% in root and 74% in leaf. APX activity in leaf was higher than in root.

The abiotic stresses like heavy metals lead to molecular damage to plant cells by generating reactive oxygen species (ROS) [10]. Although Cd does not generate ROS directly, it generates oxidative stress by interrupting the antioxidant defense system [24]. Produced these ROS mainly include GPX, APX and CAT. These antioxidant enzymes balance the ROS production and destruction. Cd also inhibits Calvin cycle enzymes and hence accumulated reduced coenzymes will not be able to accept electrons from PSI. In our experiment the activities of catalase, ascorbate peroxidase and superoxide dismutase were measured. Our results showed that 15, 30, 45 and 60 mg kg^{-1} Cd concentrations led to a significant increase in the antioxidant enzyme activity (SOD; CAT; APX) in both root and leaf of strawberry (Figures 4, 5 and 6). Our results are in agreement with previous studies that have observed findings of Gill et al. [22] who reported that activities of SOD, CAT and APX were found increased in the leaves of garden gress plant with increased dose of Cd treatment.

Effect of Cd on mineral concentration of strawberry

Increasing of Cd concentrations affected content of mineral elements in Camarosa cultivar. In leaves, K, Mg and Mn content was found higher than in root, but Fe, Cu and Zn content was found higher in root with increasing Cd concentrations (Table 1). In leaves, contents of essential elements (Ca, Mg, Fe, Mn, Cu and Zn) were found statistically significant according to Cd concentrations except K while Mg, Fe, Mn and Zn was found statistically significant in root based on Cd concentrations. Initially K, Ca, Mg Fe and Mn contents were tending to increase when compare with control then a slight decrease was observed in both root and leaf at 60 mg kg^{-1} Cd concentration. With increasing Cd concentration Zn, Cu and Mn content was observed decrease in both root and leaf except Cu in root. Nada et al. [23] found that in leaf and root of almond, Cd addition reduced the concentration of macronutrients such as Ca, Mg and K in leaves and in root. Liu et al. [4] report that the interactions of Cd and Fe, Cu and Zn are synergetic in uptake and translocation from root to shoot by rice plants. Yang et al. [21] also

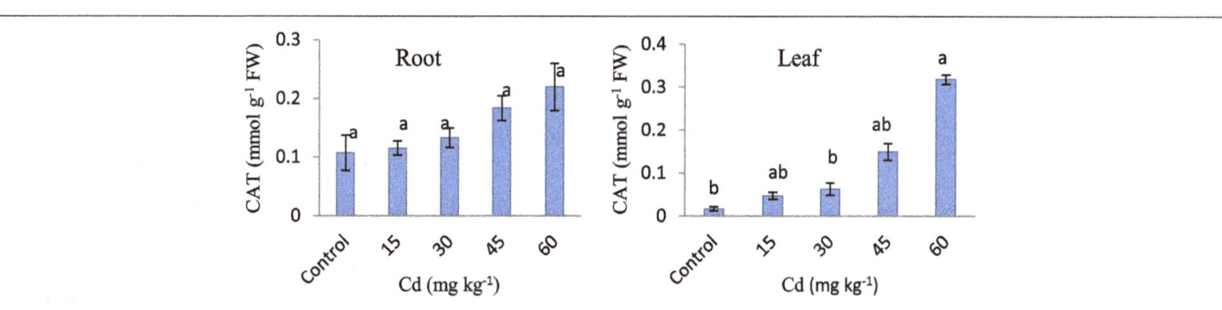

Figure 5 CAT activity in strawberry plant exposed to different Cd applications. Same letters are not significantly different according to Duncan test ($p \leq 0.05$).

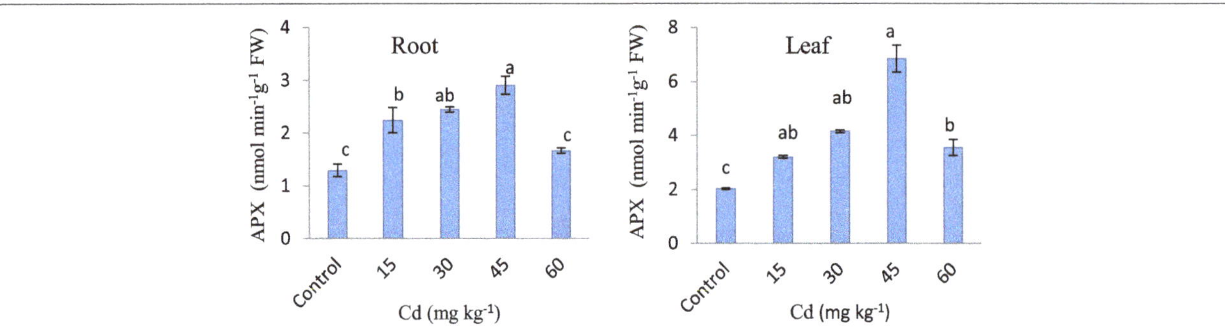

Figure 6 APX activity in strawberry plant exposed to different Cd applications. Same letters are not significantly different according to Duncan test (p ≤ 0.05).

reported that a decrease macronutrient (K and P) contents in *Potamogeten crispus*.

Conclusions

The results suggested that increasing Cd concentrations had negative effect on chlorophyll content and nearly decrease 30% in leaves. The roots accumulate about higher 70% Cd than leaves of strawberry. Results indicated that MDA and antioxidant enzymes (SOD, CAT and APX) content are considered to be indicator in determining Cd tolerance in plant. Strawberry plants affected with increased Cd concentrations. Lipid peroxidation content and antioxidant enzyme activities increased with Cd concentrations.

Table 1 Effect of Cd applications on macro-micro nutrient elements concentrations of root and leaf of strawberries Camarosa cultivars (mg kg^{-1}DW)

Cd applications		Control	15	30	45	60
	Root	3291^{a}	3608^{a}	4698^{a}	4679^{a}	3784^{a}
K	Leaf	18655^{a}	15325^{a}	16339^{a}	18846^{a}	14921^{a}
	Root/Leaf	0.17	0.23	0.28	0.24	0.25
	Root	17144^{a}	17972^{a}	18769^{a}	16295^{a}	15695^{a}
Ca	Leaf	11842^{c}	15489^{c}	21105^{ab}	22714^{a}	16302^{bc}
	Root/Leaf	1.45	1.16	0.89	0.72	0.96
	Root	4038^{a}	4134^{a}	4215^{a}	3257^{b}	2930^{b}
Mg	Leaf	4779^{b}	6037^{ab}	6759^{ab}	7704^{a}	6646^{ab}
	Root/Leaf	0.85	0.68	0.62	0.42	0.44
	Root	980^{b}	1333^{ab}	1653^{a}	1439^{ab}	1326^{ab}
Fe	Leaf	203^{b}	191^{b}	251^{a}	190^{b}	184^{b}
	Root/Leaf	4.82	6.96	6.59	7.58	7.19
	Root	38^{c}	38^{c}	60^{b}	81^{a}	31^{c}
Mn	Leaf	105^{d}	148^{c}	202^{b}	230^{ab}	261^{a}
	Root/Leaf	0.37	0.26	0.30	0.35	0.12
	Root	15.49^{a}	16.38^{a}	17.10^{a}	18.04^{a}	20.65^{a}
Cu	Leaf	10.99^{ab}	8.63^{bc}	8.60^{bc}	13.39^{a}	6.45^{c}
	Root/Leaf	1.41	1.9	1.99	1.35	3.2
	Root	206^{a}	52.87^{c}	72.94^{b}	40.30^{d}	35.63^{d}
Zn	Leaf	22^{a}	17.69^{b}	17.49^{b}	19.32^{ab}	18.90^{ab}
	Root/Leaf	9.34	2.98	4.17	2.08	1.88

Same letters in the same line are not significantly different according to Duncan test (p≤0.05).

Methods

Plant materials and pot experiment

The experiment was carried out in the greenhouse of Yuzuncu Yil University during growing period (from middle May to end of July). The experiment was conducted by using frigo plants of Strawberry (*Fragaria x ananassa* cv. Camarosa) in pot experiment. Four frigo plants were planted into every pot (72x20x17cm) that was filled with peat (4 kg) [25]. Initial stages of grown, plants were fed by adding nutrient solution to the pots. The nutrition solutions contained N 200, Mg 49, K 208, P 37, Ca 167, Mn 1.16, Fe 1.53, Zn 0.09, B 0.46, Cu 0.03 and Mo 0.02 mg/l. Flower buds were cut of early stage of plant's growth. After the plants had four or five leaves about 4 weeks, cadmium applications were started. Cadmium was added to pots at concentration of 0, 15, 30, 45 and 60 mg kg^{-1} in the form of $CdSO_4$*8 H_2O four equal times with watering during growth period. In harvest, 12 plants were harvested to every application, plants were sectioned into roots and leaves and this section was stored at −80°C until antioxidant analyze. Also for macro−micro analysis, fresh root and leafs dried in an oven (80°C) and dried parts were ground and stored until analyze.

Chlorophyll determination

Chlorophyll a and chlorophyll b, 0.5 g fresh leaves were extracted in 80% acetone and were determined spectrophotometrically by Lichtentaler formula [26].

Lipid peroxidation content

MDA content, a product of lipid peroxidation, was used to gauge the level of lipid peroxidation [27]. A leaf sample (0.5 g) was homogenized in trichloro acetic acid,

TCA (10 ml; 0.1%). The homogenate was centrifuged (15 000 g; 5 min) and supernatant was collected. To aliquot (1.0 ml) of the supernatant, 4 ml of 0.5% thiobarbituric acid (TBA) in TCA (20%) was added. The mixture was heated at 95°C for half an hour and then quickly cooled in an ice bath. After centrifugation (10 000 g; 10 min), the absorbance of the supernatant was recorded at 532 nm. The value for non-specific absorption at 600 nm was subtracted. The MDA content was calculated by its extinction coefficient of 155 mM^{-1} cm^{-1} and expressed as nmol MDA per gram fresh weight.

Preparation of extracts and determination of antioxidant enzymes

For the analysis of antioxidant enzyme, 1 g fresh tissue from fourth leaves and the roots was homojenized in 5 ml cold 0.1 M 0.1 M Na-phospat, 0.5 mM Na-EDTA and 1 mM ascorbic acid (pH: 7.5). Samples were centrifuged at 18 000 g for 30 min at a temperature 4°C. Then Catalase activity immediately was determined and the supernatant was stored at −20°C until determined for SOD.

CAT activity was determined using the modified Aebi [28] method, by measurement of the decrease in absorbance at 240 nm for 2 min, in a solution containing H_2O_2 (10 mM) in phosphate buffer (pH 7.0; 50 mM). Enzyme activity was defined as the consumption of 1 μmol H_2O_2 per min and mL using a molar absorptivity of 39.4 mM^{-1} cm^{-1}.

SOD activity was measured by monitoring the inhibition of nitroblue tetrazolioum (NBT) reduction at 560 nm as reported by Giannopolitis and Ries [29]. The reaction mixture contained phosphate buffer (pH 7; 50 mM), Na-EDTA (0.1 mM), riboflavin (75 μM), methionine (13 mM) and enzyme extract (0.1-0.2 ml). Reaction was carried out in test tubes at 25°C under fluorescent lamp (40 W) with irradiance of 75 μmol m^{-2} s^{-1}. The reaction was allowed to run for 10 min and stopped by switching the light off. Blanks and controls were run similarly but without irradiation and enzyme, respectively. Under the experimental condition, the initial rate of reaction, as measured by the difference in increase of absorbance at 560 nm in the presence and absence of extract, was proportional to the amount of enzyme.

APX activity was assayed according to the method of Nakano and Asada [30] by recording the decrease in ascorbate content at 290 nm, as ascorbate was oxidized. The reaction mixture contained potassium phosphate buffer (pH 7.0; 50 mM), ascorbic acid (5 mM), EDTA (0.1 mM), H_2O_2 (0.1 m M) and diluted enzyme (0.1 ml) in a total volume of 3.0 ml. The reaction was started with the addition of H_2O_2 and absorbance was recorded at 290 nm spectrophotometrically for 1 min.

Macronutrient and micronutrient determination

İn dried leaves and roots, Cd contents and others nutrient element concentrations were analyzed by an atomic absorption spectrophotometer (Varian Techtron Model AAS 1000, Varian Associates, Palo Alto, CA). The samples, which were digested in an acid solution (HCL 3%) were passed through the AAS system using different lamps, and calibrated with related minerals in different concentrations for different micronutrients.

Statistical analysis

The experiment was designed as a complete random block design and all measurements were replicated four times. The statistical analysis of the data obtained was performed using the software SPSS 22.0. The results were subjected to one-way ANOVA using the Duncan test to check for significant differences between means ($p < 0.05$). Error bars in graphs represent ± standard error.

Competing interests

The authors declare that they do not have competing interests.

Authors' contributions

FM, MG, SE and TE made a significant contribution to experiment design, acquisition of data, analysis and drafting of the manuscript. FB, MZ, HZEJ and SE have made a substantial contribution to interpretation of data, drafting and carefully revising the manuscript for intellectual content. All authors read and approved the final manuscript.

Acknowledgements

This research was supported by Yuzuncu Yil Universty of the head of scientific research (BAP), Van, Turkey, Project No.: 2010-ZF-B015.

Author details

[1]Department of Horticulture, Faculty of Agriculture and Natural Sciences, Abant Izzet Baysal University, Bolu, Turkey. [2]Department of Horticulture, Faculty of Agriculture, Ataturk University, Erzurum, Turkey. [3]Department of Horticulture, Faculty of Agriculture, Yuzuncu Yil University, Van, Turkey. [4]Department of Horticulture, Faculty of Agriculture, Ordu University, Ordu, Turkey. [5]Department of Crop Science, Faculty of Agriculture, University Putra Malaysia, 43400 Selangor, Malaysia. [6]The Patent Office, Karachi, Pakistan.

References

1. Ozbek K, Cebel N, Unver I. Extractability and phytoavailability of cadmium in Cd-rich pedogenic soils. Turk J Agric For. 2014;38:70–9.
2. Hassan M, Mansoor S. Oxidative stress and antioxidant defense mechanism in mung bean seedlings after lead and cadmium treatments. Turk J Agric For. 2014;38:55–61.
3. Zornoza P, Vazquez S, Esteban E, Fernandez-Pascual M, Carpena R. Cadmium-stress in nodulated white lupin: strategies to avoid toxicity. Plant Physiol Bioc. 2002;40:1003–9.
4. Liu J, Li K, Xu J, Liang J, Lu X, Yang J, et al. Interaction of Cd and five mineral nutrients for uptake and accumulation in different rice cultivars and genotypes. Field Crop Res. 2003;83:271–81.
5. Smeets K, Cuypers A, Lambrechts A, Semane B, Hoet P, Laerve AV, et al. Induction of oxidative stress and antioxidative mechanisms in *Phaseolus vulgaris* after Cd application. J Plant Physiol Biochem. 2005;43:437–44.
6. Mishra S, Srivastava S, Tripathi RD, Govidarajan R, Kuriakose SV, Prasad MNV. Phytochelatin Synthesis and response of antioxidants during cadmium stress in *Bacopa monnieri* L. J Plant Physiol Biochem. 2006;44:25–37.
7. Barcelo J, Poschenrieder C. Plant water relations as affected by heavy metal stress: a review. J Plant Nutr. 1990;13:1–37.

8. Chen YX, He YF, Luo YM, Yu YL, Lin Q, Wong MH. Physiological mechanism of plant roots exposed to cadmium. Chemosphere. 2003;50:789–93.
9. Hegedus A, Erdei S, Janda T, Toth E, Horvath G, Dubits D. Transgenic tobacco plants over producing alfafa aldose/aldehyde reductase show higher tolerance to low temperature and Cadmium stress. Plant Sci. 2004;166:1329–33.
10. Zhang H, Jiang Y, He Z, Ma M. Cadmium accumulation and oxidative burst in garlic (*Allium sativum*). J Plant Physiol. 2005;162:977–84.
11. Demiral T, Turkan I. Comparative lipid peroxidation, antioxidant defense systems and proline content in roots of two rice cultivars differing in salt tolerance. Environ Exp Bot. 2005;53:247–57.
12. Asada K. The water-water cycle in chloroplasts: scavenging of active oxygen's and dissipation of excess photons. Annu Rev Plant Physiol Plant Mol Biol. 1999;50:601–39.
13. Shah K, Ritambhara GK, Verma S, Dubey RS. Effect of cadmium on lipid peroxidation, superoxide anion generation and activities of antioxidant enzymes in growing rice seedlings. Plant Sci. 2001;161:1135–44.
14. Sbartai H, Rouabhi R, Sbartai I, Berrebbah H, Djebar RM. Induction of anti-oxidative enzymes by cadmium stress in tomato (*Lycopersicon esculentum*). Afr J Plant Sci. 2008;2:72–6.
15. Grant CA, Buckley WT, Bailey LD, Selles F. Cadmium accumulation in crops. Canadian J Plant Sci. 1998;78:1–17.
16. Esitken A, Yildiz HE, Ercisli S, Donmez MF, Turan M, Gunes A. Effects of plant growth promoting bacteria (PGPB) on yield, growth and nutrient contents of organically grown strawberry. Sci Hortic. 2010;124(1):62–6.
17. Torun AA, Aka Kacar Y, Erdem N, Bicen B, Serce S. *In vitro* screening of octoploid *Fragaria chiloensis* and *Fragaria virginiana* genotypes against iron deficiency. Turk J Agric For. 2014;38:169–79.
18. Treder W, Cieslinski G. Cadmium uptake and distribution in strawberry plants as affected by its concentration in soil. J Fruit Ornam Plant Res. 2000;8:127–35.
19. Qian H, Li J, Sun L, Chen W, Sheng GD, Liu W, et al. Combined effect of copper and cadmium on *Chlorella vulgaris* growth and photosynthesis-related gene transcription. Aquat Toxicol. 2009;94:56–61.
20. Fang Z, Bouwkamp JC, Solomos T. Chlorophyllase activities and chlorophyll degradation during leaf senescence in non-yellowing mutant and wild type of *Phaseolus vulgaris* L. J Exp Bot. 1998;49:503–10.
21. Yang HY, Shi GX, Xu QS, Wang HX. Cadmium effects on mineral nutrition and stress-related induces in *Potamogeton cripru*s. Russ J Plant Physl. 2011;58:253–60.
22. Gill SS, Khan NA, Tuteja N. Cadmium at high dose perturbs growth, photosynthesis and nitrogen metabolism while at low dose it up regulates sulfur assimilation and antioxidant machinery in garden cress *(Lepidium sativum L.)*. Plant Sci. 2012;182:112–20.
23. Nada E, Ferjani BA, Rhouma A, Bechir BR, Imed M, Makki B. Cadmium-induced growth inhibition and alteration of biochemical parameters in almond seedlings grown in solution culture. Acta Physiol Plant. 2007;29:57–62.
24. di Toppi S, Gabrielli R. Response to cadmium in higher plants. Environ Exp Bot. 1999;41:105–30.
25. Sahin U, Anapali O, Ercisli S. Physico-chemical and physical properties of some substrates used in horticulture. Gartenbauwissenshaft. 2002;67(2):55–60.
26. Lichtentaler HK. Chlorophyll and carotenoids pigments of photosynthetic biomembranes. Meth Enzymol. 1994;148:350–82.
27. Heath RL, Packer L. Photoperoxidation in isolated chloroplast. I. Kinetics and stoichiometry of fatty acid peroxidation. Arch Biochem Biophys. 1968;125:189–98.
28. Aebi H. Catalase *in vitro*. Methods Enzymol. 1984;105:121–6.
29. Giannopolitis CN, Ries SK. Superoxide dismutase. I. Occurrence in higher plants. Plant Physiol. 1977;59:309–14.
30. Nakano Y, Asada K. Hydrogen peroxide is scavenged by ascorbate specific peroxidase in spinach Chloroplasts. Plant Cell Physiol. 1981;22:867–80.

8

The antioxidative defense system is involved in the premature senescence in transgenic tobacco (*Nicotiana tabacum NC89*)

Yu Liu[1], Lu Wang[1], Heng Liu[1], Rongrong Zhao[1], Bin Liu[1], Quanjuan Fu[2] and Yuanhu Zhang[1*]

Abstract

Background: α-Farnesene is a volatile sesquiterpene synthesized by the plant mevalonate (MVA) pathway through the action of α-farnesene synthase. The *α-farnesene synthase 1* (*MdAFS1*) gene was isolated from apple peel (var. *white winter pearmain*), and transformed into tobacco (*Nicotiana tabacum* NC89). The transgenic plants had faster stem elongation during vegetative growth and earlier flowering than wild type (WT). Our studies focused on the transgenic tobacco phenotype.

Results: The levels of chlorophyll and soluble protein decreased and a lower seed biomass and reduced net photosynthetic rate (Pn) in transgenic plants. Reactive oxygen species (ROS) such as hydrogen peroxide (H_2O_2) and superoxide radicals ($O_2^{\cdot -}$) had higher levels in transgenics compared to controls. Transgenic plants also had enhanced sensitivity to oxidative stress. The transcriptome of 8-week-old plants was studied to detect molecular changes. Differentially expressed unigene analysis showed that ubiquitin-mediated proteolysis, cell growth, and death unigenes were upregulated. Unigenes related to photosynthesis, antioxidant activity, and nitrogen metabolism were downregulated. Combined with the expression analysis of senescence marker genes, these results indicate that senescence started in the leaves of the transgenic plants at the vegetative growth stage.

Conclusions: The antioxidative defense system was compromised and the accumulation of reactive oxygen species (ROS) played an important role in the premature aging of transgenic plants.

Keywords: Tobacco, Senescence, Transcriptome, Reactive oxygen species (ROS)

Background

α-Farnesene is a volatile plant sesquiterpene. It also acts as a semiochemical that can affect the behavior of some insect species [1, 2]. α-Farnesene can accumulate in apple peel, and the oxidative production of α-farnesene is associated with the development of scald symptoms [3, 4], a physiological disorder of apple and pear fruits [5, 6]. α-Farnesene synthesis mainly occurs via the cytosolic mevalonic acid (MVA) pathway, which is initiated by the first rate-limiting enzyme, 3-hydroxy-3-methylglutaryl-CoA reductase (HMGR) [7]. The other important step in the α-farnesene synthesis pathway involves farnesyl diphosphate synthase (FPS). FPS catalyzes the condensation of geranyl diphosphate (GPP) and isopentenyl diphosphate (IPP) to produce farnesyl diphosphate (FPP), which is the immediate precursor of sesquiterpenes [8]. In the final step of synthesis, α-farnesene synthase uses FPP as the substrate to catalyze the synthesis of α-farnesene. Pechous and Whitaker (2004) cloned the *α-farnesene synthase 1* (*AFS1*) gene of the 'Law Rome' apple and expressed it in bacteria [9]. When farnesyl diphosphate (FPP) was supplied as the substrate for the bacterially expressed recombinant enzyme, α-farnesene was synthesized.

Terpenoids play crucial roles in plant defense, growth, and development. Genetic engineering of key genes involved in the terpene synthesis pathway has generated

*Correspondence: yyhzhang@sdau.edu.cn
[1] State Key Laboratory of Crop Biology, College of Life Sciences, Shandong Agricultural University, 61 Dai Zong Street, Tai'an 271018, Shandong, People's Republic of China
Full list of author information is available at the end of the article

mutants and transgenic plants with altered growth and development. For example, the *hmg1* mutant exhibits dwarfing, early senescence, sterility, and earlier induction of the *senescence-associated 12* (*SAG12*) gene compared to wild-type (WT) plants [10]. In *Arabidopsis thaliana*, farnesyl diphosphate synthase 1 (FPS1) and farnesyl diphosphate synthase 2 (FPS2), which encode cytosolic FDP synthase, are differentially expressed [11]. Compared to WT, overexpression of *A. thaliana FPS1* results in a cell death/senescence-like phenotype, which grew less vigorously than wild-type plants, and premature induction of the *SAG12* gene [12]. Overexpression of *A. thaliana FPS2* leads to the synthesis of several novel sesquiterpenes, including E-β-farnesene, and transgenic plants show enhanced growth [13]. Of the three functional FPPS in maize, *FPPS3* is induced by herbivory, and is essential to the production of the volatile sesquiterpenes, including β-caryophyllene [14]. Sesquiterpenes play an important role in plant physiological and ecological functions such as scavenging for reactive oxygen species (ROS), stabilizing membrane structure, and inhibiting bacterial growth [15, 16]. Treatment of *Aquilaria sinensis* (Lour.) cell culture suspensions with hydrogen peroxide (H_2O_2) induces the production of sesquiterpenes, stimulates programmed cell death (PCD), and increases salicylic acid (SA) accumulation. These results indicate potential interactions between sesquiterpene synthesis and programmed cell death (PCD) during *A. sinensis* (agarwood) formation [17]. In plants, terpenes are synthesized via the mevalonate (MVA) pathway in the cytosol and peroxisomes, and the 2-C-methyl-D-erythritol 4-P/1-deoxy-D-xylulose 5-P (MEP) pathway in plastids [18]. The two pathways are capable of exchanging intermediates [19]. In the MEP pathway, disruption of the *1-deoxy-D-xylulose-5-phosphate reductoisomerase* (*DXR*) gene leads to biosynthetic deficiency of photosynthetic pigments, GAs and ABA, resulting in developmental abnormalities [20]. Monoterpenes are volatile terpenes synthesized by the MEP pathway. Monoterpenes also play key roles in plant defense and apoptosis-like cell death [21–23].

Leaf senescence is characterized by degradation of the chlorophyll, photosynthetic proteins and other macromolecules, conversion of peroxisomes into glyoxysomes, and increased production of ROS [24]. ROS are a central element of the senescence process, which is toxic to plant cells [25]. Excessive production of ROS causes necrosis via programmed cell death [26]. The chloroplast is the most important organelle in $^1O_2$, production and is also regarded as the major intracellular producer of partially reduced oxygen species such as H_2O_2 and $O_2^{\cdot-}$ [27]. Chloroplasts, which are a rich source of nitrogen (N), are the first organelles that are dismantled during senescence [28]. Nutrient signaling also plays important roles in leaf senescence. For example, leaf senescence is induced when sugar levels exceed or decrease beyond acceptable threshold levels [29, 30]. Sugar-induced senescence is particularly important in conditions of low nitrogen availability and may also play a role in light signaling [31, 32].

Overexpression of *α-farnesene synthase 1* (*MdAFS1*) in transgenic tobacco plants has been associated with an unexpected phenotype, specifically accelerated stem elongation and early flowering, compared to WT plants. Therefore, in this study, transcriptome analysis of transgenic tobacco plants was conducted to study the differentially expressed genes. We concluded that the antioxidative defense system is compromised, and accumulation of ROS activates the senescence process. In the future, we anticipate that the transcriptome database will be a valuable resource for improved understanding of the molecular basis for alterations in plant growth and development.

Results

Identification of transgenic plants

Three third generation lines were selected for identification of transgenic plants at the transcript level. The relative *MdAFS1* mRNA levels of the transgenic lines were average, higher, and lower in T3-1, T3-2 and T3-3, respectively (Fig. 1c). Expression levels of *MdAFS1* varied in different parts of the transgenic plants; it had a higher transcript level in the flowers (Fig. 1d). So, flowers were selected as experimental materials for GC–MS analysis of the terpenes. The release of farnesene (0.41 %) was detected in the flowers of transgenic plants (Fig. 1a, b). The chemical formula of farnesene is $C_{15}H_{24}$ and the molecular weight is 204. In addition to this, β-myrcene (0.34 %), linalool (5.92 %), and caryophyllene (2.98 %) were detected in WT plants. The release of linalool (2.3 %) and caryophyllene (1.71 %) was detected in the flowers of transgenic plants (Table 1).

Plant phenotype analysis

To determine growth differences in the WT, T3-1, T3-2 and T3-3, plant height, leaf area, and seed biomass were measured. At 8 weeks of age, the stem of transgenic plants had accelerated elongation with lengths reaching about 21 cm, whereas WT stems were approximately 2.0 cm in length (Fig. 2B). The transgenic plants flowered at about 10 weeks of age and had mature seeds at 18 weeks. At 18 weeks, the WT plants were still in full bloom. The heights of the tobacco plants were as follows: 108.0 cm, 81.7 cm, 84.0 cm and 83.3 cm for WT, T3-1, T3-2 and T3-3, respectively (Fig. 2C). Based on changes in height throughout the developmental process, the transgenic

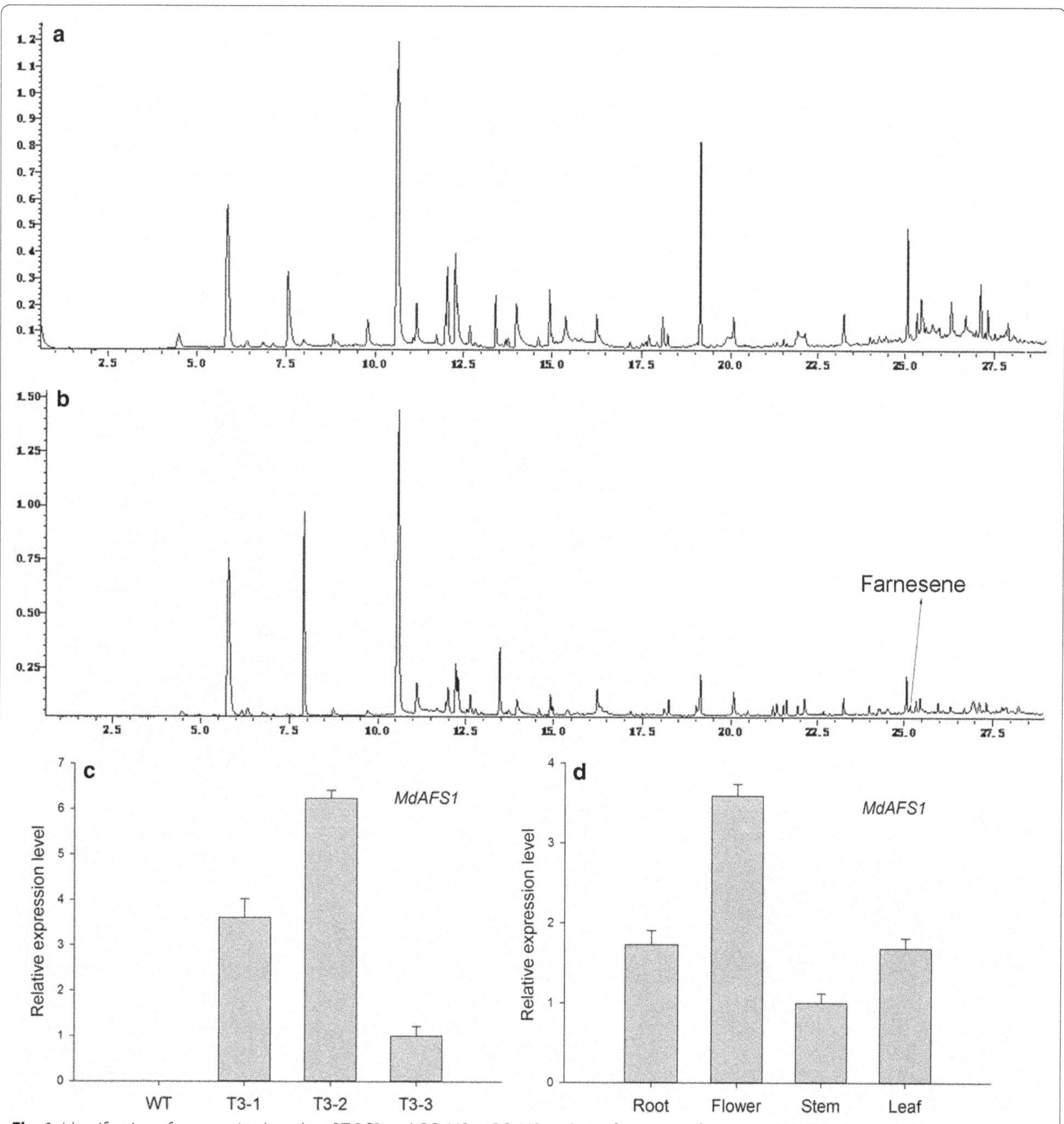

Fig. 1 Identification of transgenic plants by qRT-PCR and GC–MS. **a** GC–MS analysis of terpenes of WT plants. **b** GC–MS analysis of terpenes of transgenic plants. **c** *MdAFS1* transcript in different transgenic plants. **d** *MdAFS1* transcript in different parts of transgenic plants. The *MdAFS1* transcript level was normalized to 18S RNA expression. The standard error of the mean of three biological replicates (nested within three technical replicates)

plants had faster development (Fig. 2A). At 8 weeks old, the transgenic plants showed no change in leaf biomass, a fivefold increase in stem biomass, and a twofold increase in root biomass (Table 2). There were 9–10 leaves per transgenic plant, whereas WT plants had 8 leaves. Leaf area was measured at different locations on transgenic and WT plants. Transgenic plants showed no changes in total leaf area but the functional leaf area significantly decreased and the upper leaf area significantly increased (Table 3).

The transgenic plants have reduced seed biomass

Seed biomass was considered an important indicator of reproductive growth. One capsule of tobacco (*NC89*) contains 1500–2000 grains. Though the weight per 1000

Table 1 GC-MS analysis of tobacco terpenes

Retention time (min)	Compound formula	Compound name	Relative percentage content (%)	
			WT	T3-2
16.317	$C_{10}H_{16}$	β-Myrcene	0.34	–
19.127	$C_{10}H_{16}$	Linalool	5.92	2.3
25.064	$C_{15}H_{24}$	Caryophyllene	2.98	1.71
25.175	$C_{15}H_{24}$	Farnesene	–	0.41

"–" Indicates that this compound was not tested

grains increased by 117.1, 115.6 and 111.4 %, the number of capsules decreased by 66.7, 70.8 and 76.9 % in the T3-1, T3-2, and T3-3 lines, respectively, compared to WT plants (Table 4). Given these results, the transgenic plants produced less seed biomass than WT plants.

Physiological parameters related to senescence

Leaf senescence generally accelerates chlorophyll degradation and cell death [33, 34], which occur simultaneously with protein degradation. For these reasons, chlorophyll and soluble protein levels were measured. The T3-1, T3-2 and T3-3 plants showed a decrease in the content of chlorophyll *a* by 87.1, 83.9 and 78.2 %;

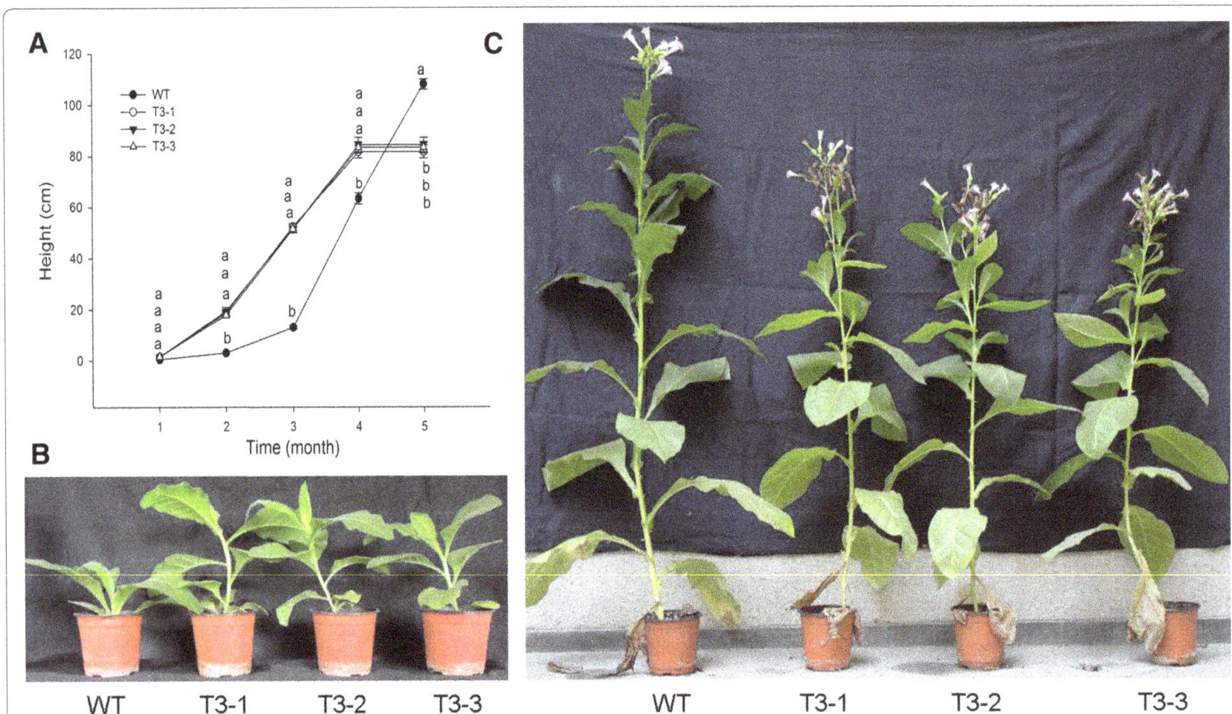

Fig. 2 Phenotypic variation of wild-type (WT) and transgenic plants at different developmental stages. **A** Variation in the height of WT and transgenic plants at different developmental stages, each *line* is the mean of five replicates. *Different letters* indicate statistically significant differences at $P \leq 0.05$, **B** eight-leaf period, and **C** filling period

Table 2 Biomass of tissues from wild-type (WT) and transgenic (T) plants

Biomass (g)	WT	T3-1	T3-2	T3-3
Root	1.73 ± 0.09b	3.26 ± 0.62a	3.06 ± 0.10a	3.65 ± 0.76a
Stem	2.19 ± 0.09b	10.84 ± 1.90a	10.76 ± 0.63a	11.46 ± 1.90a
Leaf	21.33 ± 4.33a	20.93 ± 2.82a	20.63 ± 1.52a	22.57 ± 2.61a
Total	25.25 ± 2.70b	35.03 ± 12.90a	34.46 ± 4.60a	37.67 ± 17.53a

Each line represents the mean of five replicates. *Different letters* indicate statistically significant differences at $P \leq 0.05$ in same parts of wild-type (WT) and transgenic (T) plants

Table 3 Leaf area of WT and transgenic (T) plants

Lines	1 (cm^2)	2 (cm^2)	3 (cm^2)	4 (cm^2)	5 (cm^2)	6 (cm^2)	7 (cm^2)	8 (cm^2)	9 (cm^2)
WT	28.8 ± 5.0a	62.8 ± 13.3a	82.9 ± 14.7a	158.5 ± 14.6a	154.8 ± 12.4a	112.8 ± 5.4a	77.9 ± 8.9a	25.3 ± 7.6b	
T3-1	26 ± 8.4a	59.2 ± 21.0a	97.3 ± 4.8a	96.8 ± 1.1b	132.9 ± 2.6b	103.0 ± 6.5a	89.5 ± 7.0a	56.4 ± 14.2a	18.7 ± 6.5
T3-2	30.2 ± 1.0a	67.5 ± 2.7a	89.9 ± 10.2a	124.6 ± 5.6b	133.5 ± 2.7b	111.4 ± 12.0a	86.1 ± 11.6a	53.2 ± 0.3ab	23.2 ± 2.9
T3-3	24.8 ± 3.2a	58.0 ± 8.4a	81.4 ± 1.9a	116.7 ± 11.9b	128.9 ± 6.7b	116.1 ± 1.5a	88.8 ± 6.2a	53.5 ± 10.5ab	31.1 ± 16.3

Each line represents a mean of five replicates. *Different letters* indicate statistically significant differences at $P \leq 0.05$ in same leaf position of wild-type (WT) and transgenic (T) plants

Table 4 Seed biomass of WT and transgenic (T) tobacco plants

Biomass	WT	T3-1	T3-2	T3-3
Capsule (number)	24.0 ± 1.0a	16.0 ± 1.0b	17.0 ± 3.0b	16.3 ± 2.3b
Weight per 1000 grains (mg)	76.2 ± 1.5c	89.2 ± 3.5a	88.1 ± 1.8a	84.9 ± 2.2b

Tobacco seeds dried to constant weight were used in this experiment. Values shown are means ± SE (five biological replicates is presented)

Different letters indicate statistically significant differences at $P \leq 0.05$

chlorophyll *b* decreased by 93.7, 86.3 and 80.4 %; the total chlorophyll (a + b) content decreased by 88.2, 88.3 and 84 % (Fig. 3A) and carotenoid content decreased by 70.5, 79.2 and 73.9 %, respectively (Fig. 3B). Soluble protein levels decreased significantly, by 53.8, 51.6 and 53.3 % in T3-1, T3-2 and T3-3, respectively (Fig. 3C). Chlorophyll and photosynthesis proteins are important elements in photosynthesis. The value of Pn decreased significantly by 82.7, 81.9 and 84.8 % in three transgenic lines (Fig. 3D). Water content showed a significant reduction (Fig. 3E). Malondialdehyde (MDA) is the decomposition product of membrane lipid peroxidation, and it accumulates in senescent leaves. MDA content increased by 121.7, 127.1 and 119.7 % in the T3-1, T3-2, and T3-3 lines, respectively (Fig. 3F).

Differentially expressed unigenes in leaf transcriptome data are related to senescence

The molecular mechanism underlying leaf senescence in transgenic tobacco was studied using transcriptome analysis of 8-week-old plants. De novo transcriptome assembly generated a total of 249,185 unigenes. The total number of differentially expressed transcripts ($P \leq 0.01$, ratio $\geq$2 or $\leq$0.5) was 5835, of which 2028 were upregulated and 3807 were downregulated in transgenic plants. The expression levels of six differentially expressed unigenes and three antioxidant enzyme genes were analyzed using qRT-PCR to test the reliability of the transcriptome database. The results were consistent with the transcriptome data (Additional file 1: Fig. S1). To classify the predicted functions of the unigenes, nucleotide and protein databases were used. GO and KEGG analyses showed that some differentially expressed unigenes were related to leaf senescence. For example, ubiquitin-mediated proteolysis, cell growth, and death were upregulated unigenes. However, antioxidant enzymes, nitrogen metabolism, photosynthesis, and carotenoid biosynthesis were downregulated unigenes (Additional files 2, 3). A total of 87 downregulated unigenes of the photosynthesis signaling pathway are presented in Fig. 4. All transcripts of tobacco (*N. tabacum* NC89) have been deposited to GenBank as Accession Number GDGU00000000.

Accumulation of reactive oxygen species (ROS) in transgenic plants

As indicators of oxidative stress and leaf senescence, the levels of $O_2^{\cdot-}$ and H_2O_2 in leaf tissues were measured under controlled conditions. Transgenic plants, T3-1, T3-2 and T3-3, had increases in the production rate of $O_2^{\cdot-}$ by 138.0, 147.5 and 136.6 % (Fig. 5A), and H_2O_2 levels increased by 111.5, 115.3 and 119.8 %, respectively (Fig. 5B). Histochemical staining can detect the accumulation of reactive oxygen species (ROS); deeper 3,3′-diaminobenzidine (DAB) and nitroblue tetrazolium (NBT) staining was seen in transgenic plants. These findings were consistent with the ROS content results.

Antioxidant enzyme activity and gene expression analysis

Antioxidant enzymes participated in the scavenging of ROS. However, the activity of three antioxidant enzymes in T3-1, T3-2 and T3-3 decreased significantly, APX by 52.6, 42.9 and 43.4 %; CAT by 63.7, 56.6 and 65.3 %; and SOD by 59.1, 46.4 and 44.7 %, respectively (Fig. 6). The expression of antioxidant enzyme genes were consistent with the enzyme activity, which were downregulated in transgenic plants.

Enhanced sensitivity of transgenic plants to DCMU treatment

The 3-(3,4-dichlorfenyl)-1,1-dimethylkarbonyldiamid (DCMU) molecule is an electron transfer inhibitor that acts during photosynthesis. Exogenous DCMU treatment induces production of ROS in the chloroplasts. Three week old plants were treated with 70 μM DCMU. After 10 days, all the transgenic plants had died, but the WT plants were

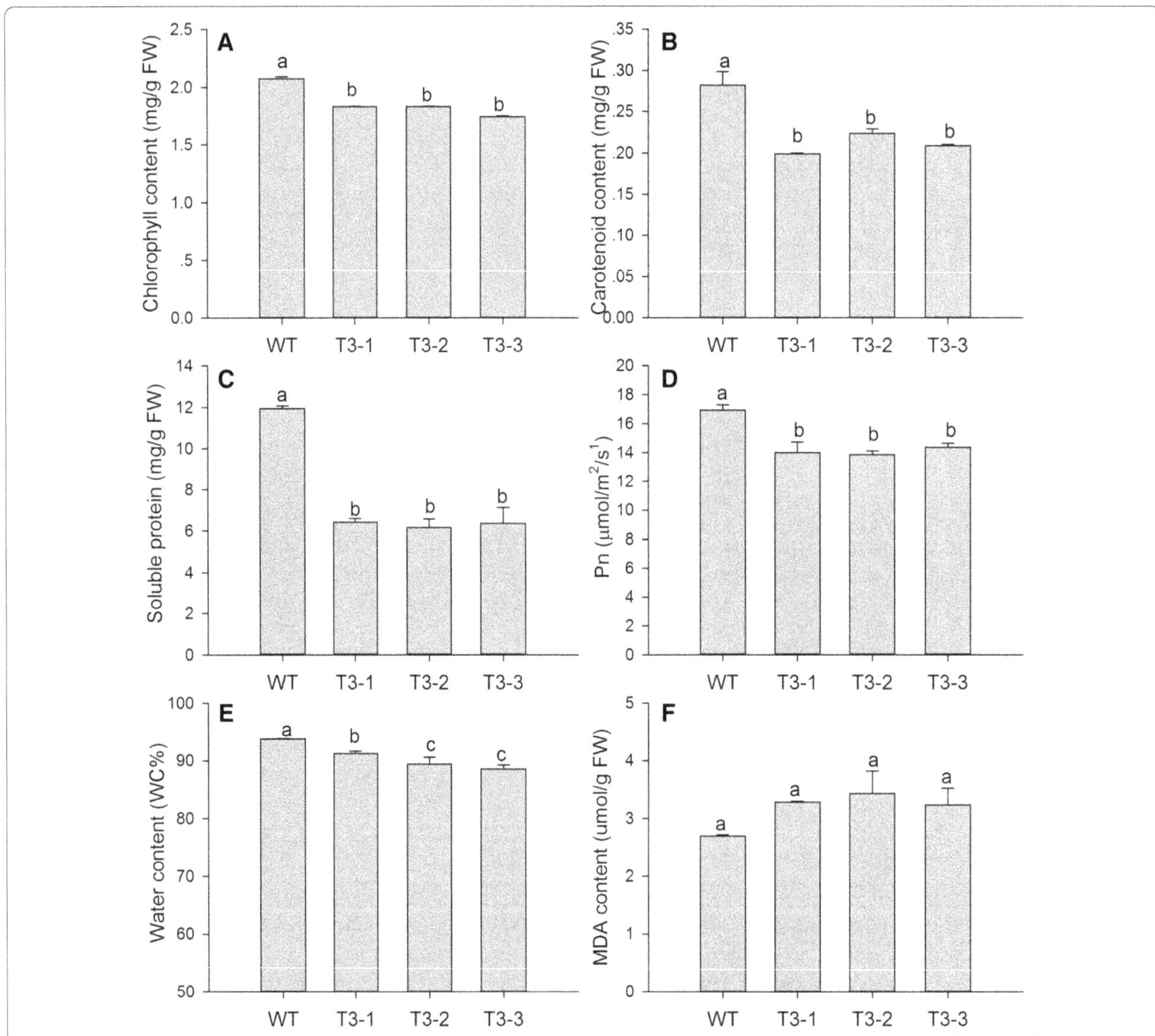

Fig. 3 Measurement of senescence-related physiological parameters in WT and transgenic plants. **A** Chlorophyll content, **B** carotenoid content, **C** soluble protein content, **D** net photosynthetic rate, **E** shows water content, and **F** signifies malondialdehyde (MDA) content. Data are mean ± SE (n = 5, five biological replicates per *line*). *Different letters* indicate statistically significant differences at $P \leq 0.05$

still alive (Fig. 7). This result indicated that the transgenic plants were more sensitive to oxidative stress.

Expression of senescence marker genes

Based on the leaf senescence database (LSD, http://www.eplantsenescence.org/) and the typical characteristics of leaf senescence, senescence-related downregulated genes were selected, including chlorophyll *a/b* binding protein (*CAB*), ribulose-1,5-bisphosphate carboxylase-oxygenase small subunit (*RBCS2B*), nitrate reductase (*Nia*), and chloroplastic glutamine synthetase (*GS2*). The upregulated genes included cysteine proteinase (*NtCP1* and *SAG12*), glutamate dehydrogenase (*GDH*), chlorophyllase (*CHL*), and *Ntdin*. qRT-PCR confirmed that *CAB*, *RBCS2B*, *Nia*, and *GS2* were downregulated and that *NtCP1*, *SAG12*, *GDH*, *CHL*, and *Ntdin* were upregulated (Fig. 8). The transcript levels of these senescence marker genes are consistent with the premature senescence characteristic.

Discussion

Transgenic plants showed a distinct premature senescence phenotype. The stems of transgenic plants rapidly elongated compared to those of WT plants. This occurred during the 7 week interval between first true leaf appearance to emergence of flower primordia (Fig. 2A).

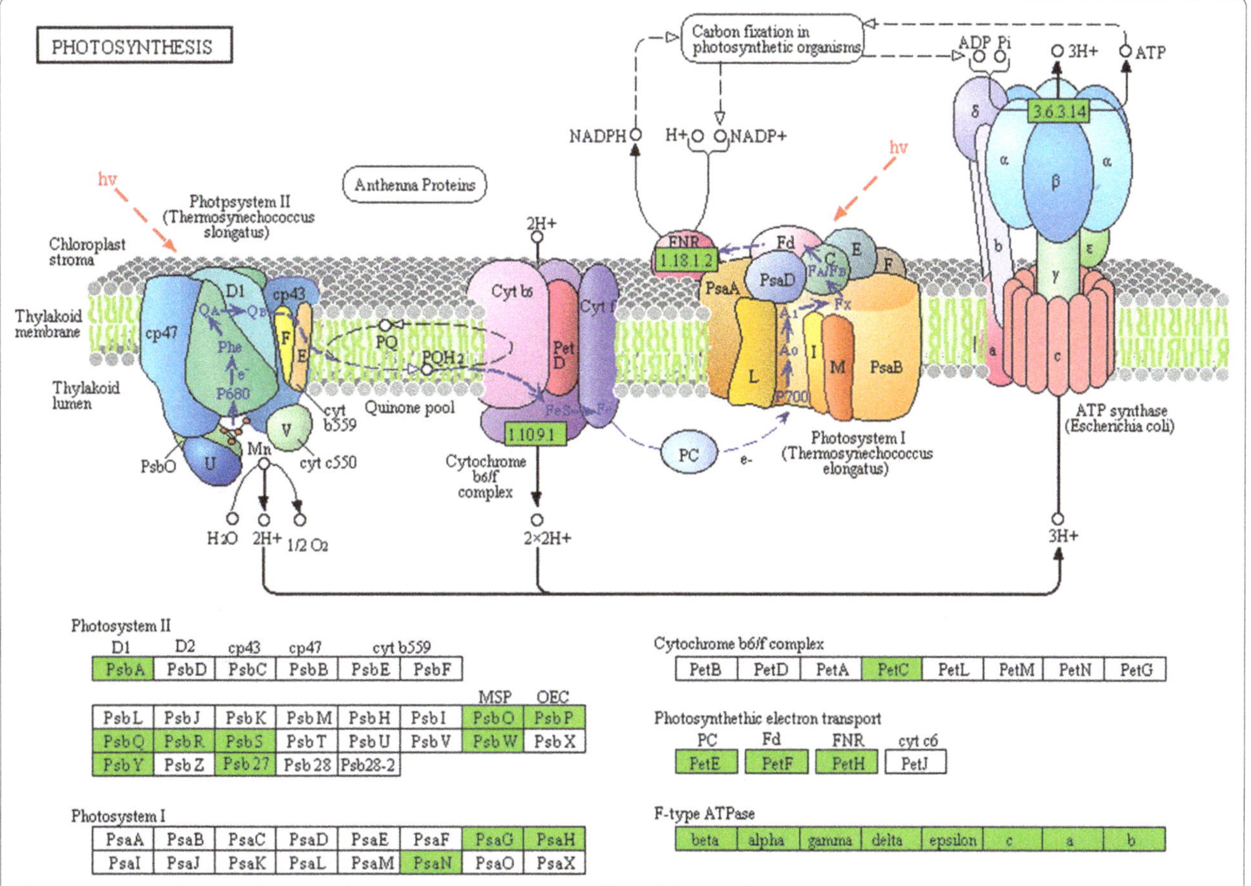

Fig. 4 Kyoto encyclopedia of genes and genomes (KEGG) pathway analysis of photosynthesis. *Green* represents downregulated unigenes ($P \leq 0.01$, ratio ≤ 0.5). *psbS* chloroplast photosystem II 22 kDa component, *psaN* photosystem I reaction center subunit, *psbY* photosystem II core complex proteins, *psbP* photosystem II oxygen-evolving enhancer protein 2, *petC* Rieske FeS precursor protein 2, *atpG* gamma subunit of ATP synthase, *petE* plastocyanin A, *psb27* photosystem II Psb27 protein, *psbO* chloroplast PsbO4 precursor, *psaH* photosystem I reaction centre subunit, *psaG* photosystem I reaction center V, *petF* ferredoxin, *atpF* ATP synthase subunit b, *psbW* photosystem II PsbW protein, *petH* ferredoxin-NADP reductase, *psbQ* photosystem II oxygen-evolving enhancer protein 3, *psbR* photosystem II 10 kDa polypeptide, *psbA* photosystem II protein D1

Transgenics had early flowering at about 50 days which was associated with a shorter vegetative period. Tobacco plants exhibited successive leaf senescence, starting from the basal leaves and advancing toward the apical leaves. At the filling period, which generally occurs in 20 week-old plants, the T3-1, T3-2 and T3-3 plants had at least two dead basal leaves, whereas WT plants only had one dead basal leaf (Fig. 2C). Changes in transgenic plant phenotype were indicative of premature senescence. Meanwhile, the transgenic plants developed new stems at the upper part of the plant. This phenomenon might be related to weak sink strength, which causes a slower progression of senescence in tobacco [35]. The new collateral delayed the death of the whole plant. Tobacco senescence is largely independent of floral development making it a model plant species for research [36]. However, leaf senescence in transgenic plants began during vegetative growth.

The main physiological purpose of leaf senescence is nutrient salvage involving the hydrolysis of macromolecules and their subsequent remobilization to other plant parts [37, 38]. We speculated that transgenic tobacco leaves activated the nutrient salvage program. Carbohydrates were temporarily stored in the stem and root tissues during vegetative growth, resulting in an increase the biomass of stems and roots (Table 2). This allocation of nutrients is beneficial to carbohydrate reserves and to the reduction of energy consumption [39], thereby remobilizing reproductive growth [28, 40]. At the reproductive stage, the premature senescent leaves of transgenic plants appeared to remobilize nutrients to the limited number of seeds, and increased the weight per 1000 grains. However, the decrease in total seed biomass is the result of senescence of transgenic plants.

Leaf senescence can be triggered by high C:N ratios [41, 42]. Glucose and sucrose repress the transcription

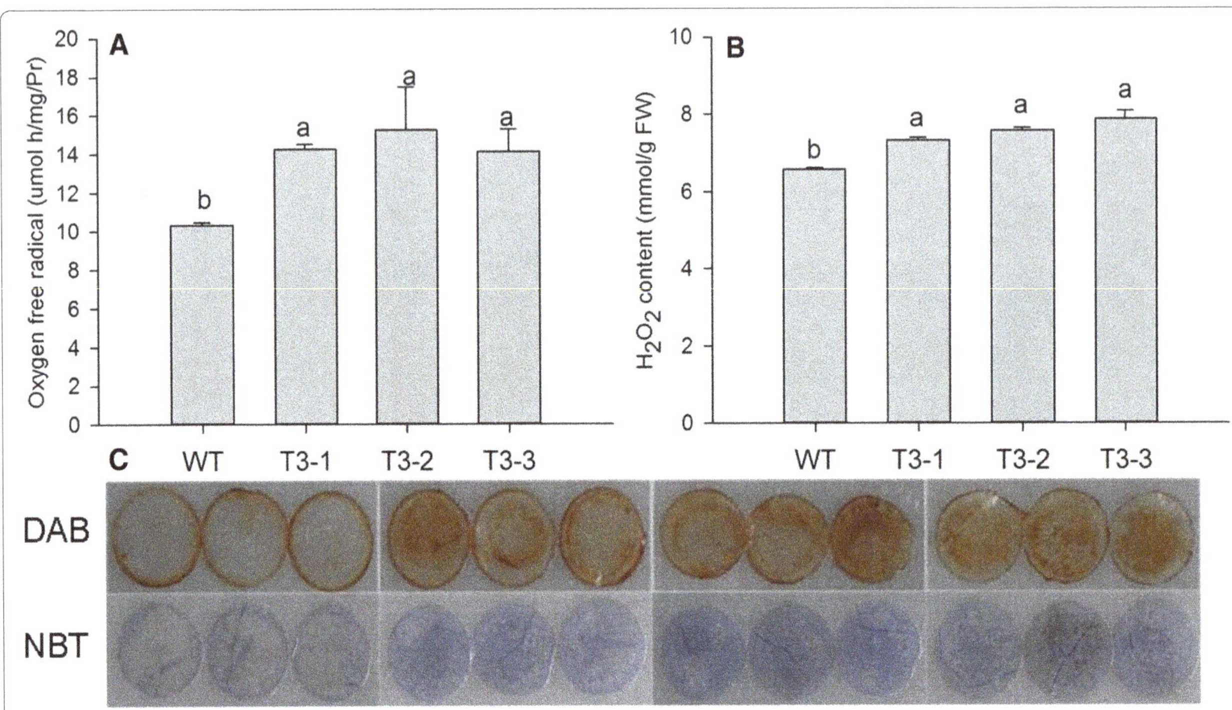

Fig. 5 H_2O_2 levels and production rate of $O_2^{\cdot-}$ in WT and transgenic plants. Data are expressed as the mean ± SE (n = 5, five biological replicates per *lines*). *Different letters* indicate statistically significant differences at $P \leq 0.05$. **a** represents the content of H_2O_2. **b** indicates production rate of oxygen free radical. **c** shows histochemical staining of H_2O_2 and $O_2^{\cdot-}$

of photosynthetic genes, probably via *hexokinase*, which acts as a sugar sensor, and by sugar phosphorylation, which mediates carbohydrate signal transduction [43]. Overexpression of *Arabidopsis hexokinase* in tomato plants induces rapid senescence [44]. In contrast, antisense or *hexokinase* mutants exhibit a delayed senescence phenotype, suggesting that *hexokinase* is involved in senescence regulation [45]. In the present study, *hexokinase* is an upregulated unigene, which is likely to be involved with inducing senescence. In nitrogen metabolism, 62 transcripts were downregulated (Additional file 2). The expression of genes involved in primary nitrogen assimilation such as *GS2* (chloroplastic glutamine synthetase) and *Nia* (nitrate reductase) were repressed; in contrast, *GDH* (glutamate dehydrogenase) mRNA accumulation was increased (Fig. 8). This suggests that primary nitrogen assimilation was suppressed, and nitrogen remobilization was activated in transgenic plants. After treatment of tobacco leaf discs with hydrogen peroxide (H_2O_2) GDH mRNA accumulated, and GS2 mRNA decreased [46]. So the expression of GDH and GS2 also suggests the accumulation of ROS in transgenic plants. In addition, nitrogen is major element of soluble protein. Soluble protein degradation was indicative of lower nitrogen content (Fig. 3C).

Leaf senescence occurs during oxidative stress, which can be induced by overproduction of ROS [47]. Plants cope with oxidative stress using antioxidative systems including antioxidative enzymes such as glutathione peroxidases, dehydroascorbate reductase, catalase (CAT), superoxide dismutase (SOD), and ascorbate peroxidase (APX), and antioxidative metabolites such as ascorbate, glutathione, tocopherol, and carotenoids [48, 49]. In our study, the lower carotenoids, polyphenols, and flavonoid level could weaken their recognized roles in protecting photosystems from oxidative stress (Fig. 3B; Additional file 4: Fig. S2). Because SOD is responsible for dismutation of $O_2^{\cdot-}$, it generates H_2O_2, which is subsequently scavenged by CAT and APX. The lower activity of antioxidative enzymes reduces effective ROS scavenging and ROS accumulate in transgenic plants (Figs. 5, 6). DCMU treatment also indicated that the transgenic plants suffered from severe oxidative stress. ROS such as $O_2^{\cdot-}$ and H_2O_2 have also been implicated as age-associated factors that trigger leaf senescence which in turn enhances membrane lipid peroxidation [50]. There was detectable, but not significant, accumulation of MDA in transgenic plants (Fig. 3F). It is possible that the transgenic plants were still undergoing vegetative growth, which prevented the membrane lipid peroxidation.

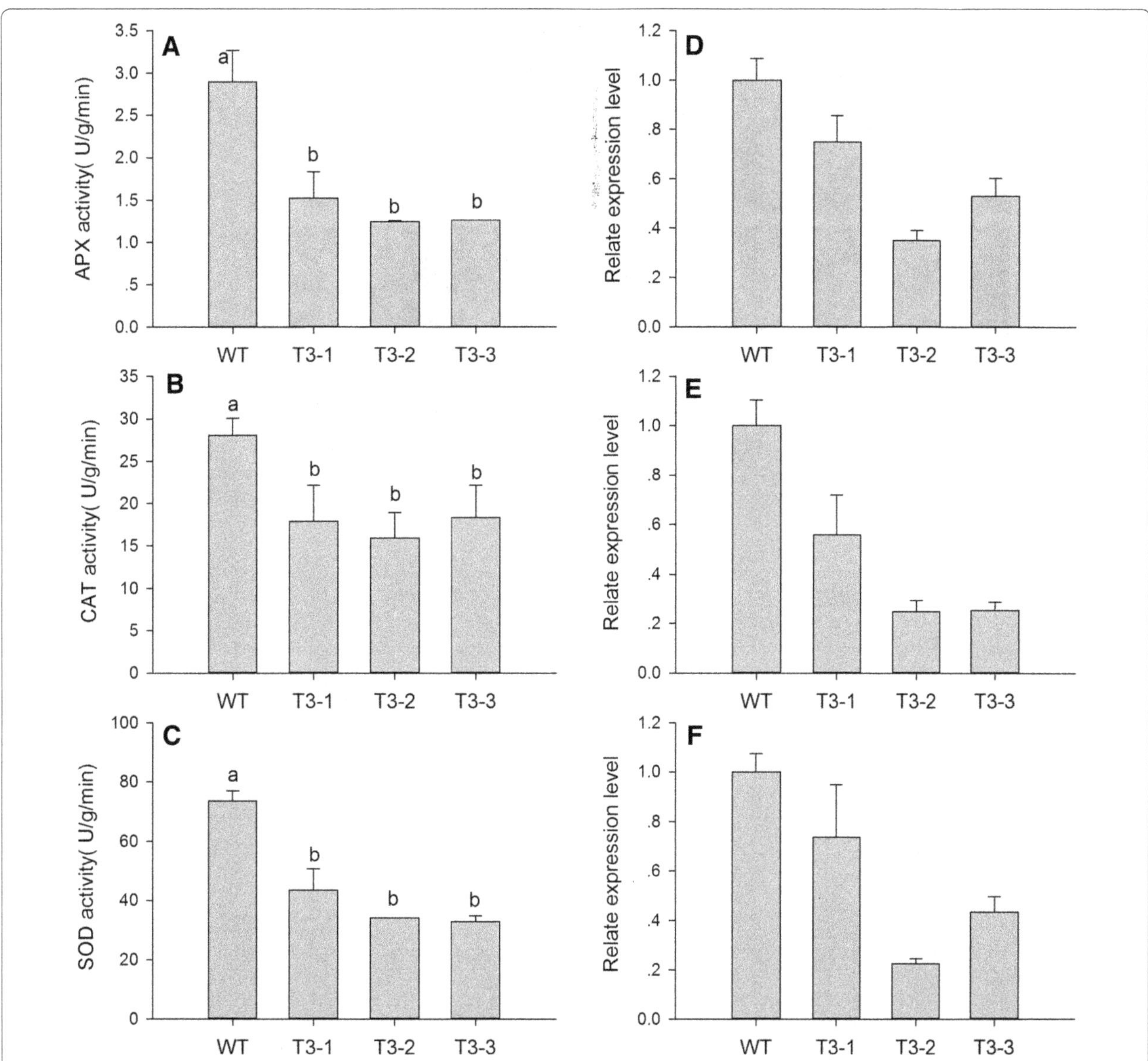

Fig. 6 Activity and expression analysis of antioxidant enzymes in the WT, T3-1, T3-2, and T3-3 tobacco plants. **A** and **D** represent ascorbate peroxidase (APX), **B** and **E** represent catalase (CAT), and **C** and **F** represent superoxide dismutase (SOD). 18S RNA (GenBank Accession Number: AJ236016) was used as a housekeeping gene. The gene names and primers used for qRT-PCR analysis are presented in Additional file 6. Five biological replicates were used for each *line* to study of antioxidant enzyme activity. The standard error of the mean of three biological replicates (nested within three technical replicates) is represented by the *error bars* in qRT-PCR analysis

Reactive oxygen species (ROS) induced plastid damage; chloroplast ROS influenced leaf senescence, and determined the viability and longevity of green tissues [51]. Downregulated genes are involved in senescence; *CAB* and *RuBisCO* small subunit, are responsible for the regulation of photosynthetic proteins. Chlorophyllase (CHL) operates the chlorophyll degradation pathway. The accumulation of *CHL* *m*RNA controls the release of ROS from the thylakoid membrane, which then initiates senescence. In addition to the above, relatively lower leaf water content led to a reduction in water potential, stomatal closure, and lower CO_2 levels in aging mesophyll tissues, thereby reducing the net photosynthetic rate (Fig. 3E; Additional file 5: Fig. S3). The degradation of chlorophyll and photosynthetic proteins in transgenic plants suppressed the net photosynthetic rate (Fig. 3). A decline in photosynthetic capacity is a major feature of senescent leaves. In photosynthetic reactions, photosystem II, photosystem I, cytochrome b6/f complex, photosynthetic electron transport, and F-type ATPase

Fig. 7 Enhanced sensitivity of transgenic plants to oxidative stress. The 3-week-old plants were treated with 70 μM DCMU for 10 days

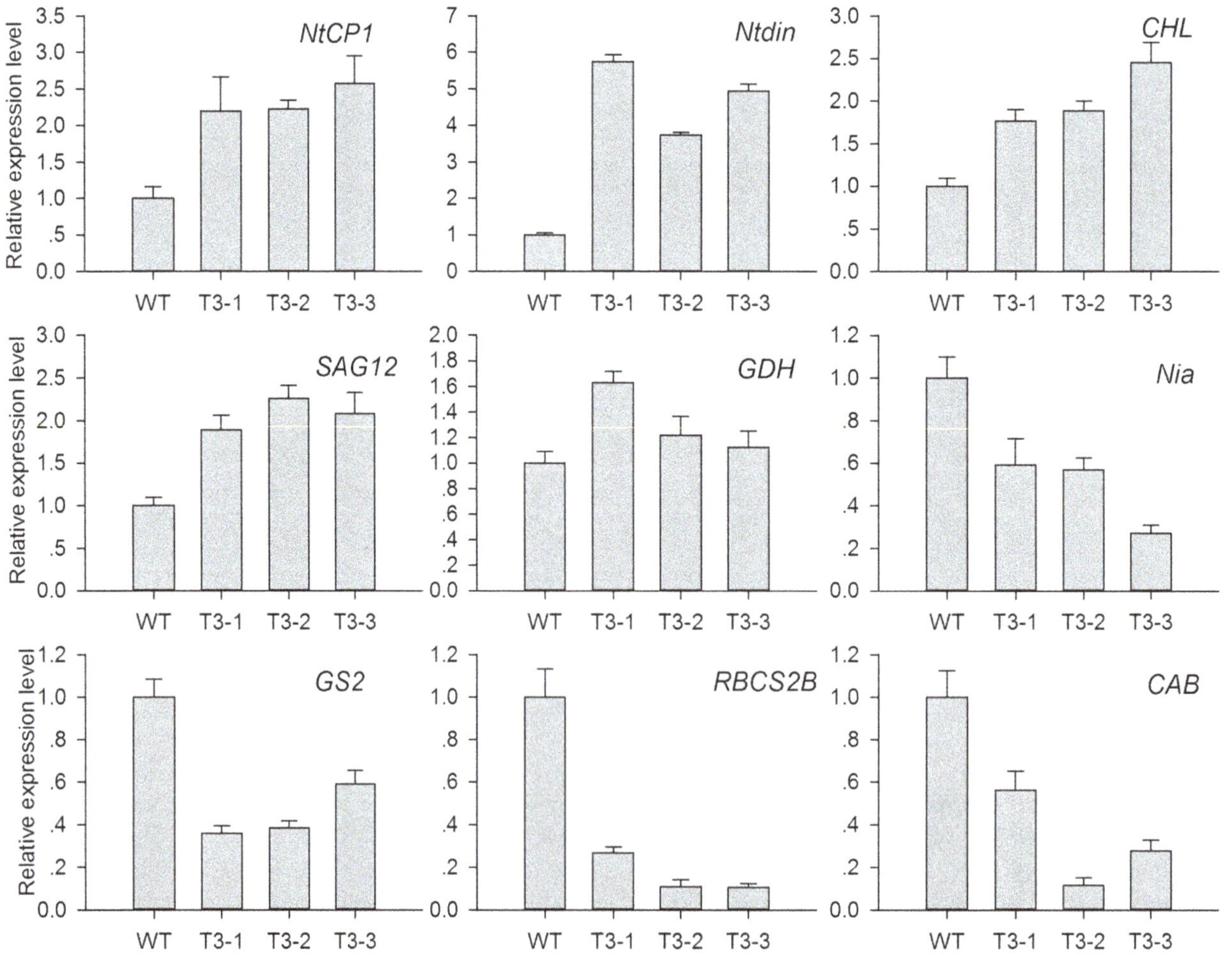

Fig. 8 Expression analysis of senescence marker genes in WT and transgenic plants. *NtCP1* cysteine protease, *SAG12* cysteine protease, *Ntdin* a tobacco senescence-associated gene, *CHL* chlorophyllase, *GDH* glutamate dehydrogenase, *Nia* nitrate reductase, *GS2* chloroplastic glutamine synthetase, *RBCS2B* RuBisCO small subunit, *CAB* chlorophyll *a/b* binding protein. 18S RNA (GenBank Accession Number: AJ236016) was used as housekeeping gene. The gene names and primers used for qRT-PCR analysis are shown in Additional file 6. The standard error of the mean of three biological replicates (nested within three technical replicates) is represented by the *error bars* in qRT-PCR analysis

distributed 87 downregulated unigenes in transgenic plants (Fig. 4). This further revealed that photosynthesis was suppressed in transgenic plants.

Proteins in plant leaves are constantly engaged in stable turnover. With plant leaf senescence, this dynamic balance is disturbed and leads to protein degradation. Protein degradation generally occurs via the ubiquitin–proteasome system. GO analysis indicated that several ubiquitin-mediated proteolysis unigenes were upregulated in transgenic plants (Additional file 2). In plants, a portion of senescence-associated proteases are localized in senescence-associated vacuoles to degrade chloroplast-derived proteins, including those encoding cysteine proteases, and vacuolar processing enzymes [52, 53]. *NtCP1* is only expressed in senescent tobacco leaves and was not induced in mature green leaves or by abiotic stress. *SAG12*, expressed in mature and senescent leaves, can also be used as a leaf senescence marker in tobacco [54]. Cysteine proteinases (*NtCP1* and *SAG12*) are important for the degradation of photosynthetic proteins such as Rubisco and Rubisco activase [55]. *NtCP1* and *SAG12* were induced indicating that senescence had started in transgenic plants.

We made a preliminary analysis of the effect of leaf senescence on the terpenoid synthesis pathway. Monoterpene production is inhibited under oxidative stress [56]. Monoterpenes, including β-Myrcene and linalool, have lower production in transgenic plants compared to wild-type plants (Fig. 1). In addition, carotenoid biosynthesis had 34 downregulated unigenes, geranylgeranyl reductase and solanesyl diphosphate synthase are downregulated in the MEP pathway (Additional file 3). The release of farnesene can reduce the supply of substrate for other terpenes in the MVA pathway such as caryophyllene and sterols. Sterols are essential for plant development and growth. This may be an additional reason for the senescence phenotype.

Conclusion

The analysis of differentially expressed unigenes, physiological and biochemical parameters related to senescence highlighted that ROS play an important role in the senescence of transgenic plants. This study also indicates that Illumina sequencing technology can be applied as a rapid method for de novo transcriptome analysis of tobacco with unavailable genomic information.

Methods

Plant materials and treatments

Seeds (*Nicotiana tabacum* NC89) of WT, T3-1, T3-2 and T3-3 were germinated on Murashige–Skoog (MS) agar medium, in closed glass dishes for 14 days at 25/20 °C day/night cycle and 16:8 (L:D) h photoperiod. Seedlings were transplanted into vermiculite and grown at 25–30 °C/20–25 °C (day/night temperature regime) under a 16:8 h photoperiod 300–700 μmol m^{-2} s^{-1} photon flux density (PFD) and 50–60 % relative humidity in a greenhouse. Leaf samples were obtained from 8-week-old plants for subsequent experiments. At 8 weeks, the transgenic plants are still in vegetative growth. So the plants were considered to be at similar developmental stages. The measurements of physiological and biochemical parameters were conducted on the youngest fully expanded leaves. Five biological repeats were used in physiological and biochemical experiments.

For reagent treatment, seeds of WT, T3-1, T3-2 and T3-3 were sown into vermiculite and grown in an illuminated incubation chamber (GXZ-260C). Three week-old WT and transgenic plants were sprayed with 70 μM 3-(3,4-dichlorfenyl)-1,1-dimethylkarbonyldiamid (DCMU).

Transformation and identification of transgenic tobacco plants

Based on the cDNA sequence of *MdAFS1* (GenBank accession number AY182241), a specific primer was designed (Additional file 6). Total RNA was isolated from apple peel of '*white winter pearmain*' using Trizol (Tiangen, Beijing, China). DNA-free total RNA was reverse-transcribed using a RevertAid First-strand cDNA Synthesis Kit (MBI Fermentas, Beijing, China) according to the manufacturer's protocol. The *MdAFS1* gene was isolated from apple (*white winter pearmain*). The open reading frame (ORF) of *MdAFS1* cDNA was inserted into the pBI122 expression vector under the control of the 35S Cauliflower mosaic virus promoter. The 35S-*MdAFS1* plasmid was introduced into *Agrobacterium tumefaciens* LBA4404, and was verified by PCR and sequencing. *N. tabacum* (NC89, saved by the current laboratory) was transformed with the resultant plasmid using the standard *Agrobacterium*- mediated method [57]. Three independent third generations lines (T3-1, T3-2, T3-3) were selected for further experimentation.

Samples and analytical conditions for GC–MS analysis

Each 1.0 g fresh tobacco flower sample was extracted with 6 ml of extraction buffer supplemented with 50 mM KCl and 10 mM $MgCl_2$ in a sealed container [58, 59]. The volatile compounds were collected with solid-phase microextraction (SPME) for 40 min at 40 °C.

Volatile compound analysis was performed with GCMS-QP2010 with a FID detector (Shimadzu, Tokyo, Japan). A Rtx-5MS fused-silica column (30 m × 0.32 mm × 0.25 μm) was used. The oven temperature was initially held for 2 min at 35 °C, ramped at 6 °C/min up to 120, 10 °C/min up to 180 and 20 °C/min up to 230 °C. Split injection (5:1) was used. The carrier gas was helium with a flow rate of 2.2 ml/min. Injector and FID detector temperature were held at 250 and 280 °C, respectively.

Qualitative method The results were analyzed using standard NIST08 gallery and spectra.

Quantitative method The results were analyzed using the normalization of peak area method.

Measurement of pigment content

Pigment was extracted from fresh leaf samples (≈0.1 g) using 80 % acetone at room temperature in darkness until the leaf tissue was completely bleached. The extract was centrifuged at 5000*g* for 5 min, and the supernatant was collected and used in spectrophotometric analysis with a spectrophotometer (UV-1780, Shimadzu, Tokyo, Japan) at an absorbance wavelength of 470, 646 and 663 nm. The chlorophyll concentration was calculated using the following formula: chlorophyll a (C_a) = 12.21*A_{663} − 2.81*A_{646}; and chlorophyll b (C_b) = 20.13*A_{646} − 5.03*A_{663}. The total chlorophyll (a + b) content (mg g^{-1} FW) was then calculated. The carotenoid concentration (C_{x*c}) was calculated using the following formula: carotenoids = (1000*A_{470} − 3.27*C_a − 104*C_b)/229.

Photosynthetic gas exchange measurements

CO_2 gas exchange was measured between 9:00 and 11:00 h on the same day using a portable photosynthesis system (CIRAS-2, PP Systems, Herts, UK). Experiments were performed under the following conditions: greenhouse temperature (25 ± 1 °C); CO_2 concentration, 390 μl l^{-1}; PFD, 1200 μmol m^{-2} s^{-1}, and relative humidity, 70–80 %. Irradiance was controlled by the automatic control function of the CIRAS-2 photosynthetic system.

Physiological assays for leaf senescence

Each 0.5 g leaf sample was homogenized in 5 ml of 50 mM sodium phosphate buffer (pH 7.8) supplemented with 1 mM EDTA and 2 % (w/v) polyvinylpyrrolidone (PVP). The homogenate was centrifuged at 12,000*g* for 20 min at 4 °C; the supernatant was immediately used for the determination of $O_2^{\cdot-}$ radical production rate, soluble protein content, and antioxidant enzyme activities. All assays were performed at 4 °C. Superoxide dismutase (SOD), catalase (CAT), and ascorbate peroxidase (APX) activities were determined as previously described [60]. Total soluble protein content was determined according to Bradford using bovine serum albumin (BSA) as the standard [61]. The assay for $O_2^{\cdot-}$ content was conducted as described by Yang [62].

H_2O_2 content was measured according to Gay and Gebicki [63]. Tobacco leaves (0.5 g) were ground with liquid nitrogen, then transferred to a centrifuge tube, to which 2 ml of cold acetone was added. After centrifugation at 10,000*g* for 10 min, 1 ml of supernatant and 3 ml of 20 % titanium tetrachloride ($TiCl_4$) were centrifuged at 4000*g* for 15 min. Twenty percentage (v/v) H_2SO_4 was added to the precipitate, dissolved, and the absorbance was determined at 410 nm.

Lipid peroxidation was determined by estimating malondialdehyde (MDA) content [64]. A 10 % solution of trichloroacetic acid (TCA) containing 0.6 % 2-thiobarbituric acid (TBA) and heated at 95 °C for 15 min, and the absorbances were determined at 450, 532 and 600 nm. All spectrophotometric analyses were performed using a spectrophotometer (UV-1780, Shimadzu, Tokyo, Japan).

Water content and seed biomass were determined using a drying oven (Yiheng, Shanghai, China). Fresh leaf samples were dried at 105 °C for 30 min, and 60 °C for 48 h. After collection, seeds were heated at 35 °C to constant weight, and the tobacco seeds were used in biomass statistics.

Histochemical staining of H_2O_2 and $O_2^{\cdot-}$

H_2O_2 and $O_2^{\cdot-}$ accumulations were detected using 3,3′-diaminobenzidine (DAB) and nitroblue tetrazolium (NBT) staining methods [65, 66], respectively. H_2O_2 reacts with DAB to form a reddish-brown stain. Tobacco leaf discs (1.4 cm diameter) were incubated in 1 mg/ml DAB solution in the dark at room temperature for 16 h. Leaf discs were boiled in ethanol (95 %) for 10 min and then cooled to room temperature. The leaf discs were then extracted with fresh ethanol and photographed. $O_2^{\cdot-}$ reacts with NBT to form a blue stain. A 0.5 mg/ml NBT solution was used in this experiment. The procedure used was similar to H_2O_2 staining.

Transcriptome analysis

The tobacco plants were grown under greenhouse conditions as described previously. Samples were collected from 8-week-old plants. To produce a transcriptome dataset with a wide coverage, RNA was extracted from pooled samples of leaves from four plants of WT and T3-2, respectively. The high-quality reads were assembled with the software package Velvet_1.2.10. The trimmed Solexa transcriptome reads were mapped onto the unique consensus sequences using Bowtie. Unigenes were compared to records in public databases, including the National Center for Biotechnology Information (NCBI, 2013), SWISS-PROT, kyoto encyclopedia of genes and genomes database (KEGG), and gene ontology (GO). Transcriptome sequencing was performed by Capital Bio Corporation (Beijing, China), using Hiseq 2000.

qRT-PCR analysis

Sampling from 8-week-old plants under the same greenhouse conditions as previously described, qRT-PCR analysis was conducted according to the MIQE guidelines [67, 68]. Total RNA was isolated using a total RNA isolation system (Tiangen, Beijing, China). First-strand cDNAs were synthesized using a First-strand cDNA Synthesis Kit (Tiangen, Beijing, China). qRT-PCR was performed on a Bio-Rad CFX96TM Real-time PCR System using SYBR Real Master Mix (Transgen, Beijing, China) using the following PCR thermal cycle conditions: predenaturation at 95 °C for 30 s; followed by 40 cycles of 95 °C for 5 s, 60 °C for 15 s, and 72 °C for 20 s. 18S RNA (GenBank Accession Number: AJ236016) was used as housekeeping gene. Three biological replicates were performed for each line, and the standard curve method was used for statistical analysis.

Statistical analysis

Statistical analyses were performed using SigmaPlot 12.0 software, Excel, and data processing system (DPS) procedures (Zhejiang University, Zhejiang, China). Differences among means were compared using Tukey's multiple range test at a 0.05 probability level.

Additional files

Additional file 1: Fig. S1. Validation of transcript levels of six candidate unigenes of the transcriptome by qRT-PCR. Upregulated unigenes of the transcriptome database: Locus_72229: *Nicotiana tabacum* glycine-rich protein precursor, Locus_4438: Ent-copalyl diphosphate synthase, Locus_712: Polyphenol oxidase; downregulated unigenes of the transcriptome database: Locus_10435: Inositol-3-phosphate synthase, Locus_5382: Ethylene-responsive transcription factor 5, Locus_93559: Calvin cycle protein. The gene names and primers used for qRT-PCR analysis are shown in Additional file. 1. The standard error of the mean of three biological replicates (nested within three technical replicates) is represented by error bars.

Additional file 2. Differentially expressed and identified unigenes in photosynthesis, nitrogen metabolism, nutrient reservoir activity, antioxidant activity, ubiquitin-mediated proteolysis, cell growth, and death.

Additional file 3. Differentially expressed and identified unigenes in secondary metabolism signaling pathways.

Additional file 4: Fig. S2. Antioxidant metabolite levels of the WT and transgenic plants.

Additional file 5: Fig. S3. The relative values of GS, Pn, Ci, in tobacco leaves.

Additional file 6. List of primers used for qRT-PCR analysis.

Abbreviations

APX: ascorbate peroxidase; CAB: chlorophyll *a/b* binding protein; CAT: catalase; CHL: chlorophyllase; DAB: 3,3′-diaminobenzidine; GDH: glutamate dehydrogenase; GO: gene ontology; GS2: chloroplastic glutamine synthetase; GC/MS: gas chromatography–mass spectrometry; H_2O_2: hydrogen peroxide; KEGG: kyoto encyclopedia of genes and genomes; LSD: leaf senescence database; MDA: malondialdehyde; MS: Murashige-Skoog; NBT: nitroblue tetrazolium; Nia: nitrate reductase; NtCP1: cysteine protease; Ntdin: a tobacco senescence-associated gene; O_2^-: superoxide anion; POD: peroxidase; qRT-PCR: quantitative real time-polymerase chain reaction; RBCS2B: RuBisCO small subunit; ROS: reactive oxygen species; SAG12: cysteine protease; SOD: superoxide dismutase; WT: wild-type.

Authors' contributions

YHZ and YL designed the research study; YL, HL, RRZ, BL, and LW conducted the study, YL and QJF analyzed the data, and YL wrote the manuscript. All authors read and approved the final manuscript.

Author details

[1] State Key Laboratory of Crop Biology, College of Life Sciences, Shandong Agricultural University, 61 Dai Zong Street, Tai'an 271018, Shandong, People's Republic of China. [2] Shandong Institute of Pomology, 66 Long Tan Road, Tai'an 271018, Shandong, People's Republic of China.

Acknowledgements

The National Natural Science Foundation of China (30970256 and 31370359) supported this study.

Competing interests

The authors declare that they have no competing interests. Compliance with ethical guidelines. This article does not contain any studies with human or animal subjects performed by any of the authors.

References

1. Kännaste A, Vongvanich N, Borg-Karlson AK. Infestation by a *Nalepella* species induces emissions of α-and β-farnesenes,(−)-linalool and aromatic compounds in Norway spruce clones of different susceptibility to the large pine weevil. Arthropod-Plant Interact. 2008;2(1):31–41.
2. Bengtsson M, Bäckman AC, Liblikas I, et al. Plant odor analysis of apple: antennal response of codling moth females to apple volatiles during phenological development. J Agric Food Chem. 2001;49(8):3736–41.
3. Bain JM, Mercer FV. The Submicroscopio Cytology of Superficial Scald, a Physiological Disease of Apples. Aust J Biol Sci. 1963;16(2):442–9.
4. Meigh DF. Volatile compounds produced by apples. I.—Aldehydes and ketones. J Sci Food Agric. 1956;7(6):396–411.
5. Whitaker BD, Solomos T, Harrison DJ. Quantification of α-farnesene and its conjugated trienol oxidation products from apple peel by C18-HPLC with UV detection. J Agric Food Chem. 1997;45(3):760–5.
6. Zubini P, Baraldi E, De Santis A, et al. Expression of anti-oxidant enzyme genes in scald-resistant 'Belfort' and scald-susceptible 'Granny Smith' apples during cold storage. J Hortic Sci Biotechnol. 2007;82(1):149–55.
7. Lange BM, Rujan T, Martin W, et al. Isoprenoid biosynthesis: the evolution of two ancient and distinct pathways across genomes. Proc Natl Acad Sci USA. 2000;97(24):13172–7.
8. Lombard J, Moreira D. Origins and early evolution of the mevalonate pathway of isoprenoid biosynthesis in the three domains of life. Mol Biol Evol. 2011;28(1):87–99.
9. Pechous SW, Whitaker BD. Cloning and functional expression of an (E, E)-α-farnesene synthase cDNA from peel tissue of apple fruit. Planta. 2004;219(1):84–94.
10. Suzuki M, Kamide Y, Nagata N, et al. Loss of function of 3-hydroxy-3-methylglutaryl coenzyme A reductase 1 (HMG1) in Arabidopsis leads to dwarfing, early senescence and male sterility, and reduced sterol levels. Plant J. 2004;37(5):750–61.
11. Keim V, Manzano D, Fernández FJ, et al. Characterization of Arabidopsis FPS isozymes and FPS gene expression analysis provide insight into the biosynthesis of isoprenoid precursors in seeds. PLoS One. 2012;7(11):e49109.
12. Masferrer A, Arró M, Manzano D, et al. Overexpression of Arabidopsis thaliana farnesyl diphosphate synthase (FPS1S) in transgenic Arabidopsis induces a cell death/senescence-like response and reduced cytokinin levels. Plant J. 2002;30(2):123–32.

13. Bhatia V, Maisnam J, Jain A, et al. Aphid-repellent pheromone E-β-farnesene is generated in transgenic Arabidopsis thaliana over-expressing farnesyl diphosphate synthase2. Ann Bot. 2015;115:581–91.
14. Richter A, Seidl-Adams I, Köllner TG, et al. A small, differentially regulated family of farnesyl diphosphate synthases in maize (*Zea mays*) provides farnesyl diphosphate for the biosynthesis of herbivore-induced sesquiterpenes. Planta. 2015;241(6):1351–61.
15. Huang M, Sanchez-Moreiras AM, Abel C, et al. The major volatile organic compound emitted from Arabidopsis thaliana flowers, the sesquiterpene (E)-β-caryophyllene, is a defense against a bacterial pathogen. New Phytol. 2012;193(4):997–1008.
16. Kempinski C, Jiang Z, Bell S, Chappell J. Metabolic engineering of higher plants and algae for isoprenoid production. Adv Biochem Eng Biotechnol. 2015;148:161–99.
17. De Ford C, Ulloa JL, Catalán CAN, et al. The sesquiterpene lactone polymatin B from *Smallanthus sonchifolius* induces different cell death mechanisms in three cancer cell lines. Phytochemistry. 2015;117:332–9.
18. Rodriguez-Concepción M, Boronat A. Elucidation of the methylerythritol phosphate pathway for isoprenoid biosynthesis in bacteria and plastids: a metabolic milestone achieved through genomics. Plant Physiol. 2002;130:1079–89.
19. Dudareva N, Andersson S, Orlova I, et al. The nonmevalonate pathway supports both monoterpene and sesquiterpene formation in snapdragon flowers. Proc Natl Acad Sci USA. 2005;102(3):933–8.
20. Xing S, Miao J, Li S, et al. Disruption of the 1-deoxy-D-xylulose-5-phosphate reductoisomerase (DXR) gene results in albino, dwarf and defects in trichome initiation and stomata closure in Arabidopsis. Cell Res. 2010;20(6):688–700.
21. Kishimoto K, Matsui K, Ozawa R, et al. Analysis of defensive responses activated by volatile allo-ocimene treatment in *Arabidopsis thaliana*. Phytochemistry. 2006;67(14):1520–9.
22. Yamasaki Y, Kunoh H, Yamamoto H, et al. Biological roles of monoterpene volatiles derived from rough lemon (Citrus jambhiri Lush) in citrus defense. J Gen Plant Pathol. 2007;73(3):168–79.
23. Chaimovitsh D, Abu-Abied M, Belausov E, et al. Microtubules are an intracellular target of the plant terpene citral. Plant J. 2010;61(3):399–408.
24. Rogers HJ. Is there an important role for reactive oxygen species and redox regulation during floral senescence? Plant Cell Environ. 2012;35(2):217–33.
25. Tewari RK, Watanabe D, Watanabe M. Chloroplastic NADPH oxidase-like activity-mediated perpetual hydrogen peroxide generation in the chloroplast induces apoptotic-like death of *Brassica napus* leaf protoplasts. Planta. 2012;235(1):99–110.
26. Van Breusegem F, Dat JF. Reactive oxygen species in plant cell death. Plant Physiol. 2006;141(2):384–90.
27. Kangasjärvi S, Neukermans J, Li S, et al. Photosynthesis, photorespiration, and light signalling in defence responses. J Exp Bot. 2012;63:1619–36.
28. Girondé A, Poret M, Etienne P, et al. A profiling approach of the natural variability of foliar N remobilization at the rosette stage gives clues to understand the limiting processes involved in the low N use efficiency of winter oilseed rape. J Exp Bot. 2015;66(9):2461–73.
29. Quirino BF, Noh YS, Himelblau E, et al. Molecular aspects of leaf senescence. Trends Plant Sci. 2000;5(7):278–82.
30. Lin CR, Lee KW, Chen CY, et al. SnRK1A-interacting negative regulators modulate the nutrient starvation signaling sensor SnRK1 in source-sink communication in cereal seedlings under abiotic stress. Plant Cell. 2014;26(2):808–27.
31. Schippers JHM, Schmidt R, Wagstaff C, et al. Living to die and dying to live: the survival strategy behind leaf senescence. Plant Physiol. 2015;169(2):914–30.
32. Wingler A, Purdy S, MacLean JA, et al. The role of sugars in integrating environmental signals during the regulation of leaf senescence. J Exp Bot. 2006;57(2):391–9.
33. Hörtensteiner S, Feller U. Nitrogen metabolism and remobilization during senescence. J Exp Bot. 2002;53(370):927–37.
34. Lin JF, Wu SH. Molecular events in senescing Arabidopsis leaves. Plant J. 2004;39(4):612–28.
35. Zavaleta-Mancera HA, Thomas BJ, Thomas H, et al. Regreening of senescent Nicotiana leaves II. Redifferentiation of plastids. J Exp Bot. 1999;50(340):1683–9.
36. Gan S, Amasino RM. Inhibition of leaf senescence by autoregulated production of cytokinin. Science. 1995;270(5244):1986–8.
37. Himelblau E, Amasino RM. Nutrients mobilized from leaves of *Arabidopsis thaliana* during leaf senescence. J Plant Physiol. 2001;158(10):1317–23.
38. Bazargani MM, Sarhadi E, Bushehri AAS, et al. A proteomics view on the role of drought-induced senescence and oxidative stress defense in enhanced stem reserves remobilization in wheat. J Proteomics. 2011;74(10):1959–73.
39. Schwachtje J, Minchin PEH, Jahnke S, et al. SNF1-related kinases allow plants to tolerate herbivory by allocating carbon to roots[J]. Proc Natl Acad Sci. 2006;103(34):12935–40.
40. Gregersen PL, Culetic A, Boschian L, et al. Plant senescence and crop productivity. Plant Mol Biol. 2013;82(6):603–22.
41. Palenchar PM, Kouranov A, Lejay LV, et al. Genome-wide patterns of carbon and nitrogen regulation of gene expression validate the combined carbon and nitrogen (CN)-signaling hypothesis in plants. Genome Biol. 2004;5(11):R91.
42. Aoyama S, Reyes TH, Guglielminetti L, et al. Ubiquitin ligase ATL31 functions in leaf senescence in response to the balance between atmospheric CO2 and nitrogen availability in Arabidopsis. Plant Cell Physiol. 2014;55:293–305.
43. Rolland F, Moore B, Sheen J. Sugar sensing and signaling in plants. Plant Cell. 2002;14(Suppl 1):S185–205.
44. Dai N, Schaffer A, Petreikov M, et al. Overexpression of Arabidopsis hexokinase in tomato plants inhibits growth, reduces photosynthesis, and induces rapid senescence. Plant Cell. 1999;11(7):1253–66.
45. Moore B, Zhou L, Rolland F, et al. Role of the Arabidopsis glucose sensor HXK1 in nutrient, light, and hormonal signaling. Science. 2003;300(5617):332–6.
46. Pageau K, Reisdorf-Cren M, Morot-Gaudry JF, et al. The two senescence-related markers, GS1 (cytosolic glutamine synthetase) and GDH (glutamate dehydrogenase), involved in nitrogen mobilization, are differentially regulated during pathogen attack and by stress hormones and reactive oxygen species in *Nicotiana tabacum* L. leaves. J Exp Bot. 2006;57(3):547–57.
47. Tewari RK, Singh PK, Watanabe M. The spatial patterns of oxidative stress indicators co-locate with early signs of natural senescence in maize leaves. Acta Physiol Plant. 2013;35(3):949–57.
48. Gill SS, Tuteja N. Reactive oxygen species and antioxidant machinery in abiotic stress tolerance in crop plants. Plant Physiol Biochem. 2010;48:909–30.
49. Sharma P, Jha AB, Dubey RS, et al. Reactive oxygen species, oxidative damage, and antioxidative defense mechanism in plants under stressful conditions. J Bot. 2012;26.
50. Khanna-Chopra R. Leaf senescence and abiotic stresses share reactive oxygen species-mediated chloroplast degradation. Protoplasma. 2012;249(3):469–81.
51. van Doorn WG, Yoshimoto K. Role of chloroplasts and other plastids in ageing and death of plants and animals: a tale of Vishnu and Shiva. Ageing Res Rev. 2010;9(2):117–30.
52. Guo Y, Cai Z, Gan S. Transcriptome of Arabidopsis leaf senescence. Plant, Cell Environ. 2004;27(5):521–49.
53. Carrión CA, Costa ML, Martínez DE, Mohr C, Humbeck K, Guiamet JJ. In vivo inhibition of cysteine proteases provides evidence for the involvement of 'senescence-associated vacuoles' in chloroplast protein degradation during dark-induced senescence of tobacco leaves. J Exp Bot. 2013;64:4967–80.
54. Beyene G, Foyer CH, Kunert KJ. Two new cysteine proteinases with specific expression patterns in mature and senescent tobacco (*Nicotiana tabacum* L.) leaves. J Exp Bot. 2006;57(6):1431–43.
55. Prins A, Van Heerden PD, Olmos E, Kunert KJ, Foyer CH. Cysteine proteinases regulate chloroplast protein content and composition in tobacco leaves: a model for dynamic interactions with ribulose-1, 5-bisphosphate carboxylase/oxygenase (Rubisco) vesicular bodies. J Exp Bot. 2008;59(7):1935–50.
56. Peñuelas J, Munné-Bosch S. Isoprenoids: an evolutionary pool for photoprotection. Trends Plant Sci. 2005;10(4):166–9.
57. Luo K, Zheng X, Chen Y, et al. The maize Knotted1 gene is an effective positive selectable marker gene for Agrobacterium-mediated tobacco transformation. Plant Cell Rep. 2006;25(5):403–9.

58. Green S, Friel EN, Matich A, et al. Unusual features of a recombinant apple α-farnesene synthase. Phytochemistry. 2007;68(2):176–88.
59. Green S, Squire CJ, Nieuwenhuizen NJ, et al. Defining the potassium binding region in an apple terpene synthase. J Biol Chem. 2009;284(13):8661–9.
60. Türkan İ, Bor M, Özdemir F, et al. Differential responses of lipid peroxidation and antioxidants in the leaves of drought-tolerant *P. acutifolius* Gray and drought-sensitive *P. vulgaris* L. subjected to polyethylene glycol mediated water stress. Plant Sci. 2005;168(1):223–31.
61. Noble JE, Bailey MJA. Quantitation of protein. Method Enzymol. 2009;463:73–95.
62. Yang S, Tang XF, Ma NN, et al. Heterology expression of the sweet pepper CBF3 gene confers elevated tolerance to chilling stress in transgenic tobacco. J Plant Physiol. 2011;168(15):1804–12.
63. Gay C, Gebicki JM. A critical evaluation of the effect of sorbitol on the ferric–xylenol orange hydroperoxide assay. Anal Biochem. 2000;284(2):217–20.
64. Rao KVM. Sresty TVS. Antioxidative parameters in the seedlings of pigeonpea (*Cajanus cajan* (L.) Millspaugh) in response to Zn and Ni stresses. Plant Sci. 2000;157(1):113–28.
65. Thordal-Christensen H, Zhang Z, Wei Y, Collinge DB. Subcellular localization of H_2O_2 in plants. H_2O_2 accumulation in papillae and hypersensitive response during the barley—powdery mildew interaction. Plant J. 1997;11(6):1187–94.
66. Cai G, Wang G, Wang L, et al. ZmMKK1, a novel group A mitogen-activated protein kinase kinase gene in maize, conferred chilling stress tolerance and was involved in pathogen defense in transgenic tobacco. Plant Sci. 2014;214:57–73.
67. Bustin SA, Benes V, Garson JA, et al. The MIQE guidelines: minimum information for publication of quantitative real-time PCR experiments. Clin Chem. 2009;55(4):611–22.
68. Bustin SA, Beaulieu JF, Huggett J, et al. MIQE precis: practical implementation of minimum standard guidelines for fluorescence-based quantitative real-time PCR experiments. BMC Mol Biol. 2010;11(1):74.

9

Soluble carbohydrate content variation in *Sanionia uncinata* and *Polytrichastrum alpinum*, two Antarctic mosses with contrasting desiccation capacities

Paz Zúñiga-González[1], Gustavo E. Zúñiga[2*], Marisol Pizarro[2] and Angélica Casanova-Katny[3,4*]

Abstract

Background: Cryptogamic vegetation dominates the ice-free areas along the Antarctic Peninsula. The two mosses *Sanionia uncinata* and *Polytrichastrum alpinum* inhabit soils with contrasting water availability. *Sanionia uncinata* grows in soil with continuous water supply, while *P. alpinum* grows in sandy, non-flooded soils. Desiccation and rehydration experiments were carried out to test for differences in the rate of water loss and uptake, with non-structural carbohydrates analysed to test their role in these processes.

Results: Individual plants of *S. uncinata* lost water 60 % faster than *P. alpinum*; however, clumps of *S. uncinata* took longer to dry than those of *P. alpinum* (11 vs. 5 h, respectively). In contrast, rehydration took less than 10 min for both mosses. Total non-structural carbohydrate content was higher in *P. alpinum* than in *S. uncinata*, but sugar levels changed more in *P. alpinum* during desiccation and rehydration (60–50 %) when compared to *S. uncinata*. We report the presence of galactinol (a precursor of the raffinose family) for the first time in *P. alpinum*. Galactinol was present at higher amounts than all other non-structural sugars.

Conclusions: Individual plants of *S. uncinata* were not able to retain water for long periods but by growing and forming carpets, this species can retain water the longest. In contrast individual *P. alpinum* plants required more time to lose water than *S. uncinata*, but as moss cushions they suffered desiccation faster than the later. On the other hand, both species rehydrated very quickly. We found that when both mosses lost 50 % of their water, carbohydrates content remained stable and the plants did not accumulate non-structural carbohydrates during the desiccation prosses as usually occurs in vascular plants. The raffinose family oligosaccarides decreased during desiccation, and increased during rehydration, suggesting they function as osmoprotectors.

Keywords: Antarctica, Antarctic vegetation, Bryophytes, Sugars

Background

Over the last decades, Antarctica has become a natural laboratory for studying plant tolerance mechanisms under extreme conditions and climate change. In the Antarctic, the development of most life forms is limited due to abiotic factors such as low temperatures, frequent cycles of freezing and thawing, high radiation, strong winds, and extreme dryness; a dryness due in part to the lack of organic soil capable of water retention, in addition to the physiological drought caused by freezing [1]. All these elements contribute to low water availability for plant growth and cellular activities which represents one of the principal limiting factors for distribution of terrestrial vegetation [2].

The Antarctic flora is poor in vascular plants, with lichens, mosses, and liverworts dominating the

*Correspondence: gustavo.zuniga@usach.cl; angecasanova@gmail.com
[2] Departamento de Biología, Facultad de Química y Biología, Universidad de Santiago, Alameda, 3363 Santiago, Chile
[3] Núcleo de Estudios Ambientales, Universidad Católica de Temuco, Casilla 15-D, Temuco, Chile
Full list of author information is available at the end of the article

landscape. Plant-lichen communities are distributed at ice-free sites along the west part of the Antarctic Peninsula and on the offshore islands of the maritime Antarctic [3]. Only a few lichen and moss species are capable of surviving the freezing temperatures and strong desiccation found further south [4].

King George Island forms part of the South Shetlands Archipelago in the maritime Antarctic, and is characterized by a semidesert landscape [5]. This island hosts 61 reported moss species located at sites that are humid, protected, and covered by relatively stable and partially organic soil [6]. *Sanionia uncinata* (Hedw.) Loeske and *Polytrichastrum alpinum* (Hedw.) G. L. Smith are frequently found on Fildes Peninsula. In predominantly bryophytic communities, *S. uncinata* grows on the borders of waterlogged areas as well as close to small water bodies, stream banks, and spots subject to melting-water runoff. *P. alpinum* grows preferentially on humid and rocky substrates and close to moraine peaks of glaciers or at dry sites [7, 8], but not in water-saturated soils. This species, together with the two native vascular plants *Deschampsia antarctica* Desv. and *Colobanthus quitensis* (Kunth.) Bartl., form the so-called herbaceous antarctic tundra [7]. In this context, *D. antarctica*, the antarctic hairgrass, is positively associated with moss beds along the Antarctic Peninsula which, have been shown to facilitate growth of *D. antarctica* seedlings in transplant experiments on Fildes Peninsula [9].

Bryophytes are characterized by a dominant gametophytic phase during their life cycle and a poorly developed vascular system. These plants are capable of easily losing and reabsorbing water through the cellular membrane. Mosses as poikilohydric organisms can rapidly adjust cellular water content in relation to air and environmental humidity [10, 11]. Their inability to maintain stable tissue water levels requires mosses to develop desiccation tolerance mechanisms, such as the total suspension of metabolic activity in order to survive water shortage [12]. Desiccation tolerance is more common in mosses than in homohydric plants (tracheophytes) [13]. The diurnal, monthly and seasonal periods of desiccation to which mosses are exposed determines their establishment and survival, especially in extreme environments such as the Antarctica [10, 13, 14]. According to Bewley [15], the following three properties of the protoplasm in cells are essential for desiccation tolerance: (1) keeping damage to a minimum during desiccation and rehydration, (2) maintaining cellular integrity during desiccation, and (3) activating repair mechanisms following rehydration. All mechanisms are ultimately focused on cellular protection and repair.

Among the mechanisms for cellular protection, soluble carbohydrate accumulation has been related to higher desiccation tolerance in plants [10, 16–18], seeds [16], angiosperm pollen [19], the gametophytes of certain mosses [11, 20, 21] and moss spores [22]. One of the reasons for this accumulation is that soluble carbohydrates contribute to cytoplasm vitrification [23], which facilitates the preservation of macromolecules and the maintenance of membrane integrity for prolonged periods [10, 11, 24, 25].

The role of sugars in the dehydration processes of higher plants has been extensively described [18]. Plants resistant to water loss accumulate soluble sugars that diminish the osmotic potential of the cell, hydrating macromolecules during desiccation stress [18]. However, mosses are poorly investigated in terms of the role of sugars in the processes of daily or seasonal dehydration and rehydration. As dominant species in many tundra communities on the ice-free soils of the maritime Antarctic, both *S. uncinata* and *P. alpinum* play fundamental ecological roles by changing soil properties [26], so understanding the functioning of these key species may also allow deeper insight into plant–plant interactions and the responses of the whole community to changes in water regime.

The present study investigated and compared the rate of water loss and uptake for *S. uncinata* which forms carpets at the wettest sites, and *P. alpinum* which grows on drier, sandy soil, forming small cushions, followed by measurements of changes in non-structural carbohydrate content and composition in both species in response to short term desiccation and rehydration. The results should not only contribute to predicting responses of the polar tundra ecosystem as a whole to climate change, but also reveal potential interactions between bryophytes and antarctic vascular plants as well as other groups of organisms such as springtails and mites.

Results

Desiccation and rehydration curves

During the first desiccation experiment which compared individual plants, *S. uncinata,* an ectohydric moss took significantly less time to completely dehydrate than the endohydric *P. alpinum* [Fig. 1a; full desiccation (D0) reached after 1.13 ± 0.34 vs. 1.8 ± 0.04 h; put stats here ($F_{(1,66)} = 63.55$, $p < 0.0001$)]. However, desiccation took much longer when discs of both mosses were used (Fig. 1b; $F_{(1,66)} = 193.2$, $p < 0.0001$). In this case, clumps of *S. uncinata* reached D0 after eleven hours while *P. alpinum* took 5 h to desiccate (Fig. 1b). During rehydration both species needed only a few minutes (<6 min) to reach the highest water tissue content (R100) (Fig. 1c). The differences observed between species ($F_{(1,66)} = 63.55$, $p < 0.0001$), the type of samples (discs or individual plants) ($F_{(1,66)} = 193.2$, $p < 0.0001$) and which treatments

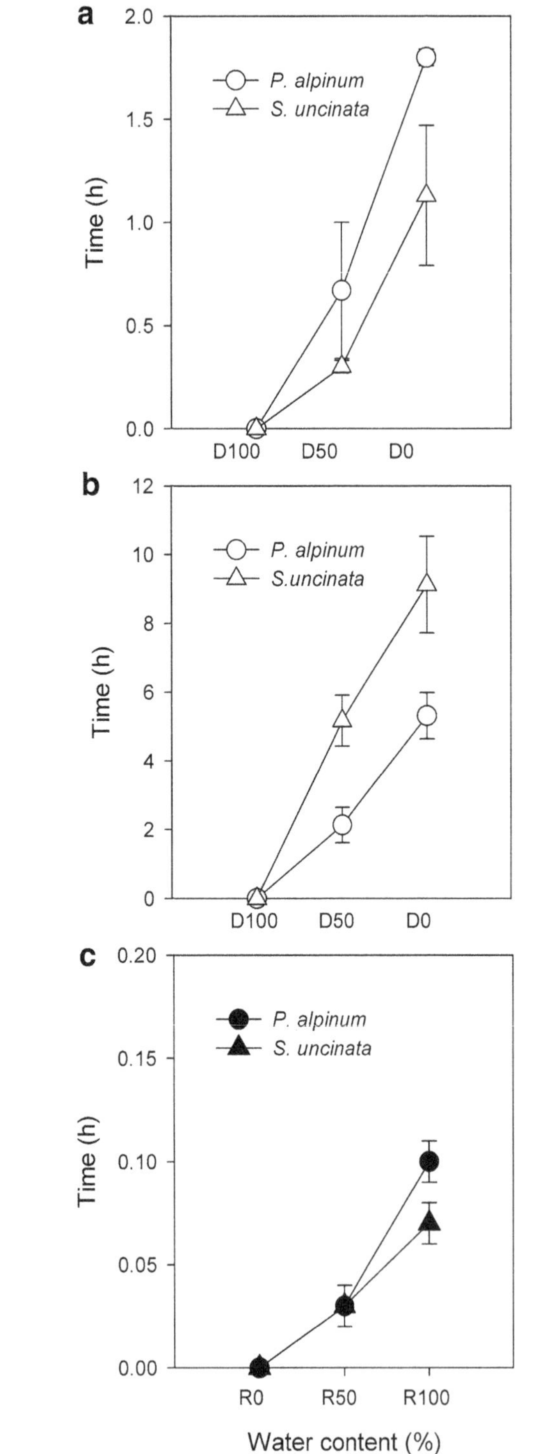

Fig. 1 Time curves (h) for water loss and uptake in Antarctic mosses. Time required to reach every water level during desiccation (from D100 to D0) of **a** individual plants or **b** disc samples and **c** rehydration (R0–R100) of *Sanionia uncinata* and *Polytrichastrum alpinum*. D100, D50, D0, R0, R50 and R100 indicated the percentage of tissue water content. Values are means (n = 4 for disc samples and n = 6 for individual plant) ± SD

($F_{(5,66)} = 332.7$, $p < 0.0001$) were statistically significant. The interaction between the variables, that is, species by which samples ($F_{(1,66)} = 95.5$, $p < 0.0001$), species by treatments ($F_{(5,66)} = 40.75$, $p < 0.0001$), and samples by treatment ($F_{(2,66)} = 245.47$, $p < 0.0001$), were also significant. Moreover, the interactions between the three variables were as well significant.

Carbohydrate content in mosses

During dehydration and rehydration assays, non-structural carbohydrate content (NSC) differed significantly with treatment ($F_{(4,30)} = 42.5$; $p < 0.0001$) and between species ($F_{(1,30)} = 186.7$; $p < 0.001$). Moreover, the interaction of both variables was statistically significant ($F_{(4,30)} = 32.2$; $p < 0.0001$). Only moss discs were used for carbohydrate analysis. On average, for the five water levels, NSC were significantly lower in *S. uncinata* (20.9 ± 1.35 mg g^{-1} DW) than in *P. alpinum* (53.49 ± 1.35 mg g^{-1} DW) (Additional file 1: Table S1). During desiccation, NSC content decreased significantly by 50 % in *S. uncinata* from fully hydrated to desicccated (D100 to D0; Additional file 1: Table S1). By full rehydration (from R0 to R100), NSC values had increased significantly by 84 % (with the highest proportion as galactinol) (Fig. 2; Additional file 1: Table S1). In this moss, NSC content was similar in range (26–24 mg g^{-1} DW) for both the initial (D100) and the final state (R100).

Significant changes in NSC content were also found in *P. alpinum* (Additional file 1: Table S1) with a similar 54 % decrease of NSC level observed during desiccation from D100 to D0. In contrast to *S. uncinata*, during full rehydration (R0 to R100), NSC content increased only slightly reaching only 58 % of the pre-desiccation value (D100; Additional file 1: Table S1).

Galactinol was the most abundant NSC in both mosses, comprising about 27 % in *S. uncinata* during all stages of dessication and ca. 37 % during rehydration, without significant changes. In *P. alpinum* however, galactinol showed significant changes due to the treatment, with higher levels during desiccation (ca. 39 % of NSC), but decreasing concentrations during the rehydration process (19 %, Additional file 1: Table S1).

The composition of carbohydrates was similar in both moss species, including sugars of the sucrose (glucose and fructose) and raffinose (stachyose and verbascose) families (Additional file 1: Table S1). We also found a series of sugar alcohols (polyols: galactinol, adonitol, arabitol, and mannitol) (Additional file 1: Table S1). Only a few soluble carbohydrates (sucrose, verbascose, adonitol, arabitol) changed significantly in quantity between treatments in *S. uncinata* (Additional file 1: Table S1; $P < 0.05$, Fig. 2). In contrast, in *P. alpinum*, 11 of the 15 analyzed sugars showed significant differences ($P < 0.001$,

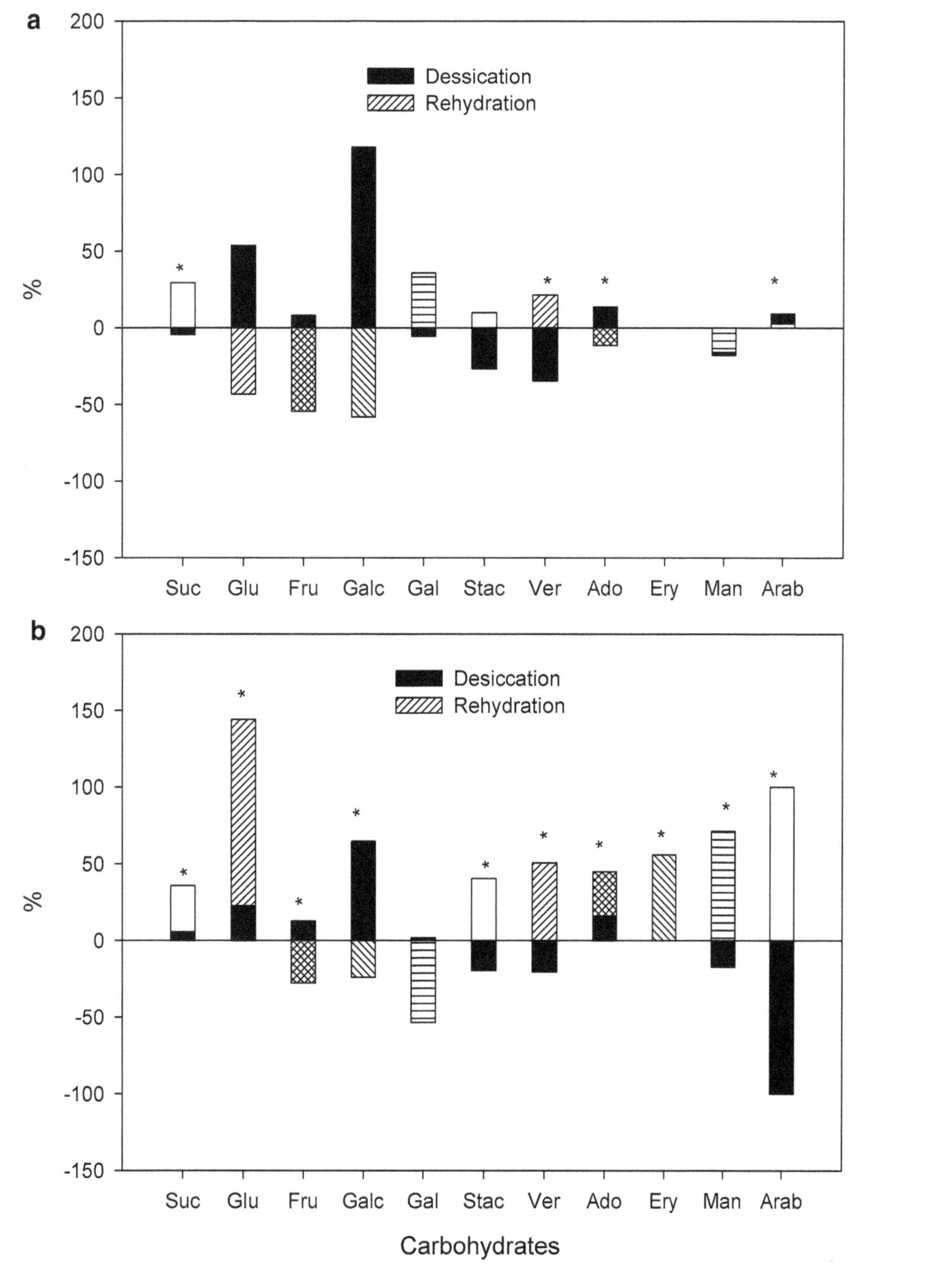

Fig. 2 Percentage of change of each soluble carbohydrate in Antarctic mosses. In **a** *S. uncinata* and **b** *P. alpinum* between the start and the end of desiccation (D100–D0) and rehydration (R0–R100). Values are percentage according to carbohydrate content showed in Additional file 1: Table S1. *Indicate significant differences in Additional file 1: Table S1. *Su* sucrose, *Glu* glucose, *Fru* fructose, *Galc* galactose, *Gal* galactinol, *Ver* verbascose, *Stac* stachyose, *Ado* adonitol, *Eri* erithritol, *Man* mannitol, *Arab* arabitol

Additional file 1: Table S1; Fig. 2). Three carbohydrates (pinitol, nystose, and kestose) were not detected (data not shown), and erythritol was only present in *P. alpinum.* In general, in both mosses the NSC content changed during desiccation and full rehydration. This effect was most siginificant when comparing starting and end point of treatments, whereas when mosses contained 50 % of water (at D50 and R50), NSC content was similar (Additional file 1: Table S1). Between the start and end of desiccation, carbohydrates varied in both species: with sucrose, fructose and glucose all declining significantly during desiccation in *P. alpinum*, while sucrose declined significantly in *S. uncinata* (Additional file 1: Table S1; Fig. 2). The disaccharide galactose increased notably more in *S. uncinata* than in *P. alpinum*, but only in the latter this change was significant (Additional file 1: Table S1; Fig. 2). Within the RFOs family (stachyose and verbascose) both diminished (Fig. 2) considerably in *S. uncinata* (ca. 27 and 35 %) but only slightly in *P. alpinum* (19 and 20 %, Fig. 2). Sugar alcohols (adonitol and arabitol) increased in *S. uncinata* and decreased in *P alpinum* during water loss (Additional file 1: Table S1). In both mosses mannitol decreased during desiccation.

The opposite was found when mosses rehydrated from R0 to R100 (Additional file 1: Table S1; Fig. 2). In *S. uncinata*, sucrose increased and concomitantly fructose and glucose were depleted (Fig. 2), along with galactose (Fig. 2). In this species during the same process, stachyose, verbascose and galactinol increased (Fig. 2). In contrast, in *P. alpinum* during rehydration, while sucrose and glucose increased, only fructose decreased (Fig. 2). At the same time, verbascose and stachyose increased, whereas galactinol decreased (Additional file 1: Table S1; Fig. 2); we also detected the production of erythritol and the increase of adonitol, mannitol and arabitol during this process (Fig. 2; Additional file 1: Table S1).

Discussion

Mosses as poikilohydric organisms are constantly subject to changes in water tissue content, with internal water maintaining equilibrium with the surrounding environment. Our studied species showed obvious differences in their response to experimental water loss and uptake: *S. uncinata* discs lost water considerably more slowly than *P. alpinum* (Fig. 1b), but individual gametophytes of each species showed the opposite pattern, with *S uncinata* desiccating faster than *P. alpinum* (Fig. 1a). The shorter water retention time of individual *S. uncinata* gametophytes as compared to *P. alpinum* can be explained by differences in functional micromorphology: *S. uncinata* is an ectohydric moss that absorbs and loses water solely through its surface as it does not possess a cuticle [8], whereas *P. alpinum* is an endohydric moss that is characterized by its rudimentary conductive tissues, analogous to the xylem and phloem of higher plants, and a thin cuticle [27, 28]. These structural differences would determine the rate at which hydric equilibrium can be achieved between tissue and relative environmental humidity. As *P. alpinum* turf is quite open, individual plants of *P. alpinum* likely show a higher capacity for water retention than *S. uncinata* which grow in dense carpets. Similarly, when comparing bryophytes growing in situ on the subantarctic Signy Island, *S. uncinata* and other ectohydric mosses such as *Schistidium antarctici* (Card.) L. Savic. and Smirn, *Calliergon sarmentosum* (Wahlenb.) Kindb, and *Chorisodontium aciphyllum* (Hook. f. and Wilson) all showed shorter desiccation times than endohydric mosses such as *P. alpinum* and *Polytrichum alpestre* [29, 30].

In the field, gametophytes of *S. uncinata*, a pleurocarpic species, form a compact carpet that reduces the exposed surface, thereby partially limiting water loss [29, 31], as can be observed by its slower desiccation rate when samples where collected as discs, keeping the agreggated form intact (Fig. 1b). According to Robinson et al. [24], the dynamics of desiccation in the field vary substantially between carpets and cushions and it is probably due to its dense growth form that *S. uncinata* discs can retain water for a longer period than *P. alpinum*. The rate at which both mosses lose and recover hydric status is not only related to structural resistance to water loss, it also determines the time available for the synthesis of compounds necessary for greater desiccation tolerance. The contrasting response to desiccation between both species can be related to carbohydrate metabolism, which changes during the treatments.

In contrast to vascular plants, we found that in these Antarctic moss species, non-structural carbohydrate (NSC) content decreases during desiccation (D100 to D0; Additional file 1: Table S1). This finding is in line with the report of Smirnoff [20] that during desiccation of three moss species, *Dicranum majus, Polytrichum formosum*, and *Tortula ruraliformis*, soluble sugars do not play an osmotic role during short-term water loss, as has been observed to occur in vascular plants [18]. In vascular plants, accumulation of soluble sugar in response to desiccation is an important mechanism for the adquisition of drought tolerance. In contrast to other reports, in our mosses NSC content decreased to ca. 50 % in *S. uncinata* and 40 % in *P. alpinum* (Additional file 1: Table S1). Moreover, we found that under laboratory conditions, fructose was higher than glucose or sucrose in both species, even though the principal and responding sugar reported in mosses under field experiments is sucrose [20, 24, 32]. Another marked difference is the presence of high levels of galactinol. This sugar alcohol was found in both mosses, a novel finding for the studied species. Interestingly, galactinol has been been linked in vascular plants to tissue viability following desiccation [33] and to drought tolerance in the desiccation-tolerant *Sporobolus stapfianus* [34]. Galactinol in *S. uncinata* represents about 27 % of all NSC during desiccation, increasing to 37 % during rehydration (Additional file 1: Table S1). In contrast, in *P. alpinum*, galactinol represents an even bigger proportion (about 39 %) of NSC during desiccation but decreases to 19 % during the rehydration process. This suggests different functional roles of galactinol during desiccation and rehydration in both species: in ectohydric *S. uncinata*, galactinol should favor water uptake during rehydration having an osmotic function, whereas in endohydric *P. alpinum* galactinol probably acts as an osmoprotector avoiding damage of membranes during water loss, while during rehydration this sugar is not necessary at high level.

Our results suggest that the biosynthetic pathway of RFOs in the examined bryophytes is active [35, 36]. In *P. alpinum* and *S. uncinata*, the presence of raffinose was not detected, however, stachyose and verbascose were found. The absence of raffinose suggested that it was depleted to form other RFOs units, especially since field experiments in the Antarctica have shown the presence of both carbohydrates during long term in situ desiccation [40]. Stachyose has been previously reported in low concentrations in Antarctic mosses [24]. In two vascular, resurrection plants, *Boea hygroscopica* and *Haberlea rhodopensis*, levels of stachyose and verbascose became significantly elevated under severe desiccation stress [35, 41]. The response to desiccation was mediated by the interplay of several groups of carbohydrates in both species. The RFO, sugar group represents a high proportion of non-structural carbohydrates in both these moss species, playing an important role during desiccation (Additional file 1: Table S1; Fig. 2), with decreasing verbascose and stachyose level during water loss. In contrast, whereas verbascose and stachyose increased in *S. uncinata* during rehydration, in *P. alpinum* this was accompanied by an increase in galactinol. It has been reported, that RFOs sugars also accumulate during desiccation in seeds of various angiosperms [37] and that they are active in higher plants exposed to cold stress [38]. Moreover, in vascular plants they have been shown to be involved in protecting membrane integrity and in cryoprotection, in addition to playing an important role as reserve sugars at low temperatures when starch cannot be used [39]. The high values of RFOs in both mosses suggest, that during full hydration (D100), verbascose and stachyose accumulate as storage sugars which are used during the water loss process, probably helping to stabilize macromolecules together with polyols.

Polyols (galactinol, mannitol, adonitol, arabitol, erithritol) play an important role in desiccation tolerance, probably acting as compatible solutes in the stabilization of macromolecules [42, 43]. The presence of polyols such as adonitol, arabitol, and mannitol has been described for other liverworts and Antarctic mosses, including *Cephaloziella exiliflora, Bryum pseudotriquetrum,* and *Grimmia antarctici* [24, 32]. However, the current report is the first to relate these sugar alcohols with processes of desiccation or rehydration. Clearly, polyols act principally in *P. alpinum*, where mannitol and arabitol have been depleted during desiccation; in contrast during water uptake, all four polyols increased considerablely, suggesting an osmotic functioning (Additional file 1: Table S1).

During water stress, carbohydrates represent a source of energy for the cell and protection for molecules, thereby decreasing the effects of water loss. In contrast to higher plants where sugars retain water through the formation of hydrogen bonds, in mosses, sugar hydrogen bonds can act as substitutes of water molecules lost during desiccation, thus maintaining the native form and activity of proteins [18, 44]. In Antarctica, mosses are not only exposed to water stress, but also to low temperatures and daily freeze–thaw cycles that impose a strong pressure on metabolism, which must continuously adjust to avoid water loss and cell damage.

It is evident that the metabolism of sugars in bryophytes is much more complex than previously assumed, especially given that recent reports have found that other moss species, grown under different conditions, are able to synthesize a series of new compounds, some of which were not previously described and which would have distinct roles in metabolic processes [45]. This creates new questions for carbohydrate metabolism in Antarctic mosses exposed to cold, freezing, and drought conditions.

Conclusions

Sanionia uncinata and *P. alpinum* presented differences in water loss and retention capacities. *S. uncinata* showed the strongest contrasting responses between plant form, with individual plants losing water rapidly while grouped discs were able to maintain a high water content over a longer period. Individual plants of *P. alpinum*, which have a rudimentary vascular system, were able to maintain water content longer than *S. uncinata*. Interestingly, both moss species showed insignificant changes in NSC contents after 50 % desiccation, only changing the level of carbohydrates during full water loss. The RFOs family of carbohydrates changed during desiccation and rehydration, and galactinol probably plays an important role during water management in both species. Differences in water loss and uptake can explain the different preferential growth sites for each moss, with *S. uncinata* growing in the flooded, sandy soil of valleys fed by run-off water from glaciers or snow banks and *P. alpinum* growing in small cushions dispersed on sandy soil without a continuous water supply. The high capacity of *S. uncinata* to maintain water for a longer time suggests that this moss species could play an important ecological role in the Antarctic tundra ecosystem, where it would provide other species with an additional water supply during drought periods; this could also partially explain the dominance of *S. uncinata* in large tundra communities of Fildes Peninsula on King George Island, as well as on other islands of the South Shetland Island Archipelago.

Methods

Plant samples

Gametophytes of *S. uncinata* (Hedw.) Loeske and *P. alpinum* (Hedw.) G.L. Smith were collected during summer 2013 at Juan Carlos Point (S62°12.03′ W058°59.66′) on Fildes Peninsula, King George Island in the South Shetland Islands Archipelago. The identification of each species was performed through microscopic analysis according to Ochyra et al. [8]. The samples were kept dry until used in desiccation and rehydration experiments. Reference specimens of each moss were deposited at the herbarium of the Universidad de Concepcion, CONC.

Experimental design

We determined first the water loss time for a) individual gametophytes of plants and b) discs consisting of various gametophytes of both species. We took discs with a punch directly from carpets (agreggated form) of *S. uncinata* or from cushions of *P. alpinum*. Disc sample sizes where similar, of 10 mm diameter (area = 78.5 mm^2), 10–15 mm in height, and 0.2–0.3 g of dry weight for *S. uncinata* and 0.30–0.35 g dry weight for *P. alpinum*. The mosses were rehydrated through submersion in distilled water for 30 h at 6–8 °C while being illuminated with photosynthetic active radiation of 100 $\mu mol/m^2\ s^{-1}$, in order to promote an active metabolism prior to desiccation. Following this, the superficially accumulated water was removed using a paper towel. Both, discs (four replicates) and individual (6 replicates) gametophytes were submerged in distilled water, dried at room temperature, and weighed during the entire process with a model M2P analytical microbalance (SARTORIUS, Germany). Given the high rate of water loss in individual gametophytes and the high sensitivity of the microbalance, the entire process of dehydration was recorded uninterruptedly using a DSC-S730 video camera (Sony, Japan). Following this, the video was reviewed, and mass was recorded every 5 min. The hydric content was calculated based on decreasing mass in mg $H_2O\ g^{-1}$ DW.

For the second experiments we used only discs of gametophytes. For the dehydration and rehydration we established three levels (a) completely hydrated, 100 % H_2O (D100); (b) moderately hydrated, 50 % H_2O (D50); and (c) dry, 0 % H_2O (D0). For this purpose, moss samples were dehydrated in a glass desiccator with desiccant agent silica gel and weighed using an analytical WTB 200 balance (RADWAG, Poland). After complete loss of water, the disc samples were rehydrated. For this, disc samples of both moss species were partially submerged in water so that hydration occurred through capillarity. The starting point of rehydration corresponded to the most desiccated treatment, (D0) but in the case of rehydration, this point was established as R0. Samples were rehydrated to reach 50 % (R50) and 100 % (R100) hydric content. For each treatment, four replicate individuals per species were used, and likewise, four tissue samples were collected for analysis of soluble sugars. Hydric content, expressed as % of H_2O, was determined by using dry weight (DW) and fresh weight (FW) of each sample according to the following formula: (FW-DW) × 100 %/(FW).

Quantifying soluble sugars

During the second experiment, we took samples for carbohydrate analysis from discs of both moss species. The extraction and quantification of total soluble sugars was performed according to Zúñiga et al. [40]. Briefly, 0.100 g of FW was taken for each desiccation and rehydration treatment, and this sample was incubated at 4 °C in 1 mL of 80 % ethanol for 96 h (4 days). For high performance liquid chromatography (HPLC) analysis, aliquots of 480 µL were concentrated (Savant DNA SpeedVac, Minn., USA) and then resuspended in 0.1 mM of calcium-EDTA buffer before being filtered (0.45 µm). A volume of 20 µL per sample was injected into an Agilent 1100 series chromatograph equipped with a 300 mm × 6.5 mm Sugar-pak I column (Waters Corp., Mass., USA) at 75 °C and with an Agilent 1100 series refractive index detector at 55 °C. The isocratic elution program consisted in a mobile phase of 0.1 mM calcium-EDTA, with a flow of 0.35 ml min^{-1} and a pressure of 38 bars per 40 min. To identify soluble carbohydrates standards of glucose, fructose, galactose, galactinol, sucrose, raffinose, stachyose, verbascose, nystose, kestose, adonitol, arabitol, erythritol, mannitol, and pinitol were used (Sigma, USA).

Statistical analysis

The time variation in responses to treatments (five water levels), sample type (individual plants or disc samples) and species (*S. uncinata* and *P. alpinum*) were analysed with ANOVA ($p < 0.05$; CI 95 %); as well as to compare the differences in carbohydrates level due to treatments and species. Thereafter, we separated the analysis of changes for each soluble carbohydrate by species, using one-way ANOVA. For a multiple comparison of measurements according to statistical differences, Tukey's test ($P < 0.05$; CI 95 %) was applied. Statistical analyses were performed using the InfoStat software [46].

Authors' contributions

PZG carried out the experiments; MP analysed the carbohydrates; ACK collected samples of mosses in Antarctica; PZG, MP, ACK and GZ, analysed and intrepreted the results. PZG and ACK wrote the manuscript. All authors read and approved the final manuscript.

Author details

[1] Laboratorio de Micología y Micorrizas, Facultad de Ciencias Naturales y Oceanográficas and Laboratorio de Investigación en Agentes Antibacterianos, Facultad de Ciencias Biológicas, Universidad de Concepción, Barrio Universitario s/n, Concepción, Chile. [2] Departamento de Biología, Facultad de Química y Biología, Universidad de Santiago, Alameda, 3363 Santiago, Chile. [3] Núcleo de Estudios Ambientales, Universidad Católica de Temuco, Casilla 15-D, Temuco, Chile. [4] Facultad de Química y Biología, Universidad de Santiago, Alameda, 3363 Santiago, Chile.

Acknowledgements

This research was funded by projects FONDECYT 1120895; INACH FR 01-12 granted to Angélica Casanova-Katny. FONDECYT 1140189 granted to Gustavo E. Zúñiga. We thank Dr. Eugenio Sanfuentes von Stowasser of the Laboratorio de Patología Forestal and Ms. Susana Casas from the Laboratory of Natural Resources of the Centro de Biotecnología de la Universidad de Concepción. Angélica Casanova Katny and Gustavo E. Zúñiga, thanks Proyectos Basales y Vicerrectoría de Investigación, Desarrollo e Innovación, Código 021543ZN_INTEXCELENC, Universidad de Santiago de Chile. We thank Dr. Sharon Robinson from Wollongong University, for comments and english correction of this paper and two anonymous reviewers for constructive criticisms. Finally we acknowledge the support of the General Directorate of Research and Postgraduate Studies of the Catholic University of Temuco, DGIPUCT Project no. CD2010-01 and MECESUP UCT 0804.

Competing interests

The authors declare that they have no competing interests.

References

1. Convey P. Antarctic climate change and its influences on terrestrial ecosystems. In: Bergstrom DM, Convey P, Huiskes AHL, editors. Trends in Antarctic terrestrial and limnetic ecosystems: Antarctica as a global indicator. Dordrecht: Springer; 2006. p. 253–72.
2. Kennedy AD. Antarctic terrestrial ecosystem repsonse to global environmental change. Ann Rev Ecol Syst. 1995;26:683–704.
3. Olech M. Plant communities on King George Island. In: Beyer L, Bölter M, editors. Ecological studies. Geoecology of Antarctic Ice-Free Coastal Landscapes, vol. 154. New York: Springer; 2002. p. 215–31.
4. Green TGA, Sancho LG, Türk R, Seppelt RD, Hogg ID. High diversity of lichens at 848S, Queen Maud Mountains, suggests preglacial survival of species in the Ross Sea region, Antarctica. Polar Biology. 2011;34:1211–20.
5. Zarzycki K. Vascular plants and terrestrial biotopes. In: Rakusa-Suszczewski S, editor. The Maritime Antarctic coastal ecosystem of Admiralty Bay. Varsovia: Polish Academy of Sciences; 1993. p. 181–7.
6. Ochyra R. The moss flora of King George Island, Antarctica. Cracovia: Polish Academy of Sciences; 1998.
7. Komárková V, Poncet S, Poncet J. Two native Antarctic vascular plants, *Deschampsia antarctica* and *Colobanthus quitensis*: a new southernmost locality and other localities in the Antarctic peninsula area. Arctic Alpine Res. 1985;17:401–16.
8. Ochyra R, Lewis R, Bednarek-Ochyra H. The illustrated moss flora of Antarctica. Cambridge: Cambridge University Press; 2008.
9. Casanova-Katny A, Cavieres LA. Antarctic moss carpets facilitate growth of Deschampsia antarctica, but not its survival. Polar Biol. 2012;35:1869–78.
10. Oliver MV, Velten J, Mishler BD. Desiccation tolerance in bryophytes: a reflection of the primitive strategy for plant survival in dehydrating habitats? Integr Comp Biol. 2005;45:788–99.
11. Proctor MC, Ligrone R, Duckett JG. Desiccation tolerance in the moss *Polytrichum formosum*: physiological and fine-structural changes during desiccation and recovery. Ann Botany. 2007;99:75–93.
12. Proctor MCF. Mosses and alternative adaptation to life on land. New Phytol. 2000;148:1–3.
13. Proctor MCF, Pence VC. Vegetative tissues: bryophytes, vascular resurrection plants, and vegetative propagules. In: Black M, Pritchard HW, editors. Desiccation and survival in plants: drying without dying. Wallingford: CABI Publishing; 2002. p. 293–318.
14. Alpert P. The discovery, scope and puzzle of desiccation tolerance in plants. Plant Ecol. 2000;151:5–17.
15. Bewley JD. Physiological aspects of desiccation tolerance. Annu Rev Plant Biol. 1979;30:195–238.
16. Sun QW, Irving TC, Leopold AC. The role of sugar, vitrification and membrane phase-transition in seed desiccation tolerance. Physiol Plant. 1994;90:621–8.
17. Vertucci CW, Farrant JM. Acquisition and loss of desiccation tolerance. In: Kigel J, Galili G, editors. Seed development and germination. New York: Marcel Dekker; 1995. p. 237–71.
18. Hoekstra FA, Golovina E, Buitink J. Mechanisms of plant desiccation tolerance. Trends Plant Sci. 2001;6:431–8.
19. Hoekstra FA, van Roekel T. Desiccation tolerance of *Papaver dubium* L. pollen during its development in the anther-possible role of phospholipid-composition and sucrose content. Plant Physiol. 1988;88:626–32.
20. Smirnoff N. The carbohydrates of bryophytes in relation to desiccation tolerance. J. Bryol. 1992;17:185–91.
21. Wasley J, Robinson SA, Popp M, Lovelock CE. Some like it wet—biological characteristics underpinning tolerance of extreme water events in Antarctic bryophytes. Funct Plant Biol. 2006;33:443–55.
22. Shortlidge EE, Rosenstiel NT, Eppley SM. Tolerance to environmental desiccation in moss sperm. New Phytol. 2012;194:741–50.
23. Buitink J, Hoekstra FA, Leprince O. Biochemistry and biophysics of tolerance systems. In: Black M, Pritchard HW, editors. Desiccation and survival in plants: Drying without dying. Wallingford: CABI Publishing; 2002. p. 293–318.
24. Robinson SA, Wasley J, Popp M, Lovelock CE. Desiccation tolerance of three moss species from continental Antarctica. Aust J Plant Physiol. 2000;27:379–88.
25. Montenegro L, Melgarejo L. Variación del contenido de azúcares totales y azúcares reductores en el musgo Pleurozium schreberi (HYLOCOMIACEAE) bajo condiciones de déficit hídrico. Acta biol Colomb. 2012;17:599–610.
26. Roberts P, Newsham KK, Bardgett RD, Farrar JF, Jones DL. Vegetation cover regulates the quantity, quality and temporal dynamics of dissolved organic carbon and nitrogen in Antarctic soils. Polar Biol. 2009;32:999–1008.
27. Chopra RN, Kumra PK. Water Relations. In: Chopra RN, Kumra PK, editors. Biology of bryophytes. New Dehli: New age international; 1988. p. 308–17.
28. Delgadillo C, Cárdenas A. Manual de Briofitas. México D.F.: Instituto de Biología UNAM; 1990.
29. Gimingham CH, Smith RIL. Growth form and water relations of mosses in the maritime Antarctic. Brit. Antarct. Surv. Bull. 1971;25:1–21.
30. Longton RE. Pattern, process and environment. In: Longton RE, editor. Biology of Polar Bryophytes and Lichens. Cambridge: Cambridge University Press; 1988. p. 66–105.
31. Gimingham CH. Quantitative community analysis and Bryophyte ecology on Signy Island. Philos T Roy Soc B. 1967;252:251–9.
32. Roser DJ, Melick DR, Ling HU, Seppelt RD. Polyol and sugar content of terrestrial plants from continental Antarctica. Antarct Sci. 1992;4:413–20.

33. Obendorf RL. Oligosaccharides and galactosyl cyclitols in seed desiccation tolerance. Seed Sci Res. 1997;7:63–74.
34. Albini FM, Murelli C, Patritti G, Rovati M, Zienna P, Finzi PV. Low-molecular weight substances from the resurrection plant *Sporobolus stapfianus*. Phytochemistry. 1994;37:137–42.
35. Albini FM, Murelli C, Finzi PV, Ferrarotti M, Cantoni B, Puliga S, et al. Galactinol in the leaves of the resurrection plant *Boea hygroscopica*. Phytochemistry. 1999;51:499–505.
36. Peterbauer T, Lahuta LB, Blöchl A, Mucha J, Jones DA, Hedley CL, et al. Analysis of the raffinose family oligosaccharide pathway in pea seeds with contrasting carbohydrate composition. Plant Physiol. 2001;127:1764–72.
37. Peterbauer T, Richter A. Biochemistry and physiology of raffinose family oligosaccharides and galactosyl cyclitols in seeds. Seed Sci Res. 2001;11:185–97.
38. Schrier AA, Hoffmann-Thoma G, van Bel AJE. Temperature effects on symplasmic and apoplasmic phloem loading and loading-associated carbohydrate processing. Aust J Plant Physiol. 2000;27:769–78.
39. Bachmann M, Matile P, Keller F. Metabolism of the Raffinose Family Oligosaccharides in leaves of *Ajuga reptans* L. (cold acclimation, translocation, and sink to source transition: discovery of chain elongation enzyme). Plant Physiol. 1994;105:1335–45.
40. Zúñiga GE, Pizarro M, Contreras RA, Kohler H. Tolerancia la desecacion en briofitas. Participacion de azucares. Cad Pesqui. 2012;24:146–54.
41. Gechev TV, Benina M, Obata T, Tohge T, Sujeeth N, Minkov I, et al. Molecular mechanisms of desiccation tolerance in the resurrection glacial relic *Haberlea rhodopensis*. Cell Mol Life Sci. 2013;70:689–709.
42. Popp M, Smirnoff N. Polyol accumulation and metabolism during water deficit. In: Smirnoff N, editor. Environment and plant metabolism: flexibility and acclimation. Oxford: Bioscientific Publishers Ltd; 1995. p. 199–214.
43. Chen TH, Murata N. Enhancement of tolerance of abiotic stress by metabolic engineering of betaines and other compatible solutes. Curr Opin Plant Biol. 2002;5:250–7.
44. Crowe JH, Crowe LM, Carpenter JF, Aurell Wistrom C. Stabilization of dry phospholipid bilayers and proteins by sugars. Biochem J. 1987;242:1–10.
45. Erxleben A, Gessler A, Vervliet-Scheebaum M, Reski R. Metabolite profiling of the moss *Physcomitrella patens* reveals evolutionary conservation of osmoprotective substances. Plant Cell Rep. 2012;31:427–36.
46. Di Rienzo J, Casanoves F, Balzarini M, Gonzalez L, Tablada M, Robledo C. InfoStat versión 2014. Grupo InfoStat, FCA, Universidad Nacional de Córdoba, Argentina.

10

Estimation of antioxidant, antimicrobial activity and brine shrimp toxicity of plants collected from Oymyakon region of the Republic of Sakha (Yakutia), Russia

Babita Paudel[1], Hari Datta Bhattarai[1], Il Chan Kim[1], Hyoungseok Lee[1], Roman Sofronov[2], Lena Ivanova[2], Lena Poryadina[2] and Joung Han Yim[1*]

Abstract

Background: Several plants are reported to be produced various biological active compounds. Lichens from the extreme environments such as high altitude, high UV, drought and cold are believed to be synthesized unique types of secondary metabolites than the other one. Several human pathogenic bacteria and fungi have been muted into drug resistant strains. Various synthetic antioxidant compounds have posed carcinogenic effects. This phenomenon needs further research for new effective drugs of natural origin. This manuscript aimed to screen new source of biological active compounds from plants of subarctic origin.

Results: A total of 114 plant species, including 80 species of higher plants, 19 species of lichens and 15 species of mosses, were collected from Oymyakon region of the Republic of Sakha (Yakutia), Russia (63°20′N, 141°42′E–63°15′N, 142°27′E). Antimicrobial, DPPH free radical scavenging and brine shrimp (Artemia salina) toxicity of all crude extract were evaluated. The obtained result was analyzed and compared with commercial standards. A total of 28 species of higher plants showed very strong antioxidant activity (DPPH IC50, 0.45-5.0 μg/mL), 13 species showed strong activity (DPPH IC50, 5-10 μg/mL), 22 species showed moderate antioxidant activity (DPPH IC50,10-20 μg/mL) and 17 species showed weak antioxidant activity (DPPH IC50 more than 20 μg/mL). Similarly, 3 species of lichen showed strong antioxidant activity, one species showed moderate and 15 species showed weak DPPH reducing activity. In addition, 4 species of mosses showed moderate antioxidant activity and 11 species showed weak antioxidant activity. Similarly, extracts of 51 species of higher plants showed antimicrobial (AM) activity against Staphylococcus aureus and 2 species showed AM activity against Candida albicans. Similarly, 11 species of lichen showed AM activity against S. aureus and 3 species showed AM activity against Escherichia coli. One species of moss showed AM activity against S. aureus. And finally, one species of higher plant Rheum compactum and one species of lichen Flavocetraria cucullata showed the toxicity against Brine shrimp larvae in 100 μg/mL of concentration.

Conclusion: The experimental results showed that subarctic plant species could be potential sources of various biologically active natural compounds.

Keywords: Antimicrobial, Antioxidant, Brine shrimp, DPPH, Lichen, Moss

* Correspondence: jhyim@kopri.re.kr
[1]Division of Life Sciences, Korea Polar Research Institute, KOPRI, Incheon 406-840, Republic of Korea
Full list of author information is available at the end of the article

Background

Reactive oxygen species (ROS) and reactive nitrogen species (NOS) are accumulated in living organisms during normal metabolic processes and exogenous stimuli such as UV-radiation, stress. ROS, such as superoxide anions (O_2^-), hydroxyl radicals (OH), hydrogen peroxide (H_2O_2), and hypocholorous acid (HOCl), and NOS such as nitric oxide radical (NO) have been associated with inflammation, cardiovascular diseases, cancer, aging-related disorders, metabolic disorders, and atherosclerosis [1]. ROS are dangerous because they can attack unsaturated fatty acids and cause membrane lipid peroxidation, decreases in membrane fluidity, loss of enzyme receptor activities, and damage to membrane proteins, ultimately leading to cell inactivation and cell death [2]. Living organisms possess a natural defense mechanism that counters the deleterious effects of ROS. Despite the existence of such a mechanism, increasing ROS accumulation over the lifetime of a cell can cause irreversible oxidative damage [3]. Thus, antioxidant agents that can slow or prevent the oxidation process by removing free radical intermediates are desired. Several strong synthetic antioxidants have already been reported [4], however most of them have been proven to be highly carcinogenic [5]. For this reason it has become very necessary to derive antioxidants from natural sources for use as supplements to human health. A wide range of natural compounds, including phenolic compounds, nitrogen compounds, and caretenoids [6-9] have antioxidant properties.

Pathogenic microbes, mainly gram-positive bacteria, pose serious threats to human. *Staphylococcus aureus* especially methicillin resistant *Staphylococcus aureus* (MRSA) species have posed serious threats to human health care settings. New alternatives for combating the spread of infection by antibiotic resistance microbes in future are necessary tools for keeping pace with the evolution of 'super' pathogens. Similarly, *Escherichia coli* and *Candida albicans* have also been considered as potential pathogen to human. In this research report we screened the antimicrobial activity of subarctic plants extracts against various bacterial and fungal pathogenic microorganisms.

Natural products have contributed significantly in development of anticancer drugs [10]. There is an urgent need of searching potential candidates of future anticancer drugs to deal with increasing number of cancer diseases in human beings. Brine shrimp toxic compounds could be a potential candidate of anticancer activity. Therefore, we have used this assay to perform primary screening of subarctic plants extracts for their toxicity.

Results and discussion

Antioxidant activity

All the tested plant extracts and the commercial standard (BHA) exhibited DPPH free radical scavenging activities in the concentration dependent manner that could be easily read by a spectrophotometer obtaining a decreased absorbance at 517 nm. BHA is a strong commercial antioxidant compound and the IC_{50} (50% inhibition) of this compound was 4.98 μg/mL in the present experiment. According to observed experimental data in term of 50% inhibition concentration (IC_{50}) (Table 1), higher plants showed strong antioxidant activity as compared to lichens and mosses. Among the higher plants, 28 species showed very strong antioxidant activity (IC_{50}, 0.45-5.0 μg/mL), 13 species showed strong activity (IC_{50}, 5–10 μg/mL), 22 species showed moderate antioxidant activity (IC_{50}, 10–20 μg/mL) and 17 species showed weak antioxidant activity (IC_{50} more than 20 μg/mL). *Rhododendron dauricum*, *Dryas grandis* and *Rhodendron redowskianum* showed the strongest antioxidant activity (IC_{50}, 0.4-0.6 μg/mL). In case of lichen, 3 species, *Thamnolia vermicularis*, *Peltigera didactyla,* and *Peltigera malacea* showed strong antioxidant activity (IC_{50}, 5.2- 6.1 μg/mL), one species of lichen *Peltigera aphthosa* showed moderate and 15 species showed weak antioxidant activity. Similarly, 4 species of mosses showed moderate and remaining 11 species showed weak antioxidant activity.

Antimicrobial activity

The observed experimental data as shown in Table 1 indicated that 51 species of higher plants showed antimicrobial (AM) activity against *S. aureus*. According to size of zone of inhibition, following four categories of antimicrobial activities against *S. aureus* were obtained. Two species of higher plants, *Empetrum nigrum and Cassiope tetragona* showed very strong AM against *S. aureus* (inhibition zone ≥ 20 mm) and moderate AM activity against *C. albicans* (inhibition zone-10 mm). Three species showed strong (inhibition zone, 15–20 mm), 32 species showed moderate (inhibition zone, 10–15 mm) and 14 species showed weak (inhibition zone ≤10 mm) AM activities. None of the tested higher plants extract showed AM activity against *E. coli* and *A. niger*. Similarly, 11 species of lichens showed various strength of AM activity against *S. aureus*. Two species of lichen, *Alectoria ochroleuca* and *Cladonia verticillata* showed very strong AM activity against *S. aureus* having zone of inhibition 28 and 27 mm respectively. One species, *Sterocaulon paschale* showed strong AM (inhibition zone, 16 mm), 5 species showed moderate and 3 species showed weak AM activity against *S. aureus*. Similarly, 3 species of lichens showed AM activity against *E. coli* out of which 2 species showed moderate and one species showed weak AM activity. Three species of lichens, *Alectoria ochroleuca*, *Cladonia amaurocraea* and *Cladonia verticillata* showed AM activity against both *S. aureus* and *E. coli.* And finally, one species of moss (*Scorpidium scorpioides*) showed AM activity against *S. aureus*. The

Table 1 Antioxidant, antimicrobial activity and brine shrimp toxicity test of plant's extracts

Symbol	Species name	DPPH-IC_{50} (μg/mL)	Antimicrobial activity-Inhibition zone (mm)[1]		
			S. aureus	*E. coli*	*C. albicans*
HP-1	*Rhododendron dauricum*	0.45 ± 0.02	-	-	-
HP-2	*Dryas grandis*	0.52 ± 0.03	-	-	-
HP-3	*Rhododendron redowskianum*	0.61 ± 0.02	-	-	-
HP-4	*Dryopteris fragrans*	1.2 ± 0.08	14 ± 1.84	-	-
HP-5	*Saxifraga bronchialis*	1.8 ± 0.07	-	-	-
HP-6	*Aconogonon tripterocarpum*	1.9 ± 0.17	-	-	-
HP-7	*Chamerion angustifolium*	2.1 ± 0.17	-	-	-
HP-8	*Salix pulchra*	2.1 ± 0.15	10 ± 0.9	-	-
HP-9	*Chamerion angustifolium*	2.2 ± 0.06	-	-	-
HP-10	*Betula divaricata*	2.5 ± 0.15	-	-	-
HP-11	*Artemisia vulgaris*	2.8 ± 0.14	-	-	-
HP-12	*Rhododendron lapponicum*	3.1 ± 0.31	15 ± 1.5	-	-
HP-13	*Andromeda polifolia*	3.2 ± 0.26	-	-	-
HP-14	*Vaccinium uliginosum*	3.4 ± 0.14	9 ± 1.06	-	-
HP-15	*Ribes triste*	3.5 ± 0.07	13 ± 1.4	-	-
HP-16	*Comarum palustre*	3.6 ± 0.18	9 ± 1.41	-	-
HP-17	*Salix reptans*	3.7 ± 0.15	-	-	-
HP-18	*Ledum palustre*	3.8 ± 0.27	-	-	-
HP-19	*Rosa acicularis*	3.9 ± 0.19	10 ± 1.8	-	-
HP-20	*Pyrola rotundifolia*	3.9 ± 0.16	-	-	-
HP-21	*Sanguisorba officinalis*	4.1 ± 0.33	12 ± 1.4	-	-
HP-22	*Carex aquatilis*	4.1 ± 0.08	-	-	-
HP-23	*Rubus matsumuranus*	4.3 ± 0.22	11 ± 1.7	-	-
HP-24	*Vaccinium vitis-idaea*	4.7 ± 0.19	8.5 ± 0.51	-	-
HP-25	*Rubus chamaemorus*	4.7 ± 0.38	11 ± 1.5	-	-
HP-26	*Veronica incana*	4.8 ± 0.14	10 ± 1.2	-	-
HP-27	*Pentaphylloides fruticosa*	4.8 ± 0.24	8.5 ± 0.74	-	-
HP-28	*Galium verum*	4.9 ± 0.34	11 ± 1.2	-	-
HP-29	*Cassiope ericoides*	5.1 ± 0.31	11 ± 1.3	-	-
HP-30	*Parnassia palustris*	5.4 ± 0.11	11 ± 0.98	-	-
HP-31	*Dracocephalum palmatum*	6 ± 0.48	10 ± 1.03	-	-
HP-32	*Orostachys spinosa*	6.2 ± 0.19	-	-	-
HP-33	*Salix tschuktschorum*	6.3 ± 0.25	-	-	-
HP-34	*Juniperus communis*	6.8 ± 0.14	11 ± 1.5	-	-
HP-35	*Ranunculus reptans*	7.3 ± 0.23	-	-	-
HP-36	*Thymus pavlovii*	7.5 ± 0.45	11 ± 1.67	-	-
HP-37	*Sparganium hyperboreum*	8.2 ± 0.43	9 ± 0.82	-	-
HP-38	*Saxifraga punctata*	8.2 ± 0.25	-	-	-
HP-39	*Pinus pumila*	9.2 ± 0.64	10 ± 0.85	-	-
HP-40	*Ribes fragrans*	9.5 ± 0.38	9 ± 1.2	-	-
HP-41	*Sedum sukaczevii*	9.8 ± 0.49	-	-	-
HP-42	*Thalictrum foetidum*	10.3 ± 0.31	10 ± 0.83	-	-
HP-43	*Sorbaria sorbifolia*	11 ± 0.88	16 ± 1.36	-	-
HP-44	*Ptarmica salicifolia*	11.1 ± 0.67	9 ± 1.3	-	-
HP-55	*Rheum compactum*	11.2 ± 0.34	18 ± 1.7	-	-
HP-46	*Artemisia lagocephala*	11.6 ± 1.1	9 ± 0.5	-	-

Table 1 Antioxidant, antimicrobial activity and brine shrimp toxicity test of plant's extracts *(Continued)*

HP-47	*Campanula rotundifolia ssp. langsdorffiana*	11.8 ± 0.71	-	-	-
HP-48	*Veratrum lobelianum*	12 ± 0.72	9 ± 0.9	-	-
HP-49	*Oxycoccus microcarpus*	12.1 ± 0.48	9 ± 0.73	-	-
HP-50	*Beckmannia syzigachne*	13 ± 0.39	10 ± 1.2	-	-
HP-51	*Empetrum nigrum*	15.1 ± 0.9	20 ± 2.2	-	10 ± 0.33
HP-52	*Euprasia hyperborea*	15.1 ± 0.45	12 ± 1.8	-	-
HP-53	*Cassiope tetragona*	15.2 ± 0.76	20 ± 1.8	-	10 ± 0.81
HP-54	*Alopecurus roshevitzianus*	18 ± 1.44	9 ± 0.8	-	-
HP-55	*Chosenia arbutifolia*	18.1 ± 1.25	14 ± 1.4	-	-
HP-56	*Dryas punctata*	18.2 ± 0.91	-	-	-
HP-57	*Achillea millefolium*	19 ± 1.33	14 ± 1.1	-	-
HP-58	*Astragalus frigidus*	19.3 ± 0.97	14 ± 1.7	-	-
HP-59	*Arctophila fulva*	19.5 ± 0.78	12 ± 1.08	-	-
HP-60	*Artemisia jacutica*	19.7 ± 1.18	10 ± 0.9	-	-
HP-61	*Huperzia selago*	19.8 ± 0.59	-	-	-
HP-62	*Equisetum arvense*	19.9 ± 1.39	11 ± 0.7	-	-
HP-63	*Equisetum arvense*	20 ± 1.2	10 ± 0.6	-	-
HP-64	*Dianthus repens*	20.2 ± 1.94	9 ± 1.02	-	-
HP-65	*Cnidium cnidiifolium*	>20	10 ± 0.8	-	-
HP-66	*Menyanthes trifoliata*	>20	12 ± 1.02	-	-
HP-67	*Lycopodium dubium*	>20	-	-	-
HP-68	*Tofieldia coccinea*	>20	-	-	-
HP-69	*Eriophorum medium*	>20	-	-	-
HP-70	*Oxyria digyna*	>20	-	-	-
HP-71	*Equisetum fluviatile*	>20	9 ± 0.83	-	-
HP-72	*Astragalus schelichovii*	>20	10 ± 1.51	-	-
HP-73	*Oxytropis adamsiana*	>20	10 ± 0.71	-	-
HP-74	*Hedysarum alpinum*	>20	10 ± 0.85	-	-
HP-75	*Calamagrostis purpurea ssp. langsdorffii*	>20	-	-	-
HP-76	*Carex saxatilis ssp. laxa*	>20	-	-	-
HP-77	*Aconitum macrorhynchum*	>20	9 ± 0.79	-	-
HP-78	*Poa botryoides*	>20	-	-	-
HP-79	*Juncus nodulosus*	>20	10 ± 0.8	-	-
HP-80	*Eleocharis acicularis*	>20	10 ± 0.7	-	-
L-91	*Thamnolia vermicularis*	5.2 ± 0.16	-	-	-
L-92	*Peltigera didactyla*	5.7 ± 0.46	-	-	-
L-93	*Peltigera malacea*	6.1 ± 0.31	-	-	-
L-94	*Peltigera aphthosa*	14.7 ± 1.03	-	-	-
L-95	*Alectoria ochroleuca*	>20	28 ± 2.38	12 ± 0.7	-
L-96	*Asahinea chrysantha*	>20	9 ± 0.69	-	-
L-97	*Cetraria laevigata*	>20	-	-	-
L-98	*Cladonia amaurocraea*	>20	10 ± 0.81	8.5 ± 0.51	-
L-99	*Cladonia arbuscula*	>20	9 ± 0.91	-	-
L-100	*Cladonia gracilis*	>20	-	-	-
L-101	*Cladonia phyllophora*	>20	-	-	-
L-102	*Cladonia stellaris*	>20	9 ± 0.7	-	-
L-103	*Cladonia stygia*	>20	10 ± 0.76	-	-
L-104	*Cladonia verticillata*	>20	27 ± 2.31	10 ± 0.49	-

Table 1 Antioxidant, antimicrobial activity and brine shrimp toxicity test of plant's extracts *(Continued)*

L-105	*Dactylina arctica*	>20	-	-	-
L-106	*Flavocetraria cucullata*	>20	13 ± 1.1	-	-
L-107	*Flavocetraria nivalis*	>20	14 ± 1.5	-	-
L-108	*Stereocaulon botryosum*	>20	13 ± 0.88	-	-
L-109	*Stereocaulon paschale*	>20	16 ± 1.6	-	-
M-100	*Sphagnum fuscum*	19.7 ± 0.98	-	-	-
M-101	*Loeskypnum badium*	19.8 ± 0.59	-	-	-
M-102	*Hylocomium splendens*	19.8 ± 1.19	-	-	-
M-103	*Polytrichastrum alpinum*	19.9 ± 0.99	-	-	-
M-104	*Scorpidium scorpioides*	>20	11 ± 1.3	-	-
M-105	*Ditrichum flexicaule*	>20	-	-	-
M-106	*Racomitrium lanuginosum*	>20	-	-	-
M-107	*Warnstorfia sarmentosa*	>20	-	-	-
M-108	*Dicranum elongatum*	>20	-	-	-
M-109	*Sphagnum lenense*	>20	-	-	-
M-110	*Sphagnum imbricatum*	>20	-	-	-
M-111	*Rhytidium rugosum*	>20	-	-	-
M-112	*Sphagnum warnstorfii*	>20	-	-	-
M-113	*Sphagnum anogstroemii*	>20	-	-	-
M-114	*Paludella squarrosa*	>20	-	-	-

HP Higher Plant, *L* Lichen, *M* Moss, (–)- no activity.
[1]None of the plant extract showed antimicrobial activity against *Aspergillus niger*.

differences in antimicrobial activities may be due to variation in antimicrobial metabolites among tested samples. AM activity of test samples is species specific. Such results clearly suggest that subarctic plants species are potential source of species specific antimicrobial active compounds.

Brine shrimp toxicity test

In the present experiment, among the tested 114 plant species, only one species of higher plant *Rheum compactum,* and one species of lichen *Flavocetraria cucullata* showed death of all tested larvae at 100 μg/mL of concentration. The remaining 112 species didn't show toxicity within 100 μg/mL concentration. In case of berberine chloride, positive control, all the *Artemia* larvae were dead at the 7 μg/mL concentration. In the negative control test all of *Artemia* larvae were alive after 24 h of experiment.

Conclusion

The observed experimental data clearly showed that most of the tested plants showed potent antioxidant activities *in vitro*. In addition, many higher plants and lichens species showed potent antimicrobial activity against human pathogenic bacteria, *Staphylococcus aureus*, *Escherichia coli* and fungi, *Candida albicans*. In addition, most of the antimicrobial and antioxidant active plants species were not toxic against *Artemia* larvae which could be an indication of being non toxic plant species. In addition, the observed data also clearly showed that several antioxidant and antimicrobial compounds could be obtained from these plants resources. Therefore, further works of isolation and characterization of antioxidant and antimicrobial compounds merit from these plants resources.

Methods

Collection and identification of Plants A total of 114 plant species, including 80 species of higher plants, 19 species of lichens and 15 species of mosses, were collected from Oymyakon region of the Republic of Sakha (Yakutia), Russia (63°20′N, 141°42′E –63°15′N, 142°27′E) in July 2010. All the plant specimens were identified by using morphological characters.

Extraction of plants

The collected samples were air dried and grinded completely. Dried and grinded samples of plants specimens (Table 1) were separately extracted in methanol–water (80:20 v/v) at room temperature (RT). The solvent was evaporated in vacuum at 45°C and the extracts were then lyophilized. The crude extracts were stored at −20°C until further use.

Biological activity evaluation

Antioxidant, antimicrobial and Brine shrimp lethality tests were performed to evaluate the biological activities of the plant extracts. To estimate the antioxidant potential of plant extracts, DPPH (1, 1-diphenyl-2-picrylhydrazyl) reducing activity was assayed as described previously [11,12]. Similarly, antimicrobial activity was assayed by disk diffusion method [13] against human pathogenic micro-organisms, *Staphylococcus aureus*, *Escherichia coli*, *Candida albicans* and *Aspergillus niger*. Brine shrimp lethality test was performed to estimate the toxicity of sample as described previously [13].

Competing interests

The authors declare that they have no competing interests.

Authors' contributions

BP and HDB designed the experiment and performed laboratory experiment of antioxidant, antimicrobial and brine shrimp toxicity test followed by writing the manuscript. ICK, HL, RS, LI and LP made sample collection and taxonomic identification of plants. JHY supervised overall experiments and field work. All authors read and approved the final manuscript.

Acknowledgements

This work was supported by a grant from the Korea Polar Research Institute, KOPRI, under the project PE13040.

Author details

[1]Division of Life Sciences, Korea Polar Research Institute, KOPRI, Incheon 406-840, Republic of Korea. [2]Institute for Biological Problems of Cryolithozone, Siberian Branch of Russian Academy of Sciences, Moscow, Russia.

References

1. Ames BN, Shigenaga MK, Hagen TM: **Oxidants, antioxidants and the degenerative diseases of aging.** *Proc Natl Acad Sci USA* 1993, **90:**7915–7922.
2. Dean RT, Davies MJ: **Reactive species and their accumulation on radical damaged proteins.** *Trends Biochem Sci* 1993, **18:**437–441.
3. Tseng TH, Kao ES, Chu CY, Chou FP, Lin Wu HW, Wang CJ: **Protective effects of dried flower extracts of *hibiscus sabdariffa* L. Against oxidative stress in rat primary hepatocytes.** *Food Chem Toxicol* 1997, **35:**1159–1164.
4. Shimizu K, Kondo R, Sakai K, Takeda N, Nagahata T, Oniki T: **Novel vitamin E derivative with 4-substituted resorcinol moiety has both antioxidant and tyrosinase inhibitory properties.** *Lipids* 2001, **36:**1321–1326.
5. Grice HC: **Safety evaluation of butylated hydroxyanisol from the prospective of effect on forest-omach and oesophageal squamous epithelium.** *Food Chem Toxicol* 1988, **26:**717–723.
6. Mei RQ, Wang YH, Du GH, Liu GM, Zhang L, Cheng YX: **Antioxidant lignans from the fruits of *broussonetia papyrifera*.** *J Nat Prod* 2009, **72:**621–625.
7. Serafini M, Testa MF, Villaño D, Pecorari M, Wieren KV, Azzini E, Brambilla A, Maiani G: **Antioxidant activity of blueberry fruit is impaired by association with milk.** *Free Radic Biol Med* 2009, **46:**769–774.
8. Sun YP, Chou CC, Yu RC: **Antioxidant activity of lactic-fermented Chinese cabbage.** *Food Chem* 2009, **115:**912–917.
9. Velioglu YS, Mazza G, Gao YL, Oomah BD: **Antioxidant activity and total phenolics in selected fruits, vegetables and grain products.** *J Agric Food Chem* 1998, **46:**4113–4117.
10. Cragg GM, Newman DJ, Snader KM: **Natural products as source of new drugs over the period 1981–2002.** *J Nat Prod* 2003, **66:**1022–1037.
11. Bhattarai HD, Paudel B, Lee HS, Lee YK, Yim JH: **Antioxidant activity of *sanionia uncinata*, a polar moss species from King George Island, Antarctica.** *Phytother Res* 2008, **22:**1635–1639.
12. Blois MS: **Antioxidant determinations by the use of a stable free radical.** *Nature* 1958, **26:**1199–1200.
13. Paudel B, Bhattarai HD, Pandey DP, Hur JS, Hong SG, Kim IC, Yim JH: **Antioxidant, antibacterial activity and Brine Shrimp toxicity test of some mountainous lichens from Nepal.** *Biol Res* 2012, **45:**387–391.

11

Chloroplast localization of *Cry1Ac* and *Cry2A* protein- an alternative way of insect control in cotton

Adnan Muzaffar[1,2], Sarfraz Kiani[1], Muhammad Azmat Ullah Khan[1], Abdul Qayyum Rao[1*], Arfan Ali[1], Mudassar Fareed Awan[1], Adnan Iqbal[1], Idrees Ahmad Nasir[1], Ahmad Ali Shahid[1] and Tayyab Husnain[1]

Abstract

Background: Insects have developed resistance against Bt-transgenic plants. A multi-barrier defense system to weaken their resistance development is now necessary. One such approach is to use fusion protein genes to increase resistance in plants by introducing more Bt genes in combination. The locating the target protein at the point of insect attack will be more effective. It will not mean that the non-green parts of the plants are free of toxic proteins, but it will inflict more damage on the insects because they are at maximum activity in the green parts of plants.

Results: Successful cloning was achieved by the amplification of *Cry2A*, *Cry1Ac*, and a transit peptide. The appropriate polymerase chain reaction amplification and digested products confirmed that *Cry1Ac* and *Cry2A* were successfully cloned in the correct orientation. The appearance of a blue color in sections of infiltrated leaves after 72 hours confirmed the successful expression of the construct in the plant expression system. The overall transformation efficiency was calculated to be 0.7%. The amplification of *Cry1Ac-Cry2A* and *Tp2* showed the successful integration of target genes into the genome of cotton plants. A maximum of 0.673 μg/g tissue of *Cry1Ac* and 0.568 μg/g tissue of *Cry2A* was observed in transgenic plants. We obtained 100% mortality in the target insect after 72 hours of feeding the 2nd instar larvae with transgenic plants. The appearance of a yellow color in transgenic cross sections, while absent in the control, through phase contrast microscopy indicated chloroplast localization of the target protein.

Conclusion: Locating the target protein at the point of insect attack increases insect mortality when compared with that of other transgenic plants. The results of this study will also be of great value from a biosafety point of view.

Keywords: Chloroplast transient peptide, *Cry1Ac*, *Cry2A*, Bt, cTP, Cry genes, Endotoxins, *GUS*, Transgenic plants

Background

The discovery of the insecticidal effects of *Bacillus thuringiensis* in the early 20th century has allowed for the development of new pest insect control methods. The Cry proteins solubilize in alkaline pH (9–12) following ingestion, and protoxins are then released. The protoxins are activated by specific enzymes in the midgut and bind to specific receptors in the microvilli of columnar cell apical membranes in lepidopteran insects [1]. The effect of Bt proteins is highly specific to certain insect species, and they are nontoxic to beneficial insects and animals [2]. Their relative safety for the environment, animals, humans, fishes, birds, and beneficial entomofauna is of great significance [3].

Transformation of these crystal protein (Bt) genes in plants, especially cotton, has been carried out for many years [4]. This limits the application of environmentally devastating pesticides. *Bacillus thuringiensis* (Bt) crystal proteins have attracted extensive attention as insecticidal molecules [5]. The reduction in pesticide application, up to 70%, has been documented in Bt cotton fields in India resulting in a saving of up to US$30 per ha in insecticide costs and an 80–87% increase in harvested cotton yield [6].

Cloning and transformation of various Bt genes have been done in higher plants but the resulting transgenic plants show lower insecticidal activity as insects develop

* Correspondence: qayyumabdul77@yahoo.com
[1]National Center of Excellence in Molecular Biology, University of the Punjab, Lahore 53700, Pakistan
Full list of author information is available at the end of the article

Bt resistance in response to the level of gene expression [7]. Low toxin levels are of huge concern nowadays. To overcome this issue, several strategies have been employed by researchers, e.g., inserting the gene into the chloroplast genome [8,9], modifying the coding sequences of the bacterial gene to plant-preferred coding sequences [10], and expressing the genes in the chloroplast using chloroplast transient peptides [11].

The new trend in transformation for localized transgene expression is chloroplast transformation [11]. This technique is very useful in expressing genes in the green parts of the plants but its application has been limited to the *Solanaceae* family [12]. Most of the work on chloroplast transformation has concentrated on tobacco because it is easy to regenerate on tissue culture media following biolistic/agrobacterium transformation [13]. However, the recalcitrant nature of cotton plants makes them impossible to regenerate on tissue culture media [14], seriously hindering the application of chloroplast transformation technology in this plant. Though the chloroplast contains its own DNA, it only codes 10% of the required protein. The rest of the proteins are imported from the cytosol to the chloroplast through specific trans-peptide (TP) signals [15] having an N-terminal extension responsible for carrying the proteins to the organelle [16]. Based on this, cotton nuclear transformation might be achieved by tagging TP at the Bt gene N-terminal to transport precursor proteins into the chloroplast [17,18].

Several reports have confirmed that lepidopteran insects develop some resistance to Bt crops with a single Cry gene. Therefore, there is a need to develop new strategies comprising multiple lines of defense to cope with this developing resistance in insects [11]. The present study focused on two aspects; first, developing resistance in plants using two genes, i.e., *Cry1Ac* and *Cry2A*, and second, achieving the benefits of chloroplast targeted expression through nuclear transformation in cotton, where tissue culture on media is impossible. A higher production of target proteins can be achieved when the genes are expressed in plant chloroplasts [17,19,20] because when the transgene is stably integrated, plastid transformation accumulates large amounts of foreign proteins (up to 46% of total leaf protein) [21]. The higher expression is the result of thousands of copies of the chloroplast genome in each plant cell, which results in high copy numbers of the functional genes [22]. Other advantages that have been seen in chloroplast transgenic plants include a 169-fold increase in transgene expression compared with nuclear transformation and a lack of transgene silencing [23]. Another advantage of chloroplast targeted engineering includes transgene stacking, i.e. simultaneous expression of multiple transgenes, thus creating multivalent vacancies in a single transformation step [22].

The present study aimed to clone *Cry1Ac* and *Cry2A* genes and transit peptides with their fusion protein, which can localize its expression in the chloroplast. This study was designed for the production of modern transgenic cotton plants with minimal biosafety concerns. The transgene is expressed only in the green tissues because the fusion-protein gene attaches to Bt on C-terminal and cTP on N-terminal resulting in higher expression levels, which enhances lepidopteran insect resistance.

Results

Construction of the plant expression vector MUZ_01

The transit peptide was isolated from Petunia (Figure 1). The construct MUX_01 was designed (Figure 2). Successful cloning was obtained by amplifying of 167 bp of *Cry2A*, 479 bp of *Cry1Ac*, and 216 bp of the transit peptide (Figure 3). The orientation was confirmed by specific primers, i.e., forward from *Cry1Ac* and reverse from *Cry2A*. An appropriate band of 805 bp and a digested product of 4.6 kb confirmed that *Cry1Ac* and *Cry2A* were successfully cloned in the correct orientation (Figure 4). The vector construction pattern is shown in a partial map (Figure 2).

Transient expression through *GUS* estimation

A total of 1000 embryos were transformed with MUZ_01 (TP-*Cry1Ac* + CryIIA) and subjected to transient expression of the *GUS* gene. The appearance of a blue color in

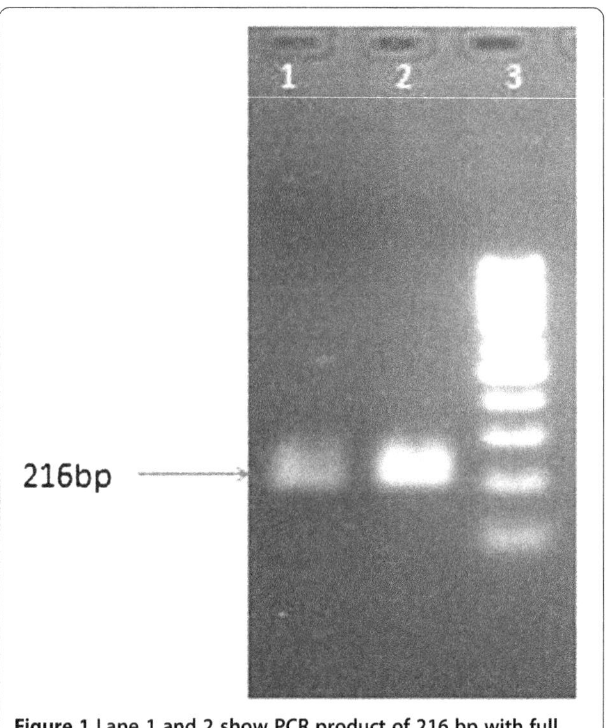

Figure 1 Lane 1 and 2 show PCR product of 216 bp with full length primers of cTP while lane 3 is 1 kb ladder.

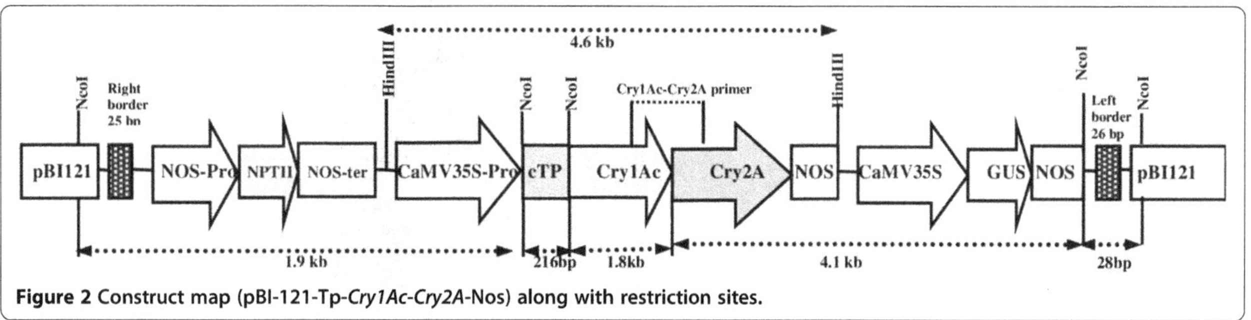

Figure 2 Construct map (pBI-121-Tp-*Cry1Ac-Cry2A*-Nos) along with restriction sites.

sections of infiltrated leaves after 72 hours confirmed the successful expression of the MUZ_01 construct vector in the plant expression system because *Cry1Ac*, *Cry2A*, and *GUS* gene expression were under the same promoter (Figures 5 and 6). A bluish green color was apparent in transgenic embryos but not in nontransgenic ones (Figure 6). Thirty plants that survived and passed screening were moved to selection free medium. In the end, seven plants survived soil acclimatization and were moved to the field. The overall transformation efficiency was calculated to be 0.7% (Table 1).

Molecular analysis of the putative transgenic plants

PCR analysis of putative transgenic plants

The amplification of 805 bp from *Cry1Ac-Cry2A* and 216 bp from *Tp2* showed successful integration of the target genes into the cotton plant genomes Figure 7. A *Tp2-Cry1Ac-Cry2A* plasmid was used as a positive control, while DNA extracted from untransformed plants was used as a negative control.

Qualitative analysis of the Bt protein in transgenic plants

The polymerase chain reaction (PCR) confirmed transgenic plants were subjected to qualitative analysis through a dipstick assay. The presence or absence of the Bt protein in transgenic plants was confirmed by the presence or absence of bands at the test position along with a control band (Figure 8).

Confirmation of MUZ_01 (Tp2- Cry1Ac-Cry2A) protein expression by ELISA

Cry1Ac and Cry2A proteins in transgenic plants were quantified by enzyme-linked immunosorbent assay (ELISA) with an Envirologix Kit (Cat # AP051, 500 Riverside Industrial Parkway Portland, Maine 04103–1486 USA). Positive and negative controls were added to the wells along with test samples. ELISA was performed according to the manufacturer's instructions and the endo-toxin (*Cry1Ac* and *Cry2A*) values were quantified as μg/g of fresh tissue [24] as shown in Figure 9. A maximum of 0.673 μg/g tissue of Cry1Ac and 0.568 μg/g tissue of Cry2A was

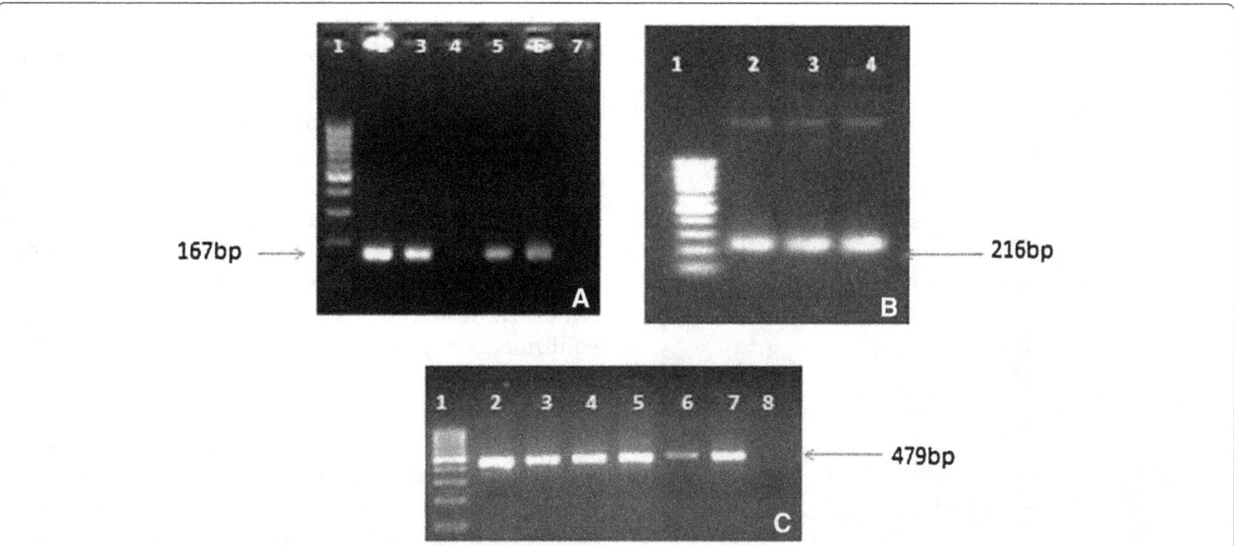

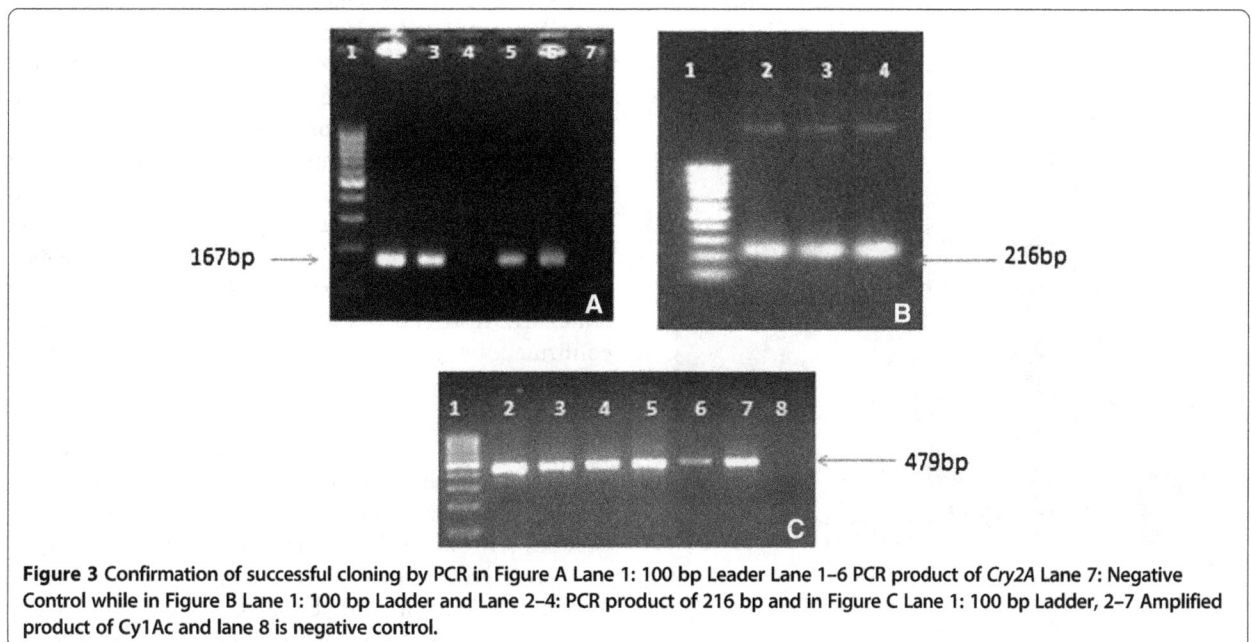

Figure 3 Confirmation of successful cloning by PCR in Figure A Lane 1: 100 bp Leader Lane 1–6 PCR product of *Cry2A* Lane 7: Negative Control while in Figure B Lane 1: 100 bp Ladder and Lane 2–4: PCR product of 216 bp and in Figure C Lane 1: 100 bp Ladder, 2–7 Amplified product of Cy1Ac and lane 8 is negative control.

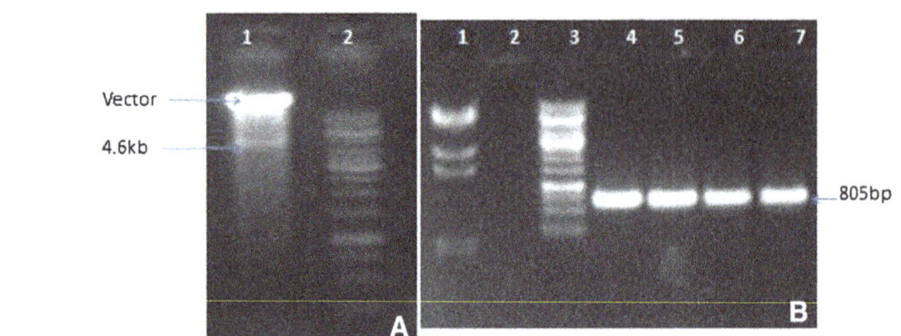

Figure 4 A Lane 1 shows the digested product of 4.6 kb and Lane 2 shows 1 kb ladder whereas in B Lane one is lamda hindi-III ladder lane 2 is negative control, lane 3 is 1 kb ladder and lane 4–6 are PCR product of 805 bp.

observed in transgenic plants, while no Bt protein expression was observed in the nontransgenic control plants.

Insect bioassay

Insect bioassays of transgenic and control plants were carried out in controlled conditions by simply using fresh leaves from transgenic cotton plants, along with nontransgenic control plants. We achieved 100% mortality in the target insect after 72 hours of feeding the 2nd instar insect larvae with transgenic plants, while 100% survival on nontransgenic leaves determined the efficacy of the *MUZ_01* gene construct against the target insect pests (Figure 10).

Phase contrast fluorescence microscopy

Fluorescein isothiocyanate (FITC) imaging of transgenic plants expressing *Cry1Ac-Cry2A* under cTP was taken at 488 nm and Chloroplast red auto-fluorescence at 580 nm excitation. Longitudinal leaf sections were labeled with primary antibody, anti-*Cry1Ac*/, anti-*Cry2A*, a secondary antibody, and FITC-conjugated IgG and observed under a phase contrast microscope (OLYMPUS DX61). 4',6-Diamidino-2-phenylindole (DAPI) was used to stain the nuclei. Cells stained with DAPI fluoresced blue, while those stained with FITC-conjugated IgG fluoresced green. The chloroplast itself gave off red auto-fluorescence, and the merged image of the transgenic leaves fluoresced yellow. In case of the control leaves, no yellow fluorescence was produced. These results indicate that Cry proteins were integrated into the chloroplast, i.e., transgenic plants under cTP, and in the case of the controls these proteins reside outside the chloroplast as indicated in Figures 11, 12, 13 and 14.

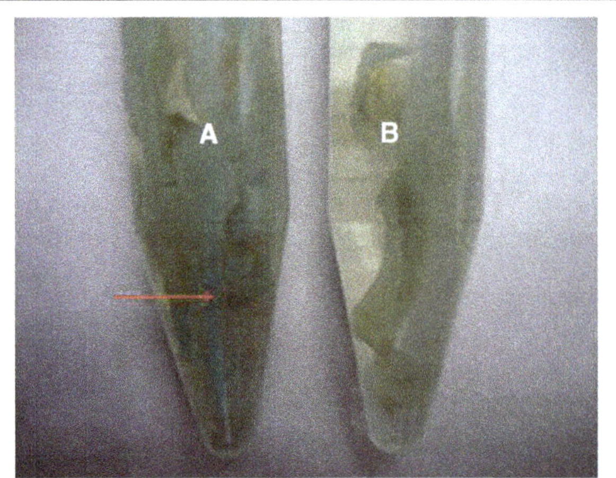

Figure 5 Gus Expression in experimental plants. **A**: transgenic plant leaves having blue-green color **B**: Non transgenic plant leaves as negative control with no color change.

Discussion

Chloroplast targeted expression of the Bt gene holds great potential for incorporating vital agronomic traits into plants. High Bt gene levels in chloroplasts permits plants to generate large quantities of crystal proteins. In the present study, two insecticidal genes, *Cry1Ac* and *Cry2A*, along with a chloroplast transit peptide were cloned in a PBI-121 vector and transformed into cotton variety MNH-786. *Cry1Ac* and *Cry2A* were selected because of their unique qualities, i.e., high expression levels and lack of competition for receptors among them.

The present study highlights the importance of cloning genes with transit peptides to demonstrate enhanced expression in cotton plants. *Cry1Ac* and *Cry2A* genes were cloned along with a chloroplast transient peptide in plant expression vector pBI-121 with the help of Hindi-III restriction sites. Successful cloning was confirmed by gene specific and orientation primers [25]. An agro-infiltration assay was used to check the efficacy of the cloned genes transient expression [11], resulting in a bluish-green color in the infiltrated region. Similar results were obtained by Ashraf, Bakhsh, and Pathi [26-28]. Transformation of the *Cry* genes with transit peptides not only makes it possible to localize transgene proteins in green parts of the plant, but it is also helpful in overcoming health and biosafety issues. Transgene protein expression was analyzed both qualitatively and

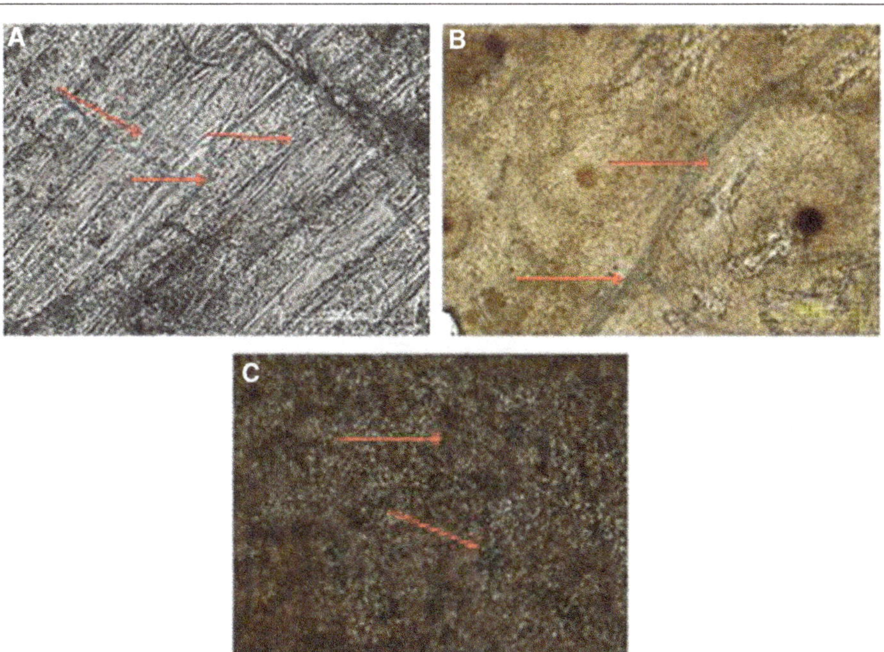

Figure 6 Transient Gus expression. A: In cotton stem section, **B**: Leaf midrib, **C**: In cotton leaf under florescent microscope. The bluish green color indicates the Gus expression.

quantitatively. Appearance of a band on the dipstick along with a control confirmed the presence of the target protein in the transgenic plant; similar results were reported by Dangat [29]. ELISA was performed to quantify the Cry proteins according to Li [30]. Maximum (0.673 ng, 0.454 ng) and minimum (0.306 ng, 0.568 ng) Cry1Ac and Cry2A proteins, respectively, were estimated in Muz-01 transgenic plants compared with control plants, which exhibited no Cry protein expression. Successful integration of Cry proteins into the chloroplast was confirmed by florescence microscopy [31] and FITC. Similar results for AtTrx-h3 expression in chloroplasts have been reported, while [32] used a similar technique for RB-60 protein expression in the cytoplasm and chloroplast. Insect bioassays revealed high mortality in the American boll worm. To check the efficacy of the transgene in the field an insect bioassay was performed. We recorded 100% mortality in the insects after feeding on the transgenic leaves, these results are comparable to those of Kiani [11].

From the above it is clear that the cloning of more than one gene, i.e., fusion genes, with transient peptides to localize the expression of these genes in the chloroplast not only increases the efficacy of the Bt proteins to kill the insects but it is also helpful in solving the biosafety concerns. On the basis of our molecular analysis we conclude that the transgenic plants with double Bt genes and a transit peptide for chloroplast expression was an excellent improvement in lepidopteron insect resistance. Our results suggest that transgenic cotton with transit peptide fusion protein genes is necessary to improve resistance against insects when compared with other genes without transit peptide fusion proteins.

Conclusions

This investigation suggests that insect resistance in cotton by modifying cotton plant genetics with gene transformation is possible. We found that protection against insects was improved by integrating some of the unique features of chloroplast transit peptides into the cotton crop. The present study was designed to produce modern transgenic cotton plants with no biosafety concerns as the transgene is only expressed in green tissues because the fusion protein gene only attaches to Bt in the C-terminal and cTP in the N-terminal. Thus, the new transgenic cotton variety exhibits greater insect resistance and enhanced Bt expression only in the green parts of plants, which will result in reduced biosafety concerns and increased cotton yield.

Table 1 Transformation efficiency of Muz-01 Construct in Cotton

Construct	Germinated embryos used	Survival (3 weeks)	Survival (8 weeks)	Shifted in soil	Transformation efficiency (TE)%
pBI-121	1000	129	36	7	0.7

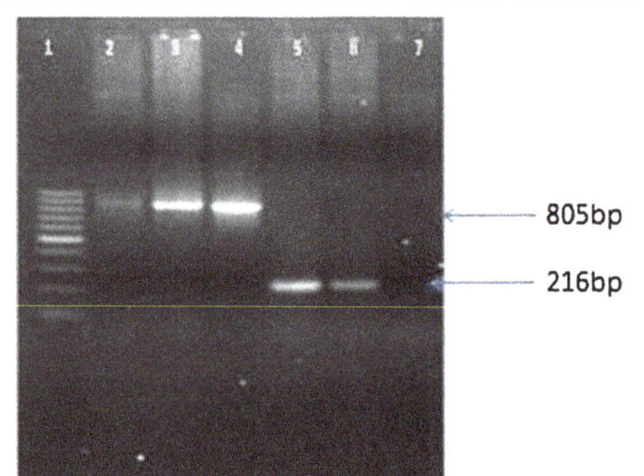

Figure 7 Confirmation of Transgenic plants by PCR with orientation (*Cry1Ac + Cry2A*) primers and Tp2 primers. Lane 1: 100 bp Ladder Lane, 2–3 PCR products with orientation primers Lane 4 positive control Lane 5–6 PCR product with Tp2 primers and Lane 7 Negative control.

Methods

Selection of plant materials

The transit peptide was first reported in petunia [Accession no. JF499829], which was locally available. For the isolation of cTP, petunia cultivar, *Grandiflora*, seeds were grown in the Center of Excellence in Molecular Biology CEMB green house at 25°C. Tobacco and cotton plants were also selected for transient expression and transformation, respectively.

Isolation of the chloroplast transit peptide (cTP)

Total RNA was extracted from petunia leaves. Oligo (dT) 18 primers and the MMLuV-RT enzyme were used for cDNA library synthesis. An NcoI restriction site was used with the forward (TTAGCCATGGATGGCACAAAT TAACAACATGG) and reverse primers (TAAGCCATGG CTGTGCTGTAGCCACTGATGC) to amplify a 216 bp fragment of the *TP* gene from the cDNA library. The amplified PCR product was cloned into a TA-vector PCR 2.1 (Invitrogen, Carlsbad, CA, United States of America). TP Sequencing was carried out with M13 primers on an ABI 310 Genetic Analyzer. Vector sequences were deleted in GeneDoc software.

Plant expression vector construction

The CaMV35S-*Cry1Ac*-NOS cassette (2476 bp) was excised from a pK2Ac vector. This excised cassette was then purified and cloned into a pTZ57 vector to overcome the NcoI constrain. TP was then digested with NcoI and ligated towards the *Cry1Ac* N-terminal. A 35S-TP-*Cry1Ac*-NOS cassette was then cloned into pBI121 using HindIII restriction sites. *Cry2A* was digested with XhoI and treated with S1 nuclease to remove single-stranded overhangs and was then ligated towards the *Cry1Ac* C-terminal, which generated TP-*Cry1Ac*-Cr2A (Muz-01 name given to this vector). The correct *Cry2A* orientation was confirmed through PCR with orientation primers. The *Cry1Ac-Cry2A* orientation primers were: forward primer (CAGCAGTGGAAATAA CATTCAGA) and reverse primer (AGCCTGTTGAG GAAGAGCTG), to give 805 bp amplification products.

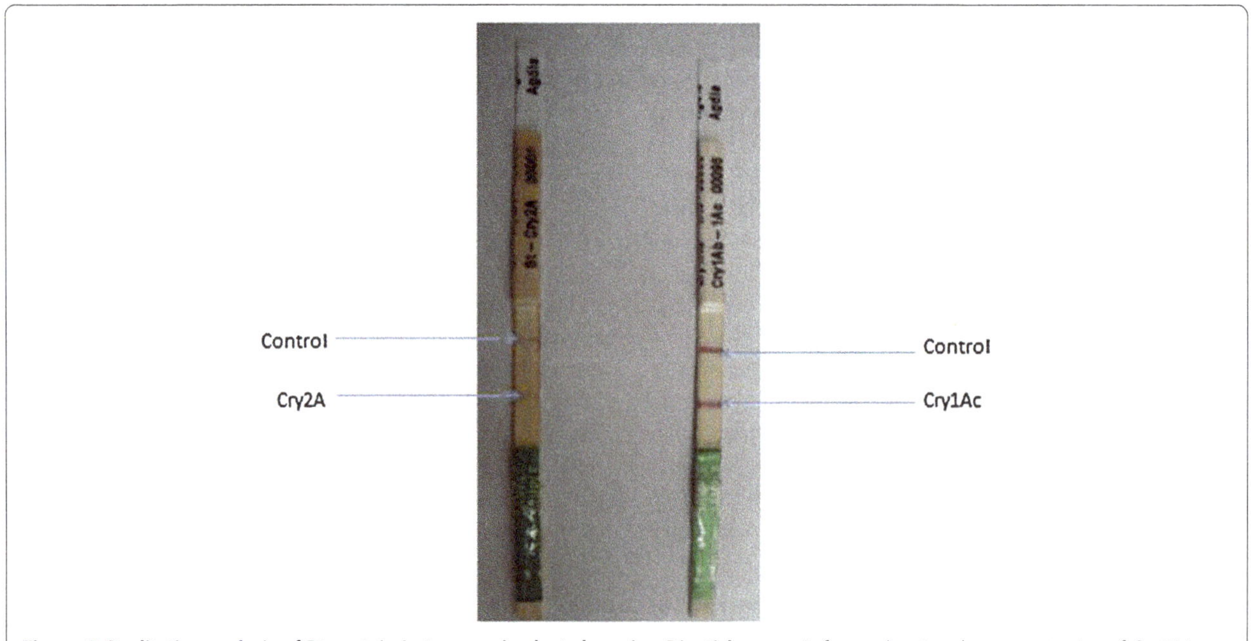

Figure 8 Qualitative analysis of Bt protein in transgenic plants by using Dipstick assay. Left row showing the concentration of Cry2A in µg/g while Cry1Ac in right row.

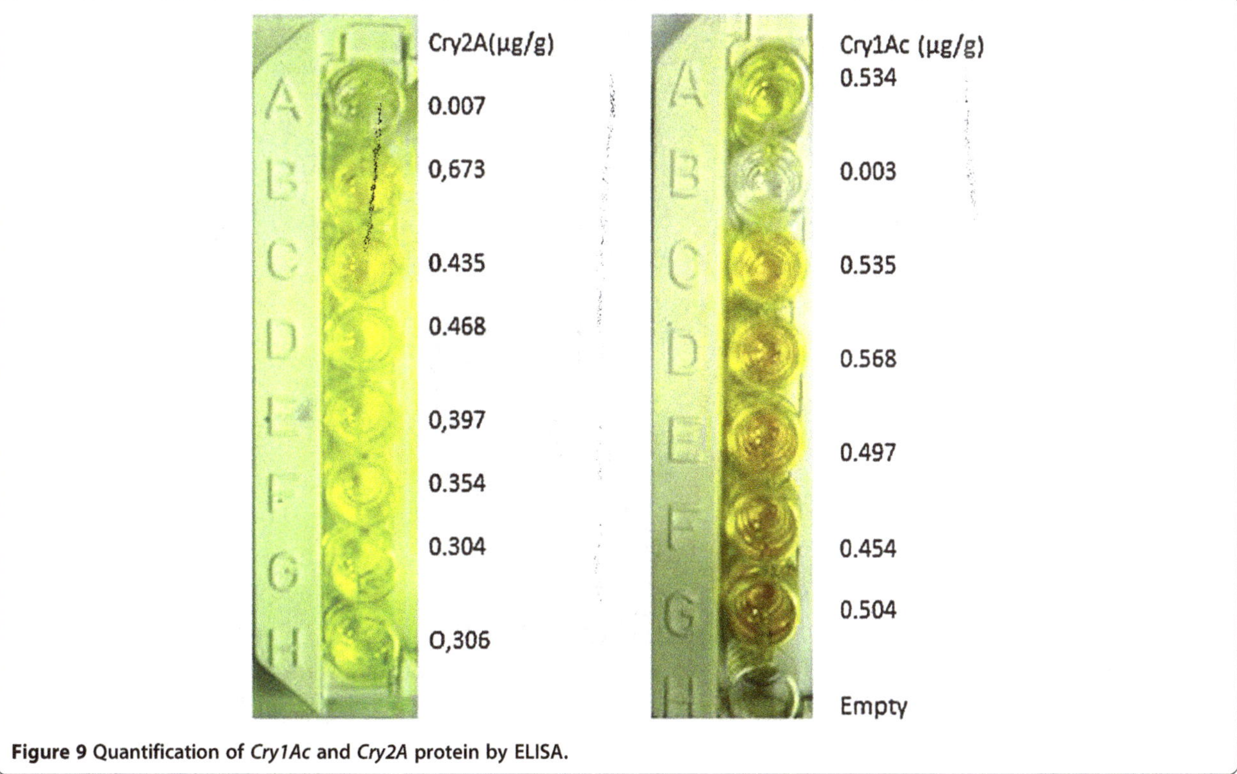

Figure 9 Quantification of *Cry1Ac* and *Cry2A* protein by ELISA.

Confirmation of successful cloning

For the confirmation of successful cloning, the construct was checked by orientation PCR and restriction-digestion. A pair of primers was designed for the orientation PCR. The forward primer (CAGCAGTGGAAATAACATT CAGA) was designed from *Cry1Ac*, while reverse primer (AGCCTGTTGAGGAAGAGCTG) was designed from *Cry2A*. Following PCR amplification with orientation, successful cloning and the correct orientation were further confirmed by digestion and ligation. The Hindi-III enzyme was used for digestion and ligation. The Hindi-III enzyme digested the complete *TP2-Cry1Ac-Cry2A*-NOS cassette, thus releasing a 4.6 kb fragment. Therefore, the restriction-digestion and orientation PCR confirmed successful cloning and that the genes were cloned in the correct orientation.

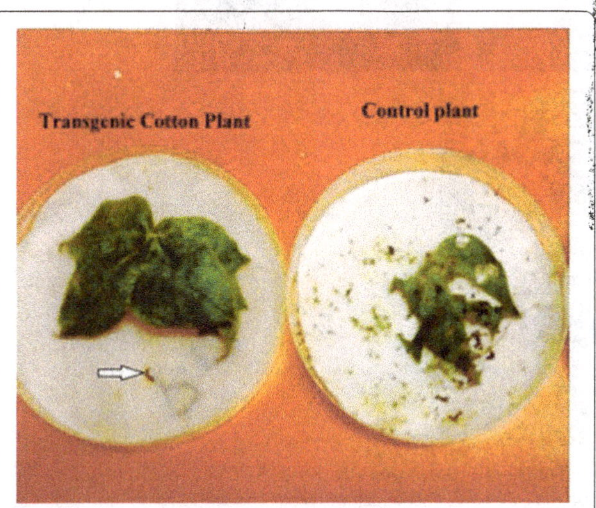

Figure 10 Bioassy with American bollworm.

GUS leaf infiltration assay

Muz-01 was transformed into *Agrobacterium tumefaciens* (LBA4404 strain) by electroporation. The efficacy of the construct vectors was confirmed by an Agrobacterium-mediated leaf infiltration *GUS* assay in both tobacco and cotton fresh leaves. The underside of the leaf was gently rubbed to remove the wax cuticle. *Agrobacterium* samples were taken with 5 mL syringes; the needle was removed, placed on the underside of the leaf, and pressed gently. Liquid diffused into mesophyllar air spaces. The infiltrated area was marked and tagged. These leaves were left for 72 hours under natural conditions and then subjected to a *GUS* assay.

Detection of *GUS* activity

GUS activity in the infiltrated leaves was detected histochemically. The infiltrated portion of the leaves was excised and incubated in *GUS* staining solution (0.08% w/v X-Gluc in 0.1 M sodium dihydrogen phosphate pH 7.0, 0.2 mM 10% Triton, and 20% methanol) at 37°C. After staining with *GUS* solution these plant tissues were immersed in fixative

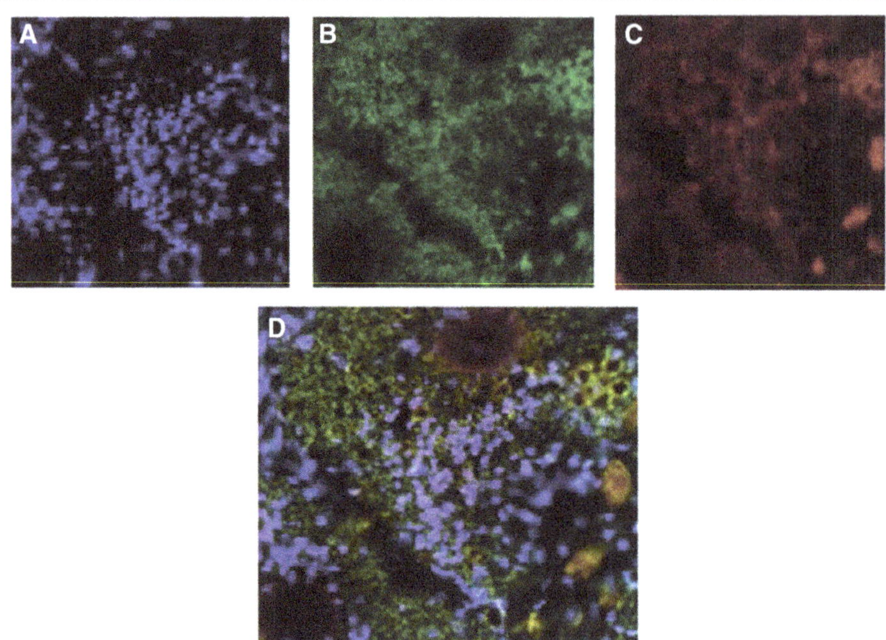

Figure 11 Phase contrast fluorescence microscopy of *Cry1Ac* transgenic plants without transit peptide. **A** = DAPI blue fluorescence. **B** = FITC green fluorescence. **C** = Chloroplast auto-fluorescence red. **D** = Merged image of I, II & III. Green red and blue colors do not merge i.e. *Cry1Ac* is outside the chloroplast.

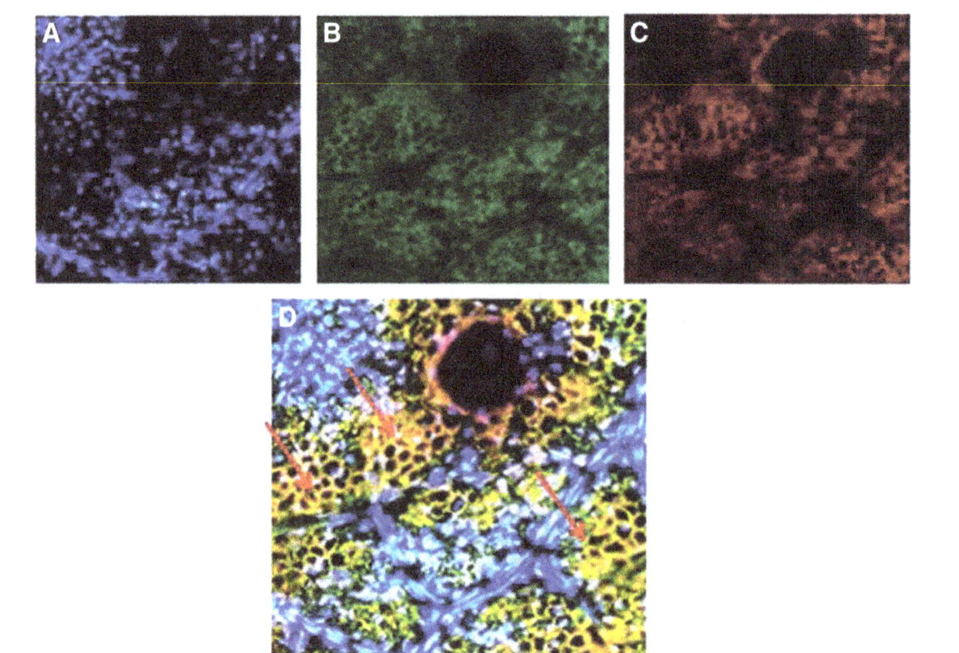

Figure 12 Phase contrast fluorescence microscopy of *Cry1Ac* transgenic plants with transit peptide. **A** = DAPI blue fluorescence. **B** = FITC green fluorescence. **C** = Chloroplast auto-fluorescence red. **D** = Merged image of A, B and C. Yellow color is produced where green and red fluorescence occurred at the same place i.e. *Cry1Ac* inside chloroplasts.

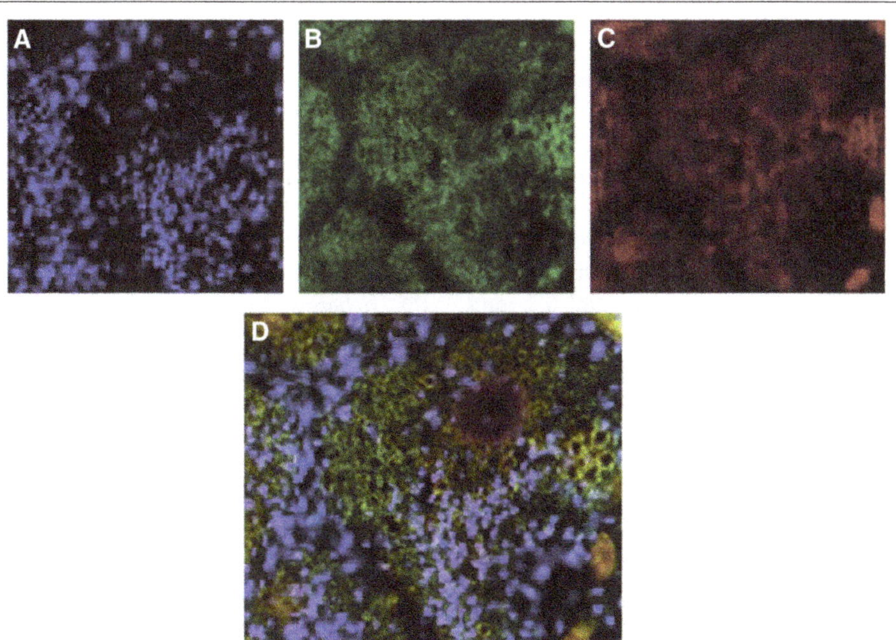

Figure 13 Phase contrast fluorescence microscopy of *Cry2A* transgenic plants without transit peptide. A = DAPI blue fluorescence. **B** = FITC green fluorescence. **C** = Chloroplast auto-fluorescence red. **D** = Merged image of I, II & III. Green red and blue colors do not merge i.e. *Cry2A* is outside the chloroplast.

solution, which consisted of formaldehyde (5%), ethanol (20%), and acetic acid (5%) for 10 minutes. To remove chlorophyll, the leaves were submerged in 70% ethanol for 48 hours. *GUS* activity was then observed by sight as well as under a florescent microscope (OLYMPUS SZX7).

Cotton construct transformation

Cotton (*G. hirsutum*) cv. MNH 786 was selected for transformation because of its high yielding potential and susceptibility to lepidopteran insects. Delinted seeds were sterilized with Tween 20 for 4 minutes and then

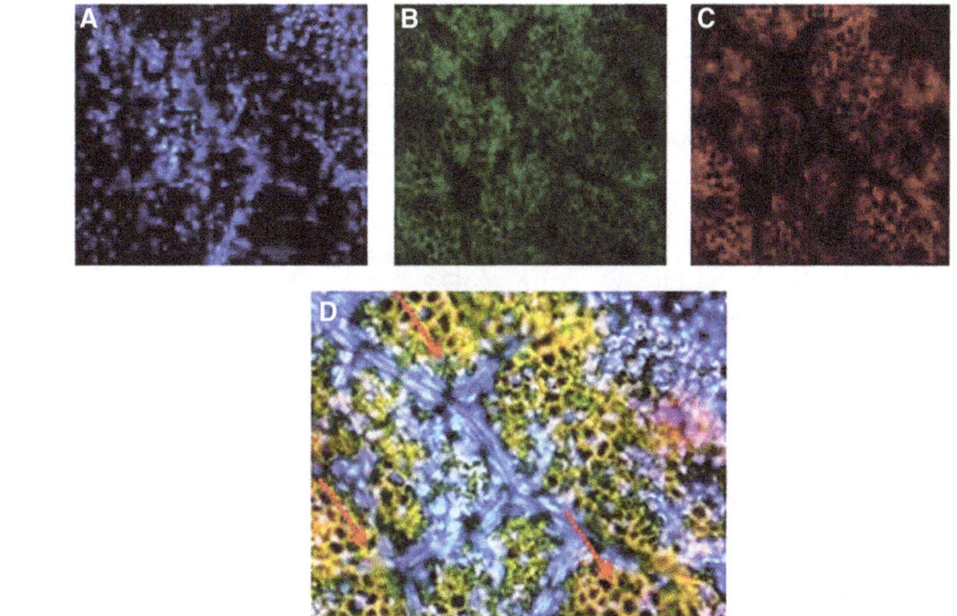

Figure 14 Phase contrast fluorescence microscopy of *Cry2A* transgenic plants with transit peptide. A = DAPI blue fluorescence. **B** = FITC green fluorescence. **C** = Chloroplast auto-fluorescence red. **D** = Merged image of A, B and C. Yellow color is produced where green and red fluorescence occurred at the same place i.e. *Cry2A* inside chloroplasts.

subjected to 0.1% HgCl2 and 0.1% sodium dodecyl sulfate. Sterilized seeds were placed in a seed germinator at 30°C overnight in the dark. Germinated seedlings were used for *Agrobacterium*-mediated transformation as used by [33] and modified by [24,34-36] at CEMB. MS medium [37] was used to culture the inoculated plants. Furthermore, 1 mg/l kinetin and 250 mg/l cefotaxime were used in the MS plates for the first 3 days, and after that the plantlets were subcultured in MS tubes containing 250 mg/l kanamycin, 0.5 mg/l benzylaminopurine, and 1 mg/l a-naphthaleneacetic acid. The putative transgenic plants were then moved to pots containing soil of equal proportions of clay, sand, and peat moss (1:1:1). Finally, the plants were moved to a greenhouse and subjected to various molecular analyses.

Genomic DNA isolation and polymerase chain reaction

Genomic DNA was isolated from the putative transgenic cotton leaves according to the method of Zhang [38]. Successful integration of the genes into the cotton genome was confirmed by PCR amplification with internal *Cry1Ac* and *Cry2A* primers.

Whole-leaf protein extraction

Whole-leaf protein was extracted from fresh transgenic cotton leaves. The leaves were crushed in liquid nitrogen. Ground leaves were put in 1.5 mL micro tubes with 400 μL of protein extraction buffer (1X). The samples were vortexed to homogenize and incubated at 4°C for 2 hours. The samples were then centrifuged for 10 min at 13,000 × *g*. The supernatant was eluted and stored in new 1.5 mL tubes and Bradford reagent was used to quantify proteins [39].

ELISA

Leaf samples were ground in liquid nitrogen with a pestle and mortar. After grinding, 300 μl of protein extraction buffer was added and it was incubated at 4°C for overnight. The next day the mixture was centrifuged at 13,000 × *g* for 10 minutes. The supernatant was eluted. Fifty microliters of *Cry1Ac* and *Cry2A* enzyme conjugate was then added to each well immediately followed by 50 μl extraction buffer, 50 μl *Cry1Ac* and *Cry2A* positive control, and 50 μl of the sample extract to the their respective wells. The contents were thoroughly mixed by moving in a rapid circular motion for 20–30 seconds. The wells were then covered with parafilm and incubated at ambient temperature for 2 hours. After incubation the cover was removed and the wells were washed three times with washing buffer. Water was removed and 100 μl of substrate was added to each well. The wells were covered with parafilm and the plate was incubated at ambient temperature for 30 minutes. After incubation, 100 μl of the stop solution was added, turning the well contents yellow. Then wavelength of the spectrophotometer was adjusted to 450 nm and the readings were recorded.

Insect bioassay

Heliothis larvae were employed for the insect bioassay [35,40] to examine whether or not the chloroplast targeted expression of the *Cry1Ac* and *Cry2A* fusion gene increased the *Heliothis* mortality. Larvae were collected from a CEMB field and used under laboratory conditions in a feeding bioassay. Leaves of both the control and transgenic cotton were placed in a petri dish, and larvae were allowed to feed on them. The leaves were examined after 48 hours.

FITCH

For immunohistochemistry, the leaves were washed with 1X PBS twice and fixed with 4% paraformaldehyde solution. The samples were then incubated with primary antibodies, Anti-*Cry1Ac* and anti-*Cry2A* in a dilution of 1:100 for 1 hour at 37°C in a humidified chamber. Incubation with primary antibodies was followed by three washes with 1X PBS. The samples were then incubated with secondary antibodies specific to each of the respective primary antibodies and stained with DAPI (Invitrogen™, CA, USA) for 1 hour at 37°C in a humidified chamber. Images were taken for each group from three separate experiments using a phase contrast microscope (OLYMPUS DX61).

Competing interests

The authors declare that they have no competing interests.

Authors' contributions

AM have made substantial contributions to conception and design, or acquisition of data, or analysis and interpretation of data; SK Helped in isolation and cloning; AQ Supervised the research; MA have been involved in drafting the manuscript; AA Helped in transformation in cotton cultivar; MF helped in Gus and FITC; AI helped in molecular analysis of the putative transgenic plants; IA Revised the manuscript; AAS have given final approval of the version to be published; and TH agree to be accountable for all aspects of the work in ensuring that questions related to the accuracy or integrity of any part of the work are appropriately investigated and resolved. All authors read and approved the final manuscript.

Acknowledgements

We are really thankful Higher Education Commission and Punjab Agriculture Research bored for supporting the research project.

Author details

[1]National Center of Excellence in Molecular Biology, University of the Punjab, Lahore 53700, Pakistan. [2]Institute of Molecular Biology, Academia Sinica, Taipei 115, Taiwan.

References

1. Kathage J, Qaim M. Economic impacts and impact dynamics of Bt (Bacillus thuringiensis) cotton in India. Proc Natl Acad Sci U S A. 2012;109:11652–6.
2. Bravo A, Gill SS, Soberon M. Mode of action of Bacillus thuringiensis Cry and Cyt toxins and their potential for insect control. Toxicon. 2007;49:423–35.

3. Jozani GRS, Komakhin RA, Piruzian ES. Comparative study of the expression of the native, modified, and hybrid cry3a genes of Bacillus thuringiensis in prokaryotic and eukaryotic cells. Russ J Genet. 2005;41:116–21.
4. Sanahuja G, Banakar R, Twyman RM, Capell T, Christou P. Bacillus thuringiensis: a century of research, development and commercial applications. Plant Biotechnol J. 2011;9:283–300.
5. Schnepf HE, Whiteley HR. Cloning and expression of the Bacillus thuringiensis crystal protein gene in Escherichia coli. Proc Natl Acad Sci U S A. 1981;78:2893–7.
6. Qaim M, Zilberman D. Yield effects of genetically modified crops in developing countries. Science. 2003;299:900–2.
7. Zhang H, Yin W, Zhao J, Jin L, Yang Y, Wu S, et al. Early Warning of Cotton Bollworm Resistance Associated with Intensive Planting of Bt Cotton in China. PLoS ONE. 2011;6:e2274-81.
8. Kota M, Daniell H, Varma S, Garczynski SF, Gould F, Moar WJ. Overexpression of the Bacillus thuringiensis (Bt) Cry2Aa2 protein in chloroplasts confers resistance to plants against susceptible and Bt-resistant insects. Proc Natl Acad Sci U S A. 1999;96:1840–5.
9. Liu CW, Lin CC, Yiu JC, Chen JJ, Tseng MJ. Expression of a Bacillus thuringiensis toxin (cry1Ab) gene in cabbage (Brassica oleracea L. var. capitata L.) chloroplasts confers high insecticidal efficacy against Plutella xylostella. Theor Appl Genet. 2008;117:75–88.
10. Sharma P, Nain V, Lakhanpaul S, Kumar PA. Synergistic activity between Bacillus thuringiensis Cry1Ab and Cry1Ac toxins against maize stem borer (Chilo partellus Swinhoe). Lett Appl Microbiol. 2010;51:42–7.
11. Kiani S, Ali A, Bajwa KS, Muzaffar A, Ashraf MA, Samiullah TR, et al. Cloning and chloroplast-targeted expression studies of insect-resistant gene with ricin fusion-gene under chloroplast transit peptide in cotton. Electron J Biotechnol. 2013;16:13–3.
12. Lössl AG, Waheed MT. Chloroplast-derived vaccines against human diseases: achievements, challenges and scopes. Plant Biotechnol J. 2011;9:527–39.
13. Verma D, Daniell H. Chloroplast Vector Systems for Biotechnology Applications. Plant Physiol. 2007;145:1129–43.
14. Liu JF, Zhao CY, Ma J, Zhang GY, Li MG, Yan GJ, et al. Agrobacterium-mediated transformation of cotton (Gossypium hirsutum L.) with a fungal phytase gene improves phosphorus acquisition. Euphytica. 2011;181:31–40.
15. Gatehouse JA. Biotechnological prospects for engineering insect-resistant plants. Plant Physiol. 2008;146:881–7.
16. Tang L, Kwon S-Y, Kim S-H, Kim J-S, Choi JS, Cho KY, et al. Enhanced tolerance of transgenic potato plants expressing both superoxide dismutase and ascorbate peroxidase in chloroplasts against oxidative stress and high temperature. Plant Cell Rep. 2006;25:1380–6.
17. Kim EH, Suh SC, Park BS, Shin KS, Kweon SJ, Han EJ, et al. Chloroplast-targeted expression of synthetic cry1Ac in transgenic rice as an alternative strategy for increased pest protection. Planta. 2009;230:397–405.
18. Rawat P, Singh AK, Ray K, Chaudhary B, Kumar S, Gautam T, et al. Detrimental effect of expression of Bt endotoxin Cry1Ac on in vitro regeneration, in vivo growth and development of tobacco and cotton transgenics. J Biosci. 2011;36:363–76.
19. Soria-Guerra RE, Alpuche-Solis AG, Rosales-Mendoza S, Moreno-Fierros L, Bendik EM, Martinez-Gonzalez L, et al. Expression of a multi-epitope DPT fusion protein in transplastomic tobacco plants retains both antigenicity and immunogenicity of all three components of the functional oligomer. Planta. 2009;229:1293–302.
20. Ruiz ON, Alvarez D, Torres C, Roman L, Daniell H. Metallothionein expression in chloroplasts enhances mercury accumulation and phytoremediation capability. Plant Biotechnol J. 2011;9:609–17.
21. De Cosa B, Moar W, Lee SB, Miller M, Daniell H. Overexpression of the Bt cry2Aa2 operon in chloroplasts leads to formation of insecticidal crystals. Nat Biotechnol. 2001;19:71–4.
22. Verma D, Daniell H. Chloroplast vector systems for biotechnology applications. Plant Physiol. 2007;145:1129–43.
23. Lee SW, Hahn TR. Light-regulated differential expression of pea chloroplast and cytosolic fructose-1,6-bisphosphatases. Plant Cell Rep. 2003;21:611–8.
24. Bakhsh A, Rao AQ, Shahid AA, Husnain T, Riazuddin S. Camv 35S is a Developmental Promoter Being Temporal and Spatial in Expression Pattern of Insecticidal Genes (Cry1ac & Cry2a) in Cotton. Aust J Basic Appl Sci. 2010;4:37–44.
25. Akhtar S, Shahid AA, Rao AQ, Bajwa KS, Muzaffar A, Latif A, et al. Genetic effects of Calotropis procera CpTIP1 gene on fiber quality in cotton (Gossypium hirsutum). Adv Life Sci. 2014;1:223–30.
26. Ashraf MA, Shahid AA, Rao AQ, Bajwa KS, Husnain T. Functional Characterization of a Bidirectional Plant Promoter from Cotton Leaf Curl Burewala Virus Using an Agrobacterium-Mediated Transient Assay. Viruses. 2014;6:223–42.
27. Bakhsh A. Expression of two insecticidal genes in Cotton. In: PhD Thesis. Lahore, Pakistan: University of the Punjab; 2010. p. 112–3.
28. Pathi KM, Tula S, Tuteja N. High frequency regeneration via direct somatic embryogenesis and efficient Agrobacterium-mediated genetic transformation of tobacco. Plant Signal Behav. 2013;8:e24354.
29. Dangat S, Rajput S, Wable K, Jaybhaye A, Patil V. A biolistic approach for transformation and expression of cry 1Ac gene in shoot tips of cotton (Gossypium hirsutum). Res J Biotechnol. 2007;2:1.
30. Li Y, Romeis J, Wang P, Peng Y, Shelton AM. A comprehensive assessment of the effects of Bt cotton on Coleomegilla maculata demonstrates no detrimental effects by Cry1Ac and Cry2Ab. PLoS One. 2011;6:e22185.
31. Park SK, Jung YJ, Lee JR, Lee YM, Jang HH, Lee SS, et al. Heat-shock and redox-dependent functional switching of an h-type Arabidopsis thioredoxin from a disulfide reductase to a molecular chaperone. Plant Physiol. 2009;150:552–61.
32. Levitan A, Trebitsh T, Kiss V, Pereg Y, Dangoor I, Danon A. Dual targeting of the protein disulfide isomerase RB60 to the chloroplast and the endoplasmic reticulum. Proc Natl Acad Sci. 2005;102:6225–30.
33. Gould JH, Magallanes-Cedeno M. Adaptation of Cotton Shoot Apex Culture to Agrobacterium-Mediated Transformation. Plant Mol Biol Report. 1998;16:283–3.
34. Bakhsh A, Siddique S, Husnain T. A molecular approach to combat spatio-temporal variation in insecticidal gene (Cry1Ac) expression in cotton. Euphytica. 2012;183:65–74.
35. Khan GA, Bakhsh A, Riazuddin S, Husnain T. Introduction of cry1Ab gene into cotton (Gossypium hirsutum) enhances resistance against Lepidopteran pest (Helicoverpa armigera). Span J Agric Res. 2011;9:296–302.
36. Rao AQ, Husnain T, Shahid AA. Impact of PHY B Gene Transformation in Physiology and Yield of Cotton. Germany: Lambert Academic Publishing 2011. [https://www.morebooks.de/store/gb/book/impact-of-phy-b-gene-transformation-in-physiology-and-yield-of-cotton/isbn/978-3-8465-0191-7].
37. Murashige T, Skoog F. A Revised Medium for Rapid Growth and Bio Assays with Tobacco Tissue Cultures. Physiol Plant. 1962;15:473–97.
38. Zhang J, Stewart J, Mac D. Economical and rapid method for extracting cotton genomic DNA. J Cotton Sci. 2000;4:193–201.
39. Bradford MM. A rapid and sensitive method for the quantitation of microgram quantities of protein utilizing the principle of protein-dye binding. Anal Biochem. 1976;72:248–54.
40. Jin S, Zhang X, Daniell H. Pinellia ternata agglutinin expression in chloroplasts confers broad spectrum resistance against aphid, whitefly, lepidopteran insects, bacterial and viral pathogens. Plant Biotechnol J. 2012;10:313-27.

12

Morphometric and phytochemical characterization of chaura fruits (*Gaultheria pumila*): a native Chilean berry with commercial potential

Evelyn Villagra[1], Carola Campos-Hernandez[1], Pablo Cáceres[1], Gustavo Cabrera[1], Yamilé Bernardo[1], Ariel Arencibia[1], Basilio Carrasco[2], Peter DS Caligari[3], José Pico[3] and Rolando García-Gonzales[1*]

Abstract

Background: For the first time, a morphometric characterization of chaura (*Gaultheria pumila*) fruits has been conducted between natural populations growing in the Villarrica National Park, Araucania Region, Chile. Chaura is a native Ericaceae from Chile that produces aromatic and tasty fruits which could be of agricultural interest.

Results: To influence the decision for a further domestication of *G. pumila*, both the fruit sizes (indicator of productivity) and the nutritional properties of the fruits have been determined from different subpopulations. Samples were a total of 74 plants and 15 fruits per plant which were randomly harvested following its natural distribution around the Villarrica volcano. Altogether, fresh weight, shape, color, diameter in the pole and the equatorial dimensions were determined as phenotypic traits of the *G. pumila* fruits. Meanwhile the total soluble solids, anthocyanin and pectin contents were calculated as nutritional traits of the Chaura fruits. Results showed a high phenotypic diversity between the sampled population with three main fruit shapes and three predominant colors. The round shapes were the most abundant, whereas a significant correlation was found among fruit size with weight and color. The highest fresh weight (597.3 mg), pole diameter (7.1 mm) and equatorial diameter (6.5 mm) were estimated in the pink color fruits.

Conclusions: The total amount of anthocyanin was higher in red fruits, while the maximum pectin content was obtained in the round white fruits. Overall results must pave the way for a further domestication and introduction of the Chaura species in the agro-productive system in Chile.

Keywords: *Gaultheria pumila*, Fruit morphometry, Fruit diversity, Pectin, Anthocyanins

Background

Gaultheria pumila is a native Chilean species from the Ericaceae family, commonly known as Chaura or Mutilla. *G. pumila* is a low bush that can reach up to 80 cm in height, depending on the environmental conditions. The species produces fleshy, flavored and aromatic fruits and inhabits the Andes mountains; its distribution ranges from the Region Metropolitana (33°26′ 16″ S; 70°39′ 01″ W) in the North to Region de Magallanes (53°9′ 45″ S; 70°55′ 21″ W) in the South [1]. Chaura berries in their natural habitat have been found for the wide variety of shapes and colors. Its natural habitat demonstrates extreme temperature variation, since in winter bushes are covered by snow for several weeks, and in summer the plants are exposed to high temperatures and high UV radiation. As a result of its adaptation, this species could be considered as an extremophile plant species because it is highly tolerant of harsh soil and environmental conditions [2]. In addition, Chaura leaves and fruits produce an intense aroma and flavor, which make it appealing for developing new food applications.

Fruit consumption is the main source of phenolics in the daily human diet, helping them to improve a large number of biological functions [3,4]. The supply of natural phenolics compounds through the human diet has been associated to a reduction of heart and brain diseases as

* Correspondence: rgarciag@ucm.cl
[1]Departamento de Ciencias Forestales, Centro de Biotecnología de los Recursos Naturales, Universidad Católica del Maule, Campus San Miguel. Av. San Miguel 3605, casilla 617, Talca, Maule Region, Chile
Full list of author information is available at the end of the article

well as cancer [5,6]. Polyphenols are a family of natural bioactive compounds with high antioxidant capacity that are capable of neutralizing and eliminating free radical species [7,8]. It has been demonstrated that polyphenolic extracts from different sources are involved in the inhibition of low density lipoproteins (LDL), confirming their potential use as nutraceuticals with positive effects on human health [5,9]. Ascorbic acid, tocopherols, tricotrienols, carotenoids, phenolic compounds and tannins can be considered among the most important plant derived antioxidants [10-12]. These compounds have been associated with anticancer activity, since they can assist complementary mechanisms of defense such as the induction of metabolizing enzymes and modulation of gene expression, cell proliferation and apoptosis [13].

Plant metabolites play a central role for protecting plants against herbivores and diseases; also they can attract pollinators and seed dispersers. Furthermore, secondary metabolites can also protect against abiotic stresses such as drought, temperature and radiation. Berry fruits, like many other species, are rich in flavonoids, phenolic acids, anthocyanin, vitamins, minerals and fiber [14,15]. Polyphenol compounds produced in berries vary their concentration across different fruit tissues, and their total content could be influenced by several factors such as plant genotype, environmental and soil conditions in orchards, fruit maturity degree, harvest practices, and fruit storage conditions after harvest among others [16].

Chaura fruits show a resistant exocarp structure with a fleshy sweet mesocarp and many seeds. The appearance of the mesocarp suggests a high content of pectin in its structure. Pectins are complex heteropolysaccharides presented in the cell wall of land plants and fruits, providing consistency and mechanical resistance to vegetable tissues. The overall structure of these macromolecules encompasses homogalacturonan blocks (1,4-linked α-D-GalA units, which can be partially methylesterified), covalently linked to type I rhamnogalacturonan blocks (repeating disaccharide [→4)-α-D-GalA-(1 → 2)-α-L-Rha-(1→] units) bearing neutral sugar side-chains [17,18]. Pectin is a worldwide food ingredient widely employed as gelling, emulsifying and stabilizing agents [19,20]. It is a soluble fiber used in jams and jellies, fruit juices, fruit drink concentrates, desserts, baking fruit preparations, dairy and delicatessen products [21].

Also, pectin is important to human health, because it has many benefits, is a compound that appears to be able to inhibit cancer metastasis and primary tumor growth in multiple types of cancer in animals [22], and it can lower cholesterol in humans [23]. Pectins are industrially produced by diluted acid hot extraction from citrus and apples peels, respectively. Citrus and apple pectin content ranges from 15 to 30% (w/w) on dry weight base. Moreover, pectin content from other species has been studied, and high variability was found among them, for example, murta pectin content (30% w/w) [24], chickpeas (67% w/w) [25], Lobeira (33.6% w/w) [26], and sweet potatoes (10.24%) [27].

In Chaura, to the best of our knowledge there is no systematic information about morphometric characteristics of chaura fruits and no studies of chemical composition of the fruit has been done before. In addition, agricultural data about crop yield, growing habits, water and fertilizer needs, as well as fruit uses is almost nonexistent. Thus, Chaura remains as an unused but seemingly quite valuable Chilean genetic resource, and more studies exploring its potential as commercial fruit should be done.

In particular, this research was aimed to describe the fruit variation of natural populations of *G. pumila* (chaura) inhabiting the Villarrica National Park in Andeans mountains of South Chile. Morphometric characterization was performed by measuring color and fruit sizes, while phytochemical characterization is done by chemical analyzing peptin content, total anthocyanins and antioxidant activity of the fruits.

Results

Morphometric characterization and fruit size variation

In order to know the main characteristics of Chaura fruits, a morphometric characterization was carried out; results are presented in Table 1. Results indicate that Chaura fruits show great phenotypic diversity and four different morphotypes can be clearly identified: pepper white, round white, red and pink. The red fruit showed the highest fresh weight value of 597.3 ± 274.4 mg, and it also had the highest value for the polar diameter (7.2 ± 2.2 mm) as well as the equatorial diameter (6.6 ± 1.7 mm). The white fruits had lower values for these three variables, except with polar diameter in pepper white. Thus, a comparative analysis of both weight and fruit size shows that there are significant differences between different morphotypes based on the two parameters analyzed, Table 1. Figure 1 shows the graphical analysis of mean from ratio polar diameter/equatorial diameter of Chaura morphotypes ($p < 0.05$).

The Principal Component Analysis significantly demonstrated the relation among fruit weight and fruit color. As shown in Figure 1B, fruits with the highest equatorial size and weight were grouped in a single group of red and pink skin fruits. The smaller fruits with the lowest weight with white skin were grouped separately. Furthermore, eight plants did not belonged to any of the established groups. The effectiveness of the PCA analysis was probed with a discriminant analysis (Wilk's lambda = 0.56; $P < 0{,}00001$), showing a 77.8% of consistency for classification based on fruit size and color.

Table 1 Morphometry studies of *G. pumila* fruits collected in South Chile (La Araucania Region) near Villarica volcano ($n = 156$)

Fruit	Fresh weight (mg)	Polar diameter (mm)	Equatorial diameter (mm)
Pepper white	223.0 ± 70.2 a	6.8 ± 2.1 a	2.5 ± 1.0 a
Round white	240.8 ± 103.9 a	4.2 ± 1.3 b	3.7 ± 1.4 b
Red	597.3 ± 274.4 b	7.2 ± 2.2 a	6.6 ± 1.7 c
Pink	456.7 ± 210.3 c	5.6 ± 2.1 c	5.5 ± 1.4 d

The evaluation of statistical significance was determined by ANOVA, followed by Tukey HSD test. Different letters represent statistical differences for each evaluated parameter ($p < 0.05$).

Total solid content (°Brix)

The values obtained by the refractometer corresponded to 8.0 ± 1.4, 8.5 ± 0.7, 9.0 ± 1.2 and 10.0 ± 1.0° Brix for fruit round white, pink, pepper white and red, respectively. These values indicate the amount of soluble sugars present in fruits, so one degree °Brix is defined by one gram of sucrose in 100 g of solution. This represents the strength of solution as percentage by weight (% w/w) and reflects the state of maturity of the fruit at harvest [28].

Anthocyanins content

Figure 2 shows a significant difference in the content of anthocyanins by morphotype. The red fruit have higher values of anthocyanins with an average of 5,942 ± 422 mg monomeric anthocyanins per 100 g of sample. The pink fruits have an average of 3,854 ± 192 mg monomeric anthocyanins per 100 g of sample, while white fruits have the lowest anthocyanin content with an average of 626.2 ± 41 mg monomeric anthocyanins per 100 g of sample. Fruits shaped like a small white pepper are those with the lowest anthocyanin content.

Pectin contents

Figure 3 shows the results of the pectin content in different Chaura fruits. For positive control, the pectin content in of orange peel (O), was used because pectin is extracted industrially from these peels. The results shown in Figure 3 are expressed in percentage of the fresh weight of fruit.

Discussion

G. pumila is an example of underutilized genetic resource where phytochemical and morphometric features including fructification and organoleptic qualities have not been determined following scientific approaches. In this way, it was determined that the Chaura wild species could produce up to 0.5 kg of fresh fruits per plant, depending on the genotype and the environmental conditions. Regarding with the fruit weight, it can be seen that together polar and equatorial diameters contribute significant for the morphotypes pepper white and round white. Pepper white fruits have a long shape characteristic, while both red and pink morphotypes have a major size. Furthermore, both red and pink morphotypes have a higher weight in comparison with white fruits (Table 1). The Principal Component Analysis has shown a consistent grouping of the *G. pumilia* variability screened in this study.

Chaura fruit collected for this study showed lower Brix values than other fruits like cranberry, which at the time of harvest has values between 12 and 14° Brix [29]. The

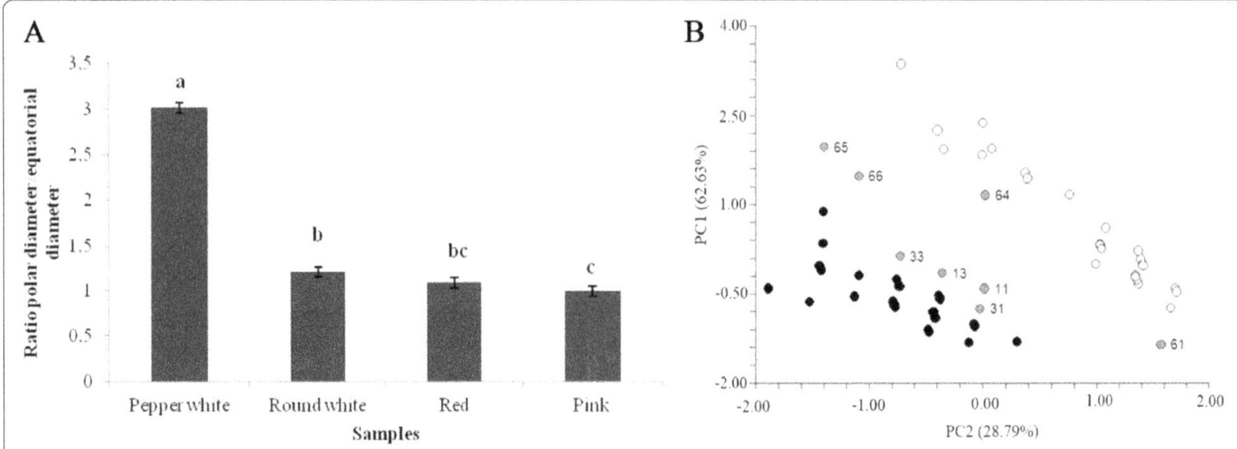

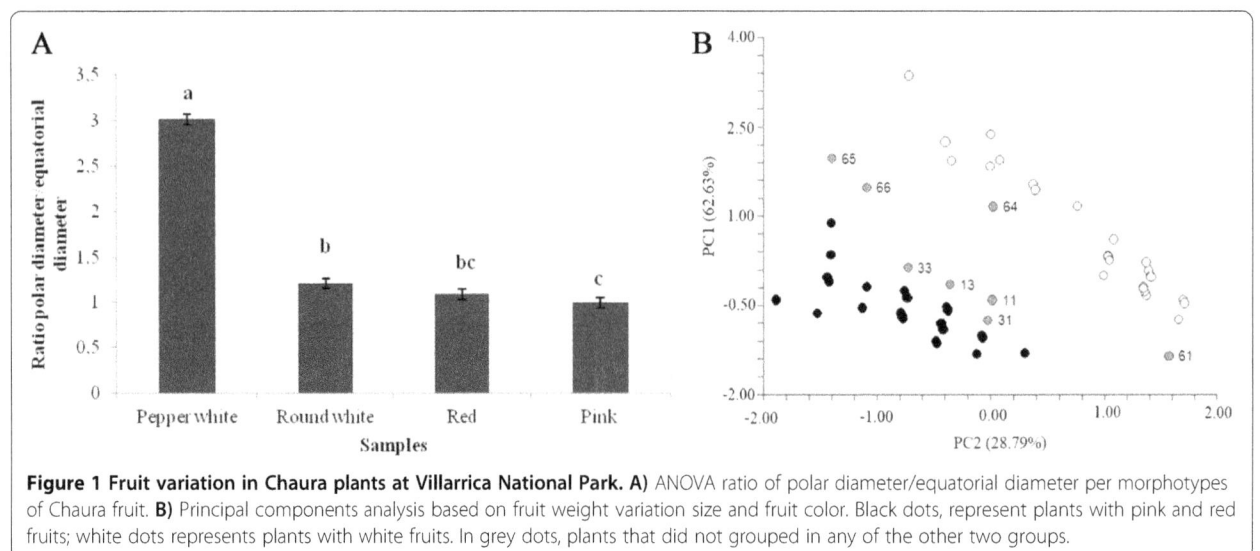

Figure 1 Fruit variation in Chaura plants at Villarrica National Park. A) ANOVA ratio of polar diameter/equatorial diameter per morphotypes of Chaura fruit. **B)** Principal components analysis based on fruit weight variation size and fruit color. Black dots, represent plants with pink and red fruits; white dots represents plants with white fruits. In grey dots, plants that did not grouped in any of the other two groups.

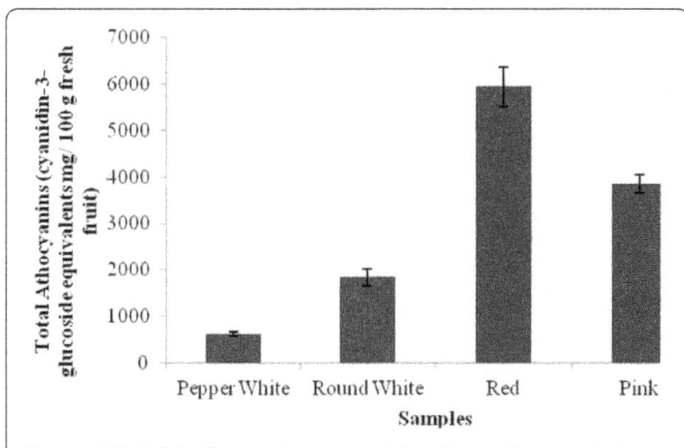

Figure 2 Total Anthocyanins content by *G. pumila* morphotype.

values obtained for Chaura vary depending on the color of the fruit: 8.5° Brix for pink fruit, 8.0° Brix for round white fruit and 9.0° Brix for pepper white fruit, while the red fruits, showed the highest value at 10.0° Brix. Despite it is necessary to collect and characterize the weather and soil conditions near the Villarrica Vulcano, where these three populations were collected, that ° Brix variation could be explained by the influence of the microclimatic conditions for each population. In general, the Villarrica National Park area near the volcano is characterized by two climatic zones. The first is more temperate: warm during the day in summer and drier, and during winter light frost or snowing occurs. Temperatures ranging between 4.0°C and 23.0°C, averaging an annual temperature of 11.5°C. In this area, the forest is very fragmented and large areas are exposed directly to sunlight. The first sampled population (El Playón) and the third sampled population (Cuesta Amarilla) are placed in this zone. The second climatic zone where the third population in this study (Centro Sky) was sampled is characterized by abundant snow and rainfall, which can reach 2,045 mm and 3,000 mm per year, largely concentrated in the months of May and August. In this zone Chaura grows mainly under the forest in a very humid environment and with no exposure to the direct sunlight radiation. Also, this population is placed in a higher altitude than the two other populations.

Another important factor could be the topography, characterized by gorges and high peaks (2,847 meters above sea level (masl)), these differences in sampling zones may influence on the ecological plasticity of the species against different environmental conditions [30] so that at harvest, the fruits do not have a homogeneous maturity. Therefore, the time to harvest the fruits, together with the microclimate generated in different sectors of sampling could also be influencing the development of the various morphotypes, as has been found before for other plant species [31-33]. However, it is remarkably necessary to correlate the phenotypic variation inside the whole studied area with the climatic conditions and for instance it is necessary to characterize in deep the climatic values for the whole area.

The fruit with intense red pigmentation had the highest content of monomeric anthocyanins, which was nine times higher than in the white fruit values. The pigmentation is directly related to the anthocyanins content because they give the red, purple and blue to the leaves, flowers and fruits [34]. Anthocyanins of berries have been reported to have potential health benefits [35]. The anthocyanin cyanidin 3-glucoside is the main active compound that has demonstrated bioactivity against carbon tetrachloride, avoiding lipid peroxidation, a process that eventually causes cell membrane to lose its physicochemical properties, culminating in cell death [36]. According to Prior (1998), anthocyanins in blueberries are mainly concentrated in the epicarp of the berry, but the amount of this chemical would increase depending on the area/volume ratio, that is related with fruit maturity.

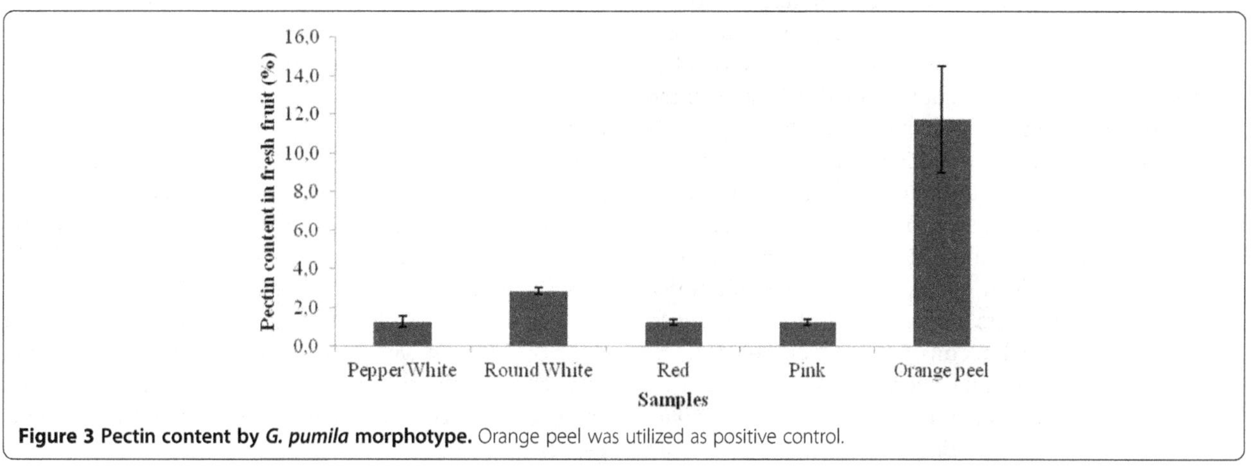

Figure 3 Pectin content by *G. pumila* morphotype. Orange peel was utilized as positive control.

Several varieties of the genus *Vaccinium* have shown values between 61.8-235 mg/100 g [35]; whereas in caneberries, values of anthocyanins have been established from 65 mg/100 g in red raspberry to 589 mg/100 g in black raspberries [6]. Strawberries have shown 10 times higher values than raspberry with 5400 mg/100 g [37]; all these cases used cyanidin 3-glucoside as the standard. Comparing these data with the results of this research, which yielded concentrations from 626.2 mg/100 g (morphotype white pepper) to 5942 mg/100 g (morphotype red), demonstrated that the fruits of Chaura, naturally high in anthocyanins, exceeded the anthocyanins values of non-native berries.

Using a standard extraction method, the pectin content from Chaura fruits and orange peel were compared. The results show that the fruits of Chaura have an important content of pectins, reaching about 3% of fresh weight pectins compared to 12% for the pectin content in the orange peel. Furthermore, no significant difference was found between the pectin content of the four morphotypes of Chaura. Although the percentage of pectin in the Chaura fruit is quantitatively lower than in the orange peel, it must be noted that the pectin content was obtained from the full fruit of the Chaura, in contrast with what happens in orange peel. The pectin in orange peel only represents about 12% of fresh weight, and the peel is not considered edible. Assuming that pectins are capable of providing health benefits in terms of lowering cholesterol [23], to use Chaura fruits alone could improve the organoleptic characteristics when used as a functional food additive [19,20]. Thus these fruits have high nutraceutical properties in addition to the antioxidant properties that might arise from the content of anthocyanins.

Conclusions

This paper deals with the first morphometric analysis and phytochemical study in *Gaultheria pumila* (Chaura) fruits from a wild species from Chile commonly known as Chaura. It was found that there is significant morphometric difference in the fruit depending on fruit color and size. Moreover, a significant difference was found in the fresh weight between unpigmented and pigmented fruit, with pink fruit showing higher fresh weight, and larger polar and equatorial diameters. In relation to Chaura phytochemistry, burgundy colored fruits had a higher concentration of anthocyanins, comparable to other berries with commercial interest, such as blueberries. In addition, high pectin content gives additional commercial value for its intrinsic ability to form gels. Both anthocyanins and pectins have proven beneficial to health. Finally, there is little information regarding native berries and bioactive compounds, which is why *G. pumila* is presented as an underutilized genetic resource, with phytochemicals and morphometric features of great interest, good fruit, and attractive organoleptic characteristics. These preliminary results could open the interest of farmers and the food industry in this species and will be precedent for future domestication and introduction of the species in the agro-productive system in Chile and the rest of the world.

Methods

Plant material

Fruits were sampled from 74 plants coming from three different wild populations grown in the area of Villarrica volcano, this volcano is placed in the Villarrica National Park, Región de la Araucanía, Chile. The coordinates of the first population (named as El Playón) are 39° 22′ 21″ S; 71° 56′ 4″ W, of the second population (named as Centro Sky) are 39° 23′ 13″ S; 71° 57′ 36″ W, and the third population (named as Cuesta Amarilla) are 39° 22′ 29″ S; 71° 56′ 3″ W (Figures 4 and 5). Fruits from plants of each geographic population were randomly harvested in February-March during the summer of 2011. Harvested fruits were stored in a cooler containing ice at 4°C, approximately, until their transport to the controlled laboratory conditions. Once in the lab, the fruits were stored in a refrigerator at 4°C, until the analysis was performed.

Fruit characterization

Morphometric characterization was conducted on fifteen fresh fruits of different shapes (pepper, full cone, and round shape), color (white, red or pink), and diameter (both pole and equator). Shape and color were evaluated visually, and a micrometer caliper (Tornado tools, Taiwan) was used to measure diameter in millimeters (mm).

Total soluble solid content (TSS)

The total soluble solid of the fruits includes organic acids, carbohydrates (sugars), and amino acids [38]. In order to determine the (TSS) of fruit samples, twenty fresh fruits were measured with a digital refractometer (Atago Master M, Atago Co Ltd., Japan) with a scale of 0–32° Brix was used. Results were expressed as degree Brix (°Brix).

Totals anthocyanins extraction

Fresh chaura fruits (2.5 g) were mixed with acidic ethanol (pH 4) at 1:10 (w/v) ratio. The mixture was sonicated for 15 min, and shaken for 60 min at room temperature. Then, the mixture was centrifuged (Boeco, U-320R, Alemania) at 4°C for 20 min at 5000 *g*. Supernatants were filtered through a funnel with glass wool, which had been washed with 3–4 mL of solvent. The extraction was repeated three times, and supernatants were pooled and stored at −20°C for further analysis. Three replications per sample were carried out.

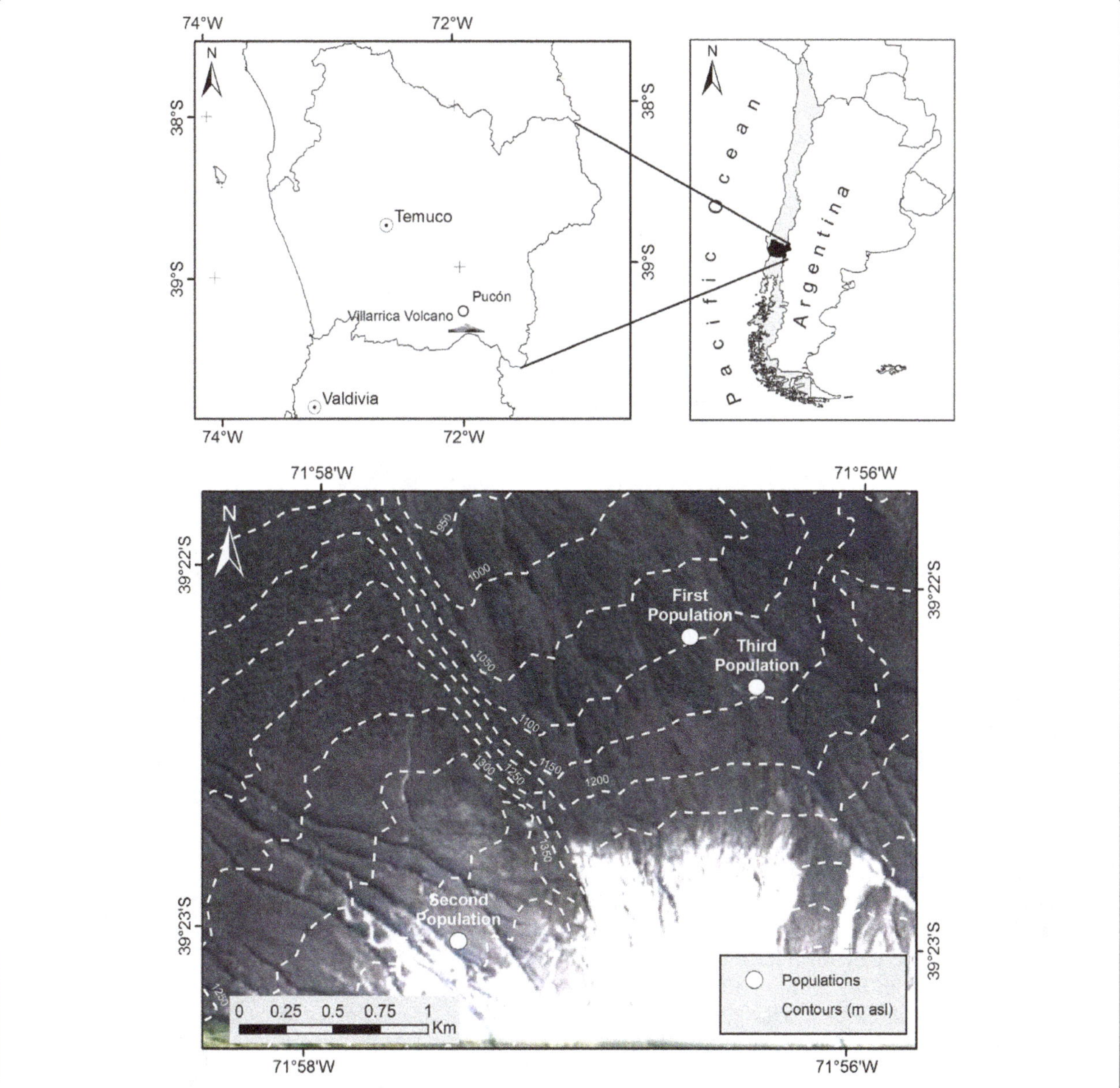

Figure 4 The population distribution of *G. pumila*. Three different wild populations in the Villarrica National Park, Región de la Araucanía, Chile. First population growing the sector locally known as El playón; Second population growing in the sector known as Centro Sky; Third population growing in the sector known as Cuesta Amarilla.

Total anthocyanin content (TAC)

The pH differential spectrophotometric method [39] was employed for quantifying the TAC in samples, using cyanidin-3-glucoside as the standard. Briefly, 200 μL of properly diluted water extract samples were mixed with 1.8 mL of either 25 mM HCl/KCl, pH 1.0 and 0.4 M acetic acid/sodium acetate, pH 4.5 solutions. The absorbance of these solutions was measured at 520 nm and 700 nm (UV mini-1240, SHIMADZU, Japan) TAC was estimated according to the following equation:

$$\text{TAC (mg/100 g fresh fruit)} = (A_M \cdot M_W \cdot V \cdot \text{Dil} \cdot 100)/(\varepsilon \cdot l \cdot m)$$

A = Absorbance of the sample
MW = Molecular weight of the standard (449.2 g mol-1)
V = Volume of the sample
Dil: Dilution of the sample
ε = Molar extinction coefficient of the standard (26 900 L cm-1 mol-1)
l = Cuvette depth
m = mass of fresh fruits

Figure 5 Flowering and *G. pumila* diversity. A) Flowering and inflorescence of Chaura. **B)** Globe shape and pink color fruits. **C)** Comparison of different morphotypes Chaura fruits, with a blueberry commercial gauge. **D)** Fructification and phenotypic diversity Chaura fruits.

To perform this analysis 2.5 g of fresh fruit sampled from different plants of each fruit type were analyzed. Three replications per sample were carried out.

Pectin extraction

5 g of frozen fruit were ground with a crusher in 50 mL water at pH 2.0, and the mixture was placed in an ultrasound bath at 80°C for 30 min. The solution was filtered hot, and then 50 mL of ethanol were added. Once the clot formed, the solution was centrifuged at 4000 rpm for 20 min, after which time the supernatant was discarded. The pectins obtained were lyophilized to determine the weight of the extract [26,40].

Statistical analysis

Descriptive statistics (mean and standard deviation), ANOVA and Turkey test (HSD) were performed using the commercial software package, Statgraphics centurion XV, version 15.01.03 (Statpoint technologies, Warrenton, VA, USA), level of significance was set at $p < 0.05$. A Principal Component Analysis was performed for fruit shape and color and grouping was confirmed by using a discriminant analysis (Wilk's lambda; $P < 0.00001$). Graphical representations of results were performed using the software Microsoft Excel from Microsoft Office 2007.

Competing interests

The authors declare that they have no competing interests.

Authors' contributions

EV, YB and GC, participated in phytochemical composition analysis, statistical analysis and manuscript writing; PC, AA, JP and CC, made field studies and manuscript writing; BC, PDSC and RG, performed the fruit analysis, statistical data processing, manuscript writing and proof reading. RG, headed the research team. All authors read and approved the final manuscript.

Acknowledgements

This project was supported partially by private funds coming from the Departamento de Ciencias Forestales at Universidad Católica del Maule and the Fondo de Innovación para la Competividad of the Región del Maule (FIC-R) through the project BIP 3303689-0. The authors would also like to thanks the Comisión Nacional de Investigación Científica y Tecnológica (CONICYT) in Chile, for partially supporting Mrs. Carola Campos-Hernández and Dr. Evelyn Villagra Quero trough the Programa Nacional de Becas de Doctorado en Chile and the Programa de Atracción e Inserción de Capital Humano (Project ID: 79112042). The authors would like to thanks to Dr. Anne Blis for her work and advices with the English style and grammar. We also thanks to the Corporación Nacional Forestal of Chile (CONAF) for giving access to the protected areas; to Mr. Luis Letelier Gálvez for the preparation of the map and to Mrs. Marjorie Seiltgens for her assistance during the sampling process.

Author details

[1]Departamento de Ciencias Forestales, Centro de Biotecnología de los Recursos Naturales, Universidad Católica del Maule, Campus San Miguel. Av. San Miguel 3605, casilla 617, Talca, Maule Region, Chile. [2]Facultad de Agronomía, e Ingeniería Forestal, Pontificia Universidad Católica de Chile, Vicuña Mackenna 4860, Macul, Santiago, Chile. [3]Instituto de Biología Vegetal y Biotecnología, Universidad de Talca, Avenida Lircay s/n, Talca, Chile.

References

1. Rodríguez R, Marticorena A, Teneb E: **Plantas vasculares de los ríos Baker y Pascua, Región de Aisén, Chile.** *Gayana Bot* 2008, **65**:39–70.
2. Gong Q, Li P, Ma S, Rupassara S, Bohnert H: **Salinity stress adaptation competence in the extremophile *Thellungiella halophila* in comparison with its relative *Arabidopsis thaliana*.** *Plant J* 2005, **44**:826–839.

3. Kähkönen M, Hopia A, Heinonen M: **Berry phenolics and their antioxidant activity.** *J Agric Food Chem* 2001, **49:**4076–4082.
4. Guerrero J, Ciampi L, Castilla A, Medel F, Schalchli H, Hormazabal E, Bensch E, Alberdi M: **Antioxidant capacity, anthocyanins, and total phenols of wild and cultivated berries in Chile.** *J Agric Food Chem* 2010, **70:**537–544.
5. Steinmetz K, Potter J: **Vegetables, fruit, and cancer, I. epidemiology.** *Can Caus Cont* 1991, **2:**325–357.
6. Wada L, Ou B: **Antioxidant activity and phenolic content of Oregon caneberries.** *J Agric Food Chem* 2002, **50:**3495–3500.
7. Arts I, Hollman P: **Polyphenols and disease risk in epidemiologic studies.** *Am J Clin Nutr* 2005, **81:**317S–325S.
8. Gil M, Tomas-Barberan F, Hess-Pierce B, Kader A: **Antioxidant capacities, phenolic compounds, carotenoids, and vitamin C contents of nectarine, peach, and plum cultivars from California.** *J Agric Food Chem* 2002, **50:**4976–4982.
9. Heinonen M, Meyer A, Frankel E: **Antioxidant activity of berry phenolics on human low-density lipoprotein and liposome oxidation.** *J Agric Food Chem* 1998, **46:**4107–4112.
10. Cao G, Sofic E, Prior R: **Antioxidant capacity of tea and common vegetables.** *J Agric Food Chem* 1996, **44:**3426–3431.
11. Seeram N: **Berry fruits: compositional elements, biochemical activities, and the impact of their intake on human health, performance, and disease.** *J Agric Food Chem* 2008, **56:**627–629.
12. Fernandez-Panchon M, Villano D, Troncoso A, Garcia-Parrilla M: **Antioxidant activity of phenolic compounds: from *in vitro* results to *in vivo* evidence.** *Crit Rev Food Sci Nutr* 2008, **48:**649–671.
13. Seeram N, Adams L, Zhang Y, Lee R, Sand D, Scheuller H, Heber D: **Blackberry, black raspberry, blueberry, cranberry, red raspberry, and strawberry extracts inhibit growth and stimulate apoptosis of human cancer cells *in vitro*.** *J Agric Food Chem* 2006, **54:**9329–9339.
14. Escribano-Bailón M, Alcalde-Eon C, Muñoz O, Rivas-Gonzalo J, Santos-Buelga C: **Anthocyanins in berries of maqui (*Aristotelia chilensis* (Mol) Stuntz).** *Phytochem Anal* 2006, **17:**8–14.
15. Ruiz A, Hermosín-Gutiérez I, Mardones C, Vergara C, Herlitz E, Vega M, Dorau C, Winterhalter P, Von Baer D: **Polyphenols and antioxidant activity of calafate (*Berberis microphylla*) fruits and other native berries from Southern Chile.** *J Agric Food Chem* 2010, **58:**6081–6089.
16. Szajdek A, Borowska E: **Bioactive compounds and health - promoting properties of berry fruits: a review.** *Plant Food Hum Nutr* 2008, **63:**147–156.
17. Ridley B, O'neill M, Mohnen D: **Pectins: structure, biosynthesis, and oligogalacturonide-related signaling.** *Phytochemistry* 2001, **57:**929–967.
18. Mohnen D: **Pectin structure and biosynthesis.** *Curr Opin Plant Biol* 2008, **11:**266–277.
19. Ngouémazong D, Tengweh F, Fraeye I, Duvetter T, Cardinaels R, Van Loey A, Moldenaers P, Hendrickx M: **Effect of de-methylesterification on network development and nature of Ca^{2+}-pectin gels: Towards understanding structure–function relations of pectin.** *Food Hydrocolloid* 2012, **26:**89–98.
20. Sun-Waterhouse D, Zhou J, Wadhwa SS: **Drinking yoghurts with berry polyphenols added before and after fermentation.** *Food Control* 2013, **32:**450–460.
21. Koubala B, Mbome L, Kansci G, Tchouanguep F, Crepeau M, Thibault J, Ralet M: **Physicochemical properties of pectins from ambarella peels (*Spondiascytherea*) obtained using different extraction conditions.** *Food Chem* 2008, **106:**1202–1207.
22. Jackson C, Dreaden T, Theobald L, Tran N, Beal T, Eid M, Gao M, Shirley R, Stoffel M, Kumar M, Mohnen D: **Pectin induces apoptosis in human prostate cancer cells: correlation of apoptotic function with pectin structure.** *Glycobiology* 2007, **17:**805–819.
23. Brouns F, Theuwissen E, Adam A, Bell M, Berger A, Rp M: **Cholesterol-lowering properties of different pectin types in mildly hyper-cholesterolemic men and women.** *Eur J Clin Nutr* 2011, **66:**591–599.
24. Taboada E, Fisher P, Jara R, Zúñiga E, Gutierrez A, Cabrera J, Gidekel M, Villalonga R, Cabrera G: **Isolation and characterization of pectic substances from murta (*Ugni molinae*) fruits.** *Food Chem* 2010, **123:**669–678.
25. Urias-Orona V, Rascon-Chu A, Lizardi-Mendoza J, Carvajal-Millan E, Gardea A, Ramírez-Wong B: **A novel pectin material: extraction, characterization and gelling properties.** *Int J Mol Sci* 2010, **11:**3686–3695.
26. Torralbo D, Batista K, Di-Medeiros M, Fernandes K: **Extraction and partial characterization of *Solanum lycocarpum* pectin.** *Food Hydrocolloid* 2012, **27:**378–383.
27. Zhang C, Mu T: **Optimisation of pectin extraction from sweet potato (*Ipomoea Batatas*, Convolvulaceae) residues with disodium phosphate solution by response surface method.** *Int J Food Sci Tech* 2011, **46:**2274–2280.
28. Das M, Kumar A, Yadav SS: *Comparative study of Brix scale and density scale of hydrometers*; 2012. AdMet 2012 Paper No MM 002. Available at http://www.metrologyindia.org/ebooks1/MM_002.pdf (accessed January 2013).
29. Figueroa D, Guerrero J, Bensch E: **Efecto de momento de cosecha y permanencia en huerto sobre la calidad en poscosecha de arándano alto (*Vaccinium corymbosum* l.), cvs. berkeley, brigitta y elliott durante la temporada 2005–2006.** *Idesia* 2010, **28:**79–84.
30. Ferreyra L, Vilardi J, Tosto D, Julio N, Saidman B: **Adaptative genetic diversity and population structure of the "algarrobo" (*Prosopis chilensis* (Molina) Stuntz) analysed by RAPD and isozyme markers.** *Eur J For Res* 2010, **129:**1011–1025.
31. Reynolds A, Wardle D: **Influence of fruit microclimate on monoterpene levels of Gewürztraminer.** *Am J Enol Vitic* 1989, **40:**149–154.
32. Hostetler G, Merwin I, Brown M, Padilla-Zakour O: **Influence of geotextile mulches on canopy microclimate, yield, and fruit composition of Cabernet franc.** *Am J Enol Vitic* 2007, **58:**431–442.
33. Saudreau M, Marquier A, Adam B, Monney P, Sinoquet H: **On the relationship between tree architecture, microclimate and fruit temperature within a tree crown.** *Acta Hort* 2008, **803:**217–224.
34. Lohachoompol V, Srzednicki G, Craske J: **The change of total anthocyanins in blueberries and their antioxidant effect after drying and freezing.** *J Biomed Biotechnol* 2004, **5:**248–252.
35. Prior R, Cao G, Martin A, Sofic E, Mcewen J, O'brien C, Lischner N, Ehlenfeldt M, Kalt W, Krewer G, Mainland C: **Antioxidant capacity as influenced by total phenolic and anthocyanin content, maturity, and variety of *Vaccinium* species.** *J Agric Food Chem* 1998, **46:**2686–2693.
36. Morazzoni P, Bombardelli E: ***Vaccinium myrtillus* L.** *Fitoterapia* 1996, **67:**3–29.
37. Paredes-López O, Cervantes-Ceja M, Vigna-Pérez M, Hernández-Pérez T: **Berries: improving human health and healthy aging, and promoting quality life-A review.** *Plant Food Hum Nutr* 2010, **65:**299–308.
38. Cano M, De Ancos B, Matallana C, Camara M, Reglero G, Tabera J: **Difference among Spanish and latin-american banana cultivars: morphological, chemical and sensory characteristics.** *Food Chem* 1997, **59:**411–419.
39. Lee J, Rennaker C, Wrolstad R: **Correlation of two anthocyanin quantification methods: HPLC and spectrophotometric methods.** *Food Chem* 2008, **110:**782–786.
40. Faravash R, Ashtiani F: **The effect of pH, ethanol volume and acid washing time on the yield of pectin extraction from peach pomace.** *Int J Food Sci Tech* 2007, **42:**1177–1187.

13

Influence of rootstocks on growth, yield, fruit quality and leaf mineral element contents of pear cv. 'Santa Maria' in semi-arid conditions

Ali Ikinci[1], Ibrahim Bolat[1], Sezai Ercisli[2] and Ossama Kodad[3*]

Abstract

Background: Rootstocks play an essential role to determining orchard performance of fruit trees. *Pyrus communis* and *Cydonia oblonga* are widely used rootstocks for European pear cultivars. The lack of rootstocks adapted to different soil conditions and different grafted cultivars is widely acknowledged in pear culture. *Cydonia* rootstocks (clonal) and *Pyrus* rootstocks (seedling or clonal) have their advantages and disadvantages. In each case, site-specific environmental characteristics, specific cultivar response and production objectives must be considered before choosing the best rootstock. In this study, the influence of three Quince (BA 29, Quince A = MA, Quince C = MC) and a local European pear seedling rootstocks on the scion yield, some fruit quality characteristics and leaf macro (N, P, K, Ca and Mg) and micro element (Fe, Zn, Cu, Mn and B) content of 'Santa Maria' pear (*Pyrus communis* L.) were investigated.

Results: Trees on seedling rootstock had the highest annual yield, highest cumulative yield (kg $tree^{-1}$), largest trunk cross-sectional area (TCSA), lowest yield efficiency and lowest cumulative yield (ton ha^{-1}) in the 10th year after planting. The rootstocks had no significant effect on average fruit weight and fruit volume. Significantly higher fruit firmness was obtained on BA 29 and Quince A. The effect of rootstocks on the mineral element accumulation (N, K, Ca, Mg, Fe, Zn, Cu, Mn and B) was significant. Leaf analysis showed that rootstocks used had different mineral uptake efficiencies throughout the early season.

Conclusion: The results showed that the rootstocks strongly affected fruit yield, fruit quality and leaf mineral element uptake of 'Santa Maria' pear cultivar. Pear seedling and BA 29 rootstock found to be more prominent in terms of several characteristics for 'Santa Maria' pear cultivar that is grown in highly calcareous soil in semi-arid climate conditions. We determined the highest N, P (although insignificant), K, Ca, Mg, Fe and Cu mineral element concentrations on the pear seedling and BA 29 rootstocks. According to the results, we recommend the seedling rootstock for normal density plantings (400 trees ha^{-1}) and BA 29 rootstock for high-density plantings (800 trees ha^{-1}) for 'Santa Maria' pear cultivar in semi-arid conditions.

Keywords: TCSA, BA-29, MA, MC, Pear seedlings, *Pyrus communis* L, *Cydonia oblonga* L

Background

Pear (*Pyrus communis* L.) is one of the major fruit in the world and grown well in temperate zones of both hemispheres. The world pear production is about 24 million tons and China is main producer shared with 68% of the world's pear production and followed by the USA (3.3%), Argentina (3.0%), Italy (2.7%) and Turkey (1.9%) [1].

In the commercial pear production, various vegetatively propagated quince and pear rootstocks and generative pear rootstocks have been used. In Turkey, the most common rootstock used for pear cultivars is wild pear seedlings with approximately 85-90% due to their tolerance to lime induced iron chlorosis, easy propagation and well graft-compatible with pear cultivars. They also grow vigorously in loamy wet soil and unfavourable conditions [2,3]. The selection of clonal quince (*C. oblonga*), such as Quince A (MA), Quince C (MC) and BA 29 in Europe, or of clonal *Pyrus communis*

* Correspondence: osama.kodad@yahoo.es
[3]Department of Pomology, National School of Agriculture, Meknes, Morocco
Full list of author information is available at the end of the article

L., such as 'Old Home' × 'Farmingdale' (OHF) in the USA or in South Africa, as substitutes for pear seedling rootstock, have clearly improved the precocity, productivity and quality of some European pear cultivars [2,3].

The rapid developments fruit tree nursery technology and rootstock research and introduction of new clonally propagated rootstocks opened in new area in fruit science [4,5]. For this reason more recently modern pear orchards with different modern training systems to start establish with use of clonal quince (*Cydonia oblonga* L.) rootstocks such as Quince A, Quince C and BA 29 in Turkey. These clonal rootstocks with dwarfing characteristics well reported to increase precocity and fruit quality, especially in the high intensity modern orchards and thus gained more importance [6-8].

Previously, several reports have been documented the relationships between various physiological parameters of pear cultivar/various rootstocks combinations [6-10]. These relationships are important from a horticultural point of view, because they provide a basis for selecting the best graft combination for particular environmental conditions and high fruit quality. Selection of an appropriate graft combination is crucial for the production of deciduous orchard species, because the scion–rootstock interaction influences water relations, leaf gas exchange, plant size, blossoming, timing of fruit set, fruit quality and yield efficiency [10-14]. Different rootstocks have also showed different mineral uptake efficiencies [15]. Leaf mineral element analysis is an effective method for fruit tree nutrient diagnosis and fertilization calculation. Similarly, symptoms of iron deficiency could be mitigated by analyses of mineral leaf composition prior to harvest [16,17]. Moreover, accurate water and fertilizer management are essential in the highly intensive orchard systems to enable the manipulation of both reproductive and vegetative development, to ensure the possibility obtaining higher fruit quality with longer storage potential and to reduce pollution and costs [7].

Southern Anatolia region in Turkey is characterized by fertile soil and semi-arid conditions favorable for growing of subtropical and temperate fruits. More recently in particular the use of clonally propagated dwarf rootstocks for temperate fruit species including pear are widespread in this region. However, the knowladge of specific rootstock effects on specific scion cultivars is of utmost importance to get maximum benefits from the enterprises.

Thus, this study is mainly focused on the effects of various clonal and seedling rootstocks on the main production traits of scion pear cv. 'Santa Maria'. Although this cultivar has already grown commercially in Southern Anatolia region, there is a need to increase the production of this fruit.

Results and discussion

As indicated in Figure 1, there were statistically significant differences among rootstocks in terms of cumulative yield and Trunk Cross Sectional Area (TCSA) ($p < 0.05$). TCSA of 'Santa Maria' pear trees were significantly affected by rootstocks ($p > 0.05$, Figure 1) and TCSA were found to be highest when 'Santa Maria' grafted on seedling rootstock and followed by BA 29, and the lowest one obtained from MA and MC (Figure 1). 'Santa Maria' pear trees grafted on the *Pyrus communis* seedling rootstock gave the highest annual yield between the years of 2008-2013, compared to the other three clonal Quince rootstocks (BA 29, MA, MC) (Figure 1). Similar to the annual yield, cumulative yield was significantly higher for 'Santa Maria' grown on seedling rootstock than the other rootstocks tested during 2008 through 2013 (Figure 1).

Cumulative yield efficiency (CYE) significantly affected by rootstocks ($p < 0.05$), with the highest was observed on MC and the lowest ones on seedling rootstock (Figure 2). There was an opposite trend between CYE and TCSA. After 10 years, the cumulative yield on BA 29 was 77.36 t ha^{-1} and MC was 68.41 t ha^{-1} considerably higher than on MA (60.69 t ha^{-1}) and 47.52 t ha^{-1} on seedling rootstock (Figure 2). Castro and Rodriguez [11] found that yield of the 'Abbe Fetel' and 'Conference' pear cultivars grafted on pear seedling was higher than the quince selections MA and BA 29. It was reported that pear cultivar 'Conference' grafted on BA 29 rootstocks had higher trunk circumference in comparison to MA and MC quince rootstocks [12]. Haak et al. [13] reported that TCSA value of 'Suvenirs' pear cultivar that is grafted onto different *Pyrus* and *Cydonia* rootstocks was the highest on *Pyrus* rootstock and the lowest on on MC rootstock, 5 years after plantation. Sotiropoulos [14] reported that production efficiency of 'William's BC' was highest when grafted on PI 27 (local quince seedlings), intermediate on MA and and lowest on *P. communis.*

Rootstock had a significant effect on tree size, as reflected by TCSA measurements. From planting of trees up to 10 years, although, trees on vigorous seedling rootstocks can have higher yield than those on dwarfing ones due to their greater size, this superiority may not hold for yield efficiency which is production per unit of growth. Yield efficiency does not seem to be clearly related to rootstock vigour [18]. Wertheim [19] reported that MC and BA 29 rootstocks showed higher yield efficiency than OH (Old Home) 11, OH 20, OH 33 and OHF (Old Home x Farmingdale) 333 rootstocks. As regards to yield efficiency the encountered data agree with Loreti et al. [12] and Giacobbo et al. [20] who analyzed different pear rootstocks and verified that the high yield efficiency is not always directly related to high production, once the rootstocks that increased production did not improve yield efficiency. These results are in

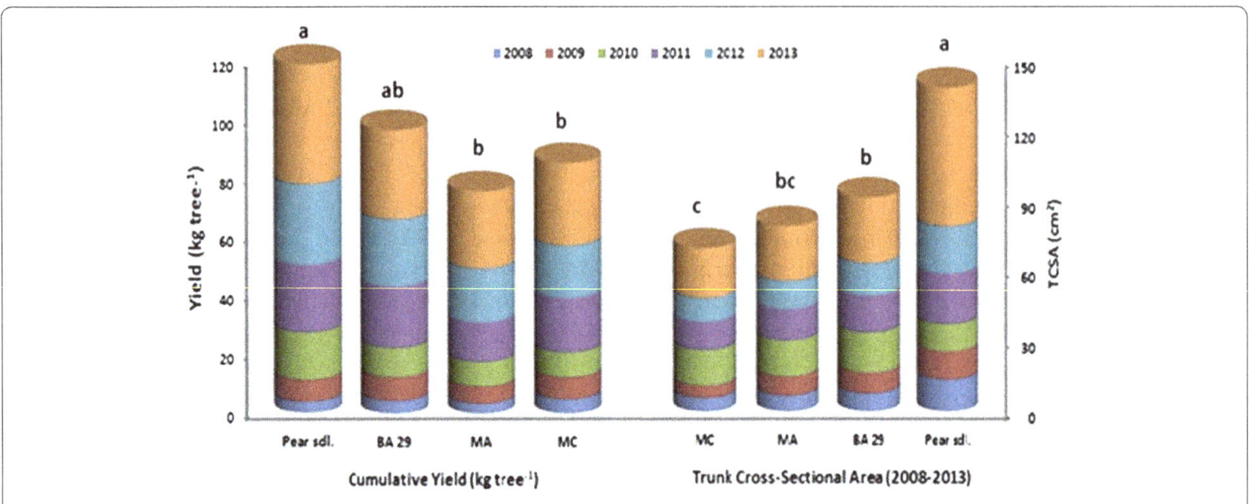

Figure 1 Cumulative yield (kg tree^{-1}) and trunk cross-sectional area (TCSA) of 'Santa Maria' pear cultivars grown on different rootstocks. Different letters denote significant differences between means, according to Duncan's multiple range test, $P< 0.05$.

general agreement with the findings of the researchers mentioned above.

There seemed to be no significant differences between rootstocks in terms of average fruit size and average fruit volume of 'Santa Maria' pear (Table 1). However, trees on MC had higher fruit size and average fruit volume than those on MA, BA 29 and seedling rootstock. Fruit flesh firmness, Soluble Solid Content (SSC) and Titratable Acidity (TA) were significantly affected by rootstocks ($p < 0.05$). Trees on BA 29 and MA had higher fruit flesh firmness than those on MC and seedling rootstocks. Fruit SSC was the highest on seedling rootstock, whereas fruit TA concentration was the highest on BA 29 rootstock (Table 1).

In our study, mean fruit weight value was between 265 and 290 g and mean fruit volume was 291-325 cm^3 in 'Santa Maria' trees on all rootstocks (Table 1). Previous studies on pear reported mean fruit weight value as 147 and 190 g in 'Santa Maria' pear cultivar [21,22]. Fruit weight values that we obtained from 'Santa Maria' cultivar are approximately 100 g higher than the fruit weight values obtained in previous studies. In this study the highest mean fruit weight and fruit volume values were obtained from MC rootstock, while the lowest values were obtained from the trees on seedling. Rootstock can effect directly or indirectly pear fruit size and weight. This effect has been shown by some *Pyrus* rootstocks (OHF 33, OHF 333), which resulted in smaller fruits than the quince rootstocks (BA 29, MC) with which they were compared in spite of lower fruit densities [14]. Wertheim [19] showed that quince rootstocks can produce larger fruit than *Pyrus* rootstocks. In this study,

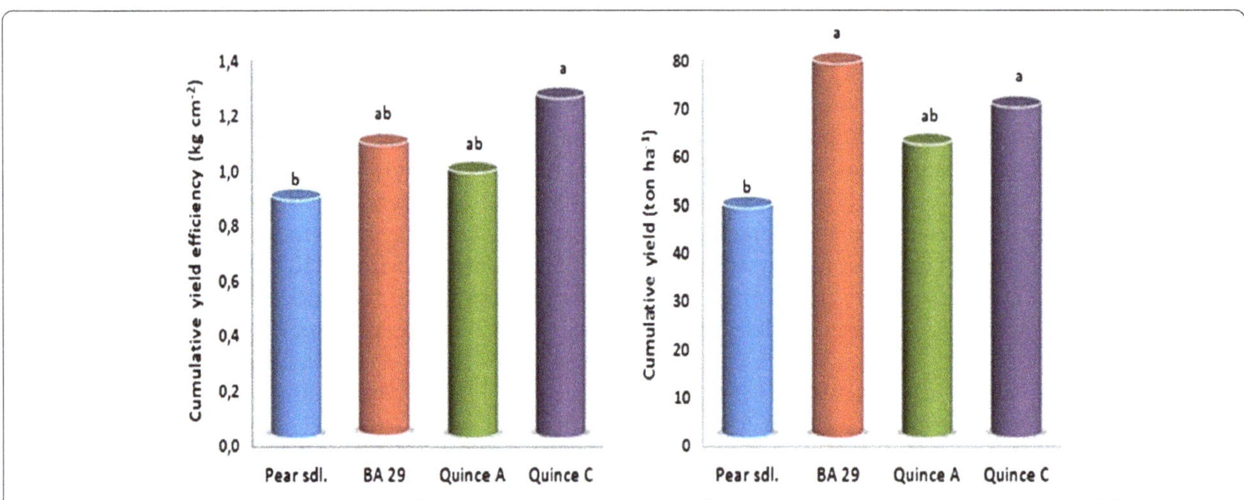

Figure 2 Cumulative yield efficiency (kg cm^{-2}) and cumulative yield (ton ha^{-1}) of 'Santa Maria' pear cultivars grown on different rootstocks. Different letters denote significant differences between means, according to Duncan's multiple range test, $P< 0.05$.

Table 1 Some quality characteristics of "Santa Maria" fruit, as influenced by rootstocks (2012-2013)

Rootstock	Average fruit weight (g)	Average fruit volume (cm^3)	Fruit flesh firmness (lb)	SSC (%)	TA (%)
Pear seedling	265.49	291.20	18.98b	15.60a	0.20c
BA 29	277.55	301.63	21.32a	14.60b	0.26a
MA	279.20	316.80	20.76a	14.90ab	0.24ab
MC	290.37	325.80	19.47ab	15.25ab	0.22bc
Significance	ns	Ns	*	**	**

ns, *and **Nonsignificant or significant at $P \leq 0.05$ or 0.01, respectively.

rootstocks did not have a statistically significant effect on fruit size. In our study, we obtained higher cumulative yield values from the trees on seedling rootstock, which is a strong rootstock, while correspondingly lower average fruit weight values were obtained from these rootstocks. The fact that the trees on seedling, which gave a higher yield and more fruit number can be considered as one of the reasons for lower mean fruit weight when compared to other rootstocks.

In this research, the highest fruit firmness (21.32 and 20.76 lb, respectively) was recorded on BA 29 and MA rootstocks whereas, the lowest (18.98 lb) on seedling rootstock (Table 1). Erdem and Öztürk [21] found fruit firmness as 18.6-21.36 lb in 'Santa Maria' pear cultivar. Fruit firmness values we obtained in our study are similar to the values reported by this researchers. Fruit firmness is one of the most important maturity and quality parameters. Fruit firmness decreases as the maturity level of fruits increase. Nutrient elements taken from the soil or given to the plant from the leaves can reduce fruit firmness.

It was found that the trees on *Pyrus* seedling rootstock had the highest SSC (15.60%) and the lowest (0.20%) TA content (Table 2). In contrast, Sotiropoulos [14] reported that SSC of fruits of 'William's BC' pear cultivar grafted on BA 29 and MA were significantly higher in comparison to *Pyrus* seedling.

SSC content increases with increasing fruit ripening, whereas the longer ripening period TA content shows the decrease [23]. Ozturk et al. [21] reported SSC and TA contents for 'Santa Maria' cultivar as 12.50% and 0.48% respectively. These differences in chemical composition of the same pear cultivar may be due to different soil type and fertility, different genetic crop loads, tree age, differences in rootstocks, differences in ecology, fertilization, irrigation level and differences in the harvest period.

SSC content of the fruits on *Cydonia oblonga* species rootstocks was determined to be lower than the fruits on *Pyrus* seedling rootstocks (Table 1). Some authors reported that Fe chlorosis can affect some growth parameters such as skin color, fruit firmness, titratable acidity, soluble solid content, organic acids, carbohydrates composition, vitamins and phenolics compounds [24,25]. This may be linked to the photosynthetic activity of plants, because CO_2 assimilation of chlorotic leaves and carbohydrates allocation in fruits were negatively affected [26].

The effect of rootstocks on leaf mineral element concentrations of 'Santa Maria' was statistically significant for N, K, Ca, Mg, Fe, Zn, Mn, Cu and B (Tables 2 and 3). The concentration of Ca, Mg, Mn and Cu in leaves increased from 30 to 90 days after full bloom (DAFB). Leaf P concentration decreased from 30 to 60 DAFB, and then increased thereafter, whereas leaf N, K, Fe, Zn, and B concentrations decreased from 30 to 90 DAFB. 'Santa Maria' on seedling rootstock had the highest leaf N (2.05%), K (1.53%), Fe (50.56 $\mu g\ g^{-1}$), Cu (21.80 $\mu g\ g^{-1}$), and

Table 2 Seasonal variation in N, P, K, Ca, and Mg foliar concentration of Santa Maria' pear cultivar grown on different rootstocks

Rootstock	N (%)			P (%)			K (%)			Ca (%)			Mg (%)		
	30 DAFB	60 DAFB	90 DAFB	30 DAFB	60 DAFB	90 DAFB	30 DAFB	60 DAFB	90 DAFB	30 DAFB	60 DAFB	90 DAFB	30 DAFB	60 DAFB	90 DAFB
Pear sdl.	2.38 a	2.14 a	2.05 a	0.18	0.17	0.17	1.94 a	1.88 a	1.53 a	1.66 a	1.84 a	1.88 a	0.40 a	0.45 a	0.47 a
BA 29	2.25 ab	2.11 ab	1.98 ab	0.22	0.15	0.18	1.89 a	1.85 a	1.42 ab	1.51 ab	1.64 a	1.67 a	0.38 a	0.44 a	0.45 a
Quince A	2.19 b	1.99 c	1.87 b	0.18	0.13	0.15	1.53 b	1.39 b	1.27 b	1.34 b	1.37 b	1.46 b	0.37 ab	0.40 ab	0.42 ab
Quince C	2.17 b	2.03 bc	1.97 ab	0.19	0.11	0.14	1.58 b	1.45 b	1.44 ab	1.11 c	1.30 b	1.38 b	0.22 b	0.34 b	0.36 b
Significance	**	**	**	Ns	Ns	Ns	***	***	**	***	***	***	**	*	*

Values are the mean for 2012 and 2013.
Norms: Heckman [36] N (2.20-2.80%), P (0.11-0.25%), K (1.00-2.00%), Ca (1.00-1.50%) and Mg (0.25-0.50%).
ns, *, **and ***Nonsignificant or significant at $P \leq 0.05$, 0.01 or 0.001, respectively.

B (16.43 ppm) concentrations at the 90 DAFB than other rootstocks. The highest leaf Ca (1.88% and 1.67%) and Mg (0.47% and 0.45%) concentrations were shown on seedling and BA 29 rootstocks, respectively. Among the rootstocks, 'Santa Maria' pear cultivar had the highest leaf Zn (26.7 and 24.6 μg g^{-1}, respectively) concentrations on MC and MA, whereas the highest leaf Mn (75.10 ppm) concentration only on MC.

In our study, we found significant differences in the mineral concentrations of leaves of 'Santa Maria' pear trees grafted on BA 29, MA, MC, and seedling rootstocks (Tables 2 and 3). Other researchers have also reported significant rootstock effects on scion leaf mineral nutrients concentrations of some fruit trees under different environmental conditions [27-29].

Based on our findings of macro elements, Ca and Mg levels generally showed an increase from 30 DAFB to 90 DAFB, while N and K concentrations generally showed a decrease (Table 2). P concentration was found to show a decrease between 30 DAFB - 60 DAFB and then to show an increase between 60 DAFB - 90 DAFB (Table 2). Belkhodja et al. [30] found that N, P and K concentrations were decreased in leaves of peach trees from 60 to 120 DAFB, whereas Ca and Mg concentrations were increased from 60 to 120 DAFB. The results we have obtained from this study's findings are in agreement with Belkhodja et al. [30]. In our study, the results that we have obtained similarly, in another study, the concentration of most nutrients in leaves decreased as the growing season progressed, with only that of Ca, Mg, and Mn showing an increase [31].

Leaf Ca and Mg concentration was higher (at the 90 DAFB) on seedling and BA 29, and lower on MA and MC (Table 2). Trees grafted on MA and MC appears to have the lowest leaf macronutrient concentration. The same effect on the leaf concentration was also found in lower vigour rootstocks in apple [29,32,33]. Several researchers have shown that scion leaves of trees on more vigorous rootstocks have higher mineral (K, Mg) content than those on size-controlling rootstocks [32,34]. Previously dwarf rootstock is rated as sensitive to Ca and K deficiencies, which is in agreement with our result [34].

It can be concluded that dwarfing rootstocks were less effective than others in terms of uptake of some macronutrients from root medium. Their nutrient uptake capacity is less due to poor root volume in the soil [35]. Differences in nutrient concentrations among rootstocks can also be explained with the structure of root systems, deviations of root cation exchange capacities, rhizosphere pH, characteristics of root exudates etc. [35,36].

Previously in pear leaf sample collection time as 35 to 70 DAFB period is reported more suitable [37], some reported that the collection time is 120 DAFB more suitable [36]. Leaf analysis results of 'Santa Maria' pear trees in 90 DAFB period showed that N, Fe, Zn and B concentrations in some rootstocks (Tables 2 and 3) were much lower than reference values [36]. N and Fe were found to be particularly lower in the leaves of the trees on MA and MC rootstocks.

Leaf Fe concentration was found to be lower than the threshold value (60 μg g^{-1}) reported by Heckman [36] starting from 30 DAFB in pear trees on *Cydonia* rootstock and starting from 60 DAFB in trees on seedling rootstock (Table 3). Leaf Fe concentration of 'Santa Maria' trees on all rootstocks has fallen well below the reference value at 90 DAFB. Pear is the leading fruit species with the most commonly Fe chlorosis seen in all fruit species [38]. Significantly high soil pH and lime ratio in soil in the orchard where the study was carried out is a major cause of the decrease of Fe concentration in 'Santa Maria' leaves starting from 30 DAFB.

Iron chlorosis increased markedly leaf K concentrations, and only slightly leaf N, Mg and Mn concentrations, whereas leaf P, Cu and Zn were not affected much by chlorosis [30]. Our pear leaf results agree with the leaf nutrient concentrations time courses obtained by Belkhodja et al. [30] in peach grown in Zaragosa (Spain).

Zn concentration in 'Santa Maria' leaves began to decrease starting from 30 DAFB on all rootstocks (Table 3).

Table 3 Seasonal variation in Fe, Zn, Mn, Cu, and B foliar concentration of Santa Maria' pear cultivar grown on different rootstocks

Rootstock	Leaf Fe (μg g^{-1})			Leaf Zn (μg g^{-1})			Leaf Mn (μg g^{-1})			Leaf Cu (μg g^{-1})			Leaf B (μg g^{-1})		
	30 DAFB	60 DAFB	90 DAFB	30 DAFB	60 DAFB	90 DAFB	30 DAFB	60 DAFB	90 DAFB	30 DAFB	60 DAFB	90 DAFB	30 DAFB	60 DAFB	90 DAFB
Pear sdl.	72.3 a	61.6 a	50.6 a	27.9 c	20.8 b	16.2 c	37.8 d	38.5 d	44.9 d	14.8 a	18.0 a	21.8 a	25.8 b	19.8 a	16.4 a
BA 29	69.2 b	60.1 a	46.7 b	32.7 b	24.9 b	21.3 b	43.9 b	46.0 c	52.2 c	15.8 a	15.9 b	17.8 b	27.4 a	7.3 c	6.7 c
Quince A	57.9 d	50.4 c	40.7 c	37.2 a	30.9 a	24.6 a	42.9 c	48.8 b	59.2 b	10.3 c	16.0 b	17.0 c	17.4 d	9.2 b	8.2 b
Quince C	64.9 c	55.7 b	41.2 c	40.8 a	32.7 a	26.7 a	45.8 a	68.8 a	75.1 a	12.9 b	14.4 c	15.7 d	21.0 c	9.5 b	7.9 b
Significance	***	***	***	**	**	**	***	***	**	***	***	***	***	***	***

Values are the mean for 2012 and 2013.
Norms: Heckman [36] Fe (60-250 μg g^{-1}), Zn (25-200 μg g^{-1}), Mn (30-100 μg g^{-1}), Cu (5-20 μg g^{-1}) and B (20-70 μg g^{-1}).
**and ***Significant at $P \leq 0.01$ or 0.001.

Zn concentration in the leaves of 'Santa Maria' on seedling and BA 29 rootstocks fell below the reference value (25-200 μg g^{-1}) at 60 DAFB. This decrease continued at 90 DAFB. It was found that Zn concentration of 'Santa Maria' leaves (26.7 μg g^{-1}) was above the reference value only on 90 DAFB date on MC rootstock. No visual sign of Zn deficiency was observed on 'Santa Maria' leaves until 90 DAFB stage in 2012 and 2013. Erdem and Ozturk [22] reported that 'Santa Maria' cultivar is more resistant to Zn deficiency than 'Akça' and 'Deveci' cultivars and that; it used the existing zinc in soil better. Swietlik [39] reported that Zn deficiency is common in fruit trees that grow in alkaline soils with high pH content. It can be stated that high pH level of the soil is a reason for Zn concentration level below the threshold value in 'Santa Maria' leaves.

Trees on seedling rootstock had higher leaf B concentrations than those on other rootstocks (Table 3). Lombard and Westwood [40] reported that pear seedling rootstocks have a higher B uptake than quince rootstocks.

Conclusion

The effects of rootstocks on fruit yield and quality and mineral element uptake of 'Santa Maria' pear cultivar showed variations. Pear seedling and BA 29 rootstock became prominent in terms of several characteristics for 'Santa Maria' pear cultivar that is grown in highly calcareous soil in semi-arid climate conditions. The trees on seedling rootstock were found to have higher values than other rootstocks in terms of annual yield, cumulative yield and TCSA value. 77.4 ton ha^{-1} yield was obtained from 10 year old 'Santa Maria' trees grafted on BA 29 rootstock at a density of 800 trees ha^{-1} in 2008-2013 periods. In the orchard used in the study, soil pH was significantly high. The highest N, P (although insignificant), K, Ca, Mg, Fe and Cu concentrations were determined in the trees on pear seedling and BA 29 rootstocks. The lowest leaf Fe concentrations in pear trees were determined in the trees on MA and MC rootstocks. Leaf Fe concentrations of the trees on these rootstocks began to decrease from 30 DAFB and began to fall below the critical threshold after this date. According to the results obtained from this study, we recommend the seedling rootstock for normal density plantings (400 trees ha^{-1}) and BA 29 rootstock for high density plantings (800 trees ha^{-1}) at the 'Santa Maria' orchards in semi-arid conditions.

Methods

Site description

The experiment was carried out at the Harran University Pome Fruit Research Station in Sanliurfa, Turkey (37°10' N, 38°59' E; alt. 520 m) during 2008-2013. Sanliurfa province has semi-arid climate features with cold and wet during the winter and very hot and dry in the summer seasons. During the experiment, the air temperatures were in average 29.9°C in summer and 9.4°C in winter, while annual precipitation ranged between 355-447 mm, mainly concentrated between the months of November and April (Figure 3). The average relative humidity is at the level of 52.2%. Relative humidity is the highest (66%) ratio in January and in July is the lowest (36%) level. The orchard was established in a calcareous (21.5% total carbonates and 10.7% active lime), alkaline and clay-loam textured soil. The physical and chemical characteristics of the soil were clay 58.5%, silt 18.5% and sand 21%, with the low level of organic matter (1.16%), pH 7.92 (in 1M KCl), and optimum concentrations of available P (80 mg kg^{-1}), K (160 mg kg^{-1}), Mg (50 mg kg^{-1}), and Fe (DTPA-extractable Fe:1.45 mg kg^{-1}) in the top soil layer (0–40 cm).

Plant material and experimental design

'Santa Maria' pear trees were planted in December 2004 with 1-year-old scions. The following rootstocks were tested: Local pear seedling (*Pyrus communis* L.), clonal MA, MC and BA 29 (*Cydonia oblonga* Mill.). The experiment was laid out in a randomized complete-block design with three blocks, each consisting of three rows of trees. There were 15 trees in each row. Each experimental plot contained seven trees in each row. Data were collected from the five central trees in each row, using the remaining trees as guards. Trees on pear seedling rootstocks (hereafter referred to as "seedling rootstocks") were planted at 5 × 5 m (400 trees ha^{-1}) and trees on the *Cydonia oblonga* variety rootstocks were planted at 5 × 2.5 m (800 trees ha^{-1}) distance and trained as a central leader system.

Cultural treatments

Irrigation of the orchard was carried out using a computerized drip irrigation system. Irrigation frequency was two times per week from May to October each season according to regional recommendations using class-A pan. Each treatment (tree) received the same total amount of water in each season. All treated trees were similarly fertigated with essential minerals using the fertigation method. No foliar application of nutrients was made to these trees. Thinning of flowers or fruitlets was not carried out during the experiment. Weed, disease, and insect control was managed using the practices that were commonly used for commercial production, and all the treatments were under the identical management. A copper spray was put on at budbreak to protect the trees from fireblight.

Data collection on growth, yield, fruit characteristics

Trunk diameter 20 cm above the graft union was measured with digital callipers in December each year. The

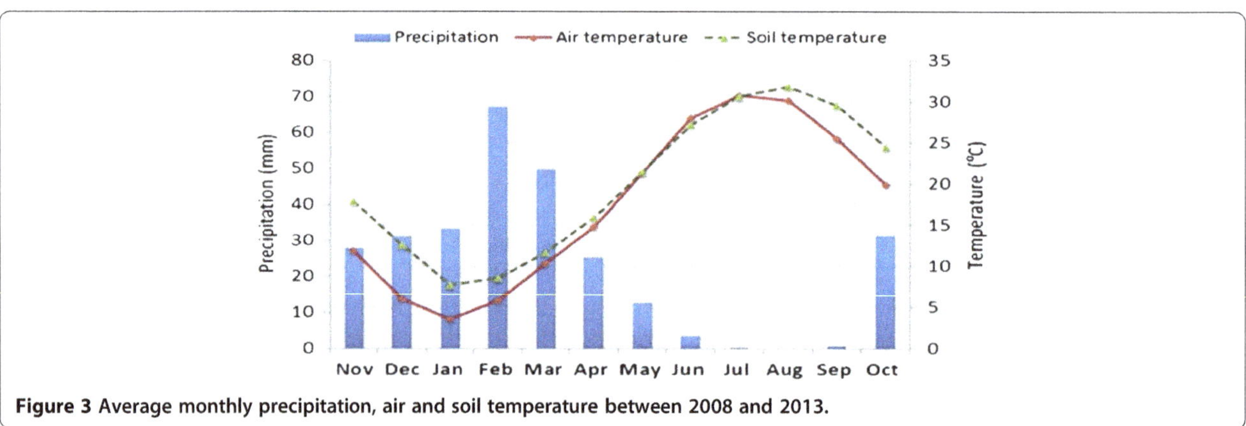

Figure 3 Average monthly precipitation, air and soil temperature between 2008 and 2013.

average of two readings (north-south and east- west) was converted to trunk cross-sectional area (TCSA) for analysis. Annual yields, yield efficiency (yield/TCSA), cumulative yield and cumulative yield efficiency (cumulative yield/TCSA in 2013) were calculated. Cumulative fruit yield efficiency (CYE) was expressed as kg cm^{-2} [41].

Fruit yield was determined each year by harvesting five central trees from each plot in September. Fruit firmness, soluble solids concentration (SSC), and titratable acidity (TA) of fruits at harvest were determined using a randomly selected sample of 20 fruits for each plot. Fruit yield per tree and average fruit weight were measured at fruit harvest in September. Fruit firmness was measured individually on two opposite faces of peeled fruits by using Effegi penetrometer (model. FT–327; McCormick Fruit Tech, Yakima, WA) with an 8 mm diameter tip and expressed in terms of lb force. The SSC was determined with an Atago Palette Series Model PR-101a digital refractometer (Atago Co. Ltd., Tokyo, Japan) at 22°C in the juice squeezed from the fruit homogenate (expressed as °Brix). TA was determined by titrating the fruit homogenate with 0.1 N NaOH to pH 8.1. The TA results represented malic acid content expressed as a percentage. All analyses were performed according to standard methods [42].

Data collection on leaf mineral elements content

Leaf mineral concentrations were determined in 2012 and 2013. Leaf sampling was done at 30, 60 and 90 days after full bloom (DAFB). Each leaf sample consisted of 50 new but fully developed midterminal leaves from current-year shoots at 150 cm above the ground in the tree canopy [28,36]. Collected leaves were immediately packed into polyethylene bags and transported to the laboratory in a portable refrigerator. The leaf samples were washed in tap water, 0.1 mol L^{-1} of HCl and deionized water then dried in a forced air drying oven at 65°C for 48 h to constant weight. Leaves were ground to pass a 40 mesh screen and stored in an oven at 60°C until analysis. One g of dried ground leaf sample dry ashed at 550°C for 5 h. The ash was then dissolved in 0.1 N HCl. Analyses were performed by a colorimetric method for P (phospho-vanadate reaction), Atomic Emission Spectrometry for K and Na, and Atomic Absorption Spectrometry (Perkin-Elmer 1100 B, Norwalk, CT) for Ca, Mg, Fe, Mn, Zn and Cu. Nitrogen was determined by the Kjeldahl procedure. Leaf boron (B) concentration was determined by spectrophotometry using the Azomethine-H method, in extracts obtained from leaf ashes (oven digestion) according to procedure described by Kacar [43]. Each determination was replicated three times. The results were expressed on a dry matter basis: % for macro (N, P, K, Ca and Mg) and mg kg^{-1} for microelements (Fe, Mn, Cu, Zn and B).

Statistical analysis

Analyses of variance were performed on all the data collected. Percentage data were subjected to arcsine transformation before analysis, to provide a normal distribution. Differences between the means were ascertained with Duncan's multiple range tests, using the SAS software package (SAS Institute, Cary, NC). The mean values for the combinations labeled with the same letters do not significantly differ at the significance level $\alpha = 0.05$.

Competing interests

The authors declare that they have no competing interests.

Authors' contributions

AI and IB, made a significant contribution to experiment design, acquisition of data, analysis and drafting of the manuscript. SE and OK have made a substantial contribution to interpretation of data, drafting and carefully revising the manuscript for intellectual content. All authors read and approved the final manuscript.

Author details

[1]Horticulture Department, Harran University, Agriculture Faculty, 63330 Sanliurfa, Turkey. [2]Horticulture Department, Ataturk University, Agriculture Faculty, 25240 Erzurum, Turkey. [3]Department of Pomology, National School of Agriculture, Meknes, Morocco.

References

1. FAOSTAT: *Food and Agriculture Organization (FAO) of the United Nations, Rome, Italy*; 2012. Available at: http://faostat.fao.org/site/567/DesktopDefault.aspx (Accessed: 24 April 2014).
2. Ercisli S, Esitken A, Orhan E, Ozdemir O: **Rootstocks used for temperate fruit trees in Turkey: an overview.** *Sodininkyste ir Darzininkyste* 2006, **25**:27–33.
3. Gunen Y, Misirli A: **Rootstock usage in pear (*Pyrus*spp.) growing.** *Anadolu* 2004, **14**:111–127.
4. Koc A, Bilgener S: **Morphological characterization of cherry rootstock candidates selected from Samsun Province in Turkey.** *Turk J Agric For* 2013, **37**:575–584.
5. Sarropoulou V, Dimassi-Theriou K, Therios I: **In vitro rooting and biochemical parameters in the cherry rootstocks CAB-6P and Gisela 6 using L-methionine.** *Turk J Agric For* 2013, **37**:688–698.
6. Jacobs JN, Cook NC: **The effect of rootstock cultivar on the yield and fruit quality of 'Packham's Triumph', 'Doyenne du Comice', 'Forelle', 'Flamingo' and 'Rosemarie' pears.** *S Afr J Plant Soil* 2003, **20**:25–30.
7. Stassen PJC, North MS: **Nutrient distribution and requirement of 'Forelle' pear trees on two rootstocks.** *Acta Hortic* 2005, **671**:493–500.
8. Lewko J, Scibisz K, Sadowski A: **Performance of two pear cultivars on six different rootstocks in the nursery.** *Acta Hortic* 2007, **732**:227–231.
9. Sugar D, Basile SR: **Performance of 'Comice' pear on quince rootstocks in Oregon, USA.** *Acta Hortic* 2011, **909**:215–218.
10. Bosa K, Jadczuk-Tobjasz E, Kalaji M, Majewska M, Allakhverdiev SI: **Evaluating the effect of rootstocks and potassium level on photosynthetic productivity and yield of pear trees.** *Russ J Plant Physl* 2014, **61**(2):231–237.
11. Castro HR, Rodriguez RO: **The behaviour of quince selections as pear rootstocks for 'Abbe Fetel' and 'Conference' pear cultivars in the Rio Negro Valley, Argentina.** *Acta Hortic* 2002, **596**:363–368.
12. Loreti F, Massai R, Fei C, Cinelli F: **Performance of 'Conference' cultivar on several quince and pear rootstocks: Preliminary results.** *Acta Hortic* 2002, **596**:311–318.
13. Haak E, Kviklys D, Lepsis J: **Comparison of Cydonia and Pyrus rootstocks in Estonia, Latvia and Lithuania.** *Sodininkyste ir Darzininkyste* 2006, **25**:322–326.
14. Sotiropoulos TE: **Performance of the pear (*Pyrus communis*) cultivar 'William's Bon Chretien' grafted on seven rootstocks.** *Aust J Exp Agr* 2006, **46**:701–705.
15. Tagliavini M, Bassi D, Marangoni B: **Growth and mineral nutrition of pear rootstocks in lime soil.** *Sci Hortic* 1993, **54**:13–22.
16. Sanz M, Pascual J, Machin J: **Prognosis and correction of iron chlorosis in peach trees: Influence on fruit quality.** *J Plant Nutr* 1997, **20**:1567–1572.
17. Tagliavini M, Marangoni B: **Major nutritional issues in deciduous fruit orchards of Northern Italy.** *HortTechnology* 2000, **12**:26–31.
18. Sugar D, Powers KA, Basile S: **Effect of rootstock on fruit characteristics and tree productivity in seven red-fruited pear cultivars.** *Fruit Varieties J* 1999, **53**:148–154.
19. Wertheim S: **Rootstocks for European pear: a review.** *Acta Hortic* 2002, **596**:299–307.
20. Giacobbo CL, GazollaNeto A, Pazzin D, Francescatto P, Fachinello JC: **The assessment of different rootstocks to the pear tree cultivar 'Carrick'.** *Acta Hortic* 2010, **872**:353–358.
21. Ozturk G, Basim E, Basim H, Emre RA, Karamursel OF, Eren I, Isci M, Kacal E: **Development new resistatnt pear cultivars by croos breeding against *Erwinia amylovora*: First fruit observation.** In *Proceedings of 6th National Horticultural Congress, 4-8 October 2011, Sanliurfa, Turkey.* 2011:1–9.
22. Erdem H, Ozturk H: **Effect of foliar applied zinc on yield, mineral element contents and biochemical properties of pear varieties grafted to BA 29 rootstock.** *SDU J Fac Agric* 2012, **7**:93–106.
23. Ozkaya O, Dundar O, Kuden A: **Storage capacity of 'Angelina' plum cv. Grown in Adana.** In *Proceedings 3rd Horticulture Crops Storage and Marketing Symposium 6-9 September 2005, Antakya, Turkey*; 2005:406–408.
24. Tagliavini M, Rombola AD: **Iron deficiency and chlorosis in orchard and vineyard ecosystems.** *Eur J Agron* 2001, **15**:71–92.
25. Alvarez-Fernandez A, Abadia J, Abadia A: **Iron deficiency, fruit yield and fruit quality.** In *Iron Nutrition in Plants and Rhizospheric Microorganisms*, Volume 4. Edited by Barton LL, Abadia J. Dordrecht, Netherlands: Springer; 2006:85–101.
26. Sorrenti G, Toselli M, Marangoni M: **Use of compost to manage Fe nutrition of pear trees grown in calcareous soil.** *Sci Hortic* 2012, **136**:87–94.
27. Giorgi M, Capocasa F, Scalzo J, Murri G, Battino M, Mezzetti B: **The rootstock effects on plant adaptability, production, fruit quality, and nutrition in the peach.** *Sci Hortic* 2005, **107**:36–42.
28. Yin X, Bai J, Seavert CF: **Pear responses to split fertigation and band placement of nitrogen and phosphorus.** *HortTechnology* 2009, **19**:586–592.
29. Kucukyumuk Z, Erdal I: **Rootstock and cultivar effect on mineral nutrition, seasonal nutrient variation and correlations among leaf, flower and fruit nutrient concentrations in apple trees.** *Bulg J Agric Sci* 2011, **17**:633–641.
30. Belkhodja R, Morales F, Sanz M, Abadía A, Abadia J: **Iron deficiency in peach trees: effects on leaf chlorophyll and nutrient concentrations in flowers and leaves.** *Plant Soil* 1998, **203**:257–268.
31. Cheng L, Raba R: **Accumulation of macro- and micronutrients and nitrogen demand-supply relationship of 'Gala'/'Malling 26' apple trees grown in sand culture.** *J Amer Soc Hort Sci* 2009, **134**:3–13.
32. Fallahi E, Colt WM, Fallahi B, Chun I: **The importance of apple rootstocks on tree growth, yield, fruit quality, leaf nutrition and photosynthesis with an emphasis on 'Fuji'.** *HortTechnology* 2002, **12**:38–44.
33. Amiri ME, Fallahi E, Safi-Songhorabad M: **Influence of rootstock on mineral uptake and scion growth of 'Golden Delicious' and 'Royal Gala' apples.** *J Plant Nutr* 2014, **37**:16–29.
34. Abdalla OA, Khatamian H, Miles NW: **Effect of rootstocks and interstems on composition of 'Delicious' apple leaves.** *J Am Soc Hort Sci* 1982, **107**:730–733.
35. Marschner H: *Mineral nutrition of higher plants.* Secondth edition. London: Academic Pres Inc; 1996:446.
36. Heckman JR: *Leaf analysis for fruit trees.* 2001. Available at: http://www.Rce.Rutgers.Edu/Pubs/Pdfs/Fs627.Pdf. (Accessed: 24 April 2014).
37. McGinnis M, Stokes C, Cleveland B: *NCDA & CS Plant tissue analysis guide: Plant/ Waste/Solution/ Media Analysis Section. Agronomic Division.* Raleigh, North Caroline-USA: North CarolineDepartmant of Agriculture Consumer Services; 2012.
38. Sanz M, Heras L, Montanes L: **Relationship between yield and leaf nutrient contents in peach trees: Early nutritional status diagnosis.** *J Plant Nutr* 1992, **15**:1457–1466.
39. Swietlik D: **Zinc nutrition of fruit crops.** *HortTechnology* 2002, **12**:45–50.
40. Lombard PB, Westwood MN: **Pear rootstocks.** In *Rootstocks for Fruit Crops.* Edited by Rom CR, Calson RF. New York, USA: Wiley-Interscience Publication, John Wiley and Sons; 1987:145–183.
41. Stern RA, Doron I: **Performance of 'Coscia' pear (*Pyrus communis*) on nine rootstocks in the north of Israel.** *Sci Hortic* 2009, **119**:252–256.
42. A.O.A.C: *Association of Official Agricultural Chemists.* 15th edition. Arlington, VA, USA: Association of Analytical Communities; 1990:484.
43. Kacar B: *Plant and Soil Chemical Analysis III. Soil Analysis. Nobel Publication No:1387.* Ankara; 2009:467.

14

Evaluation of phenolic profile, antioxidant and anticancer potential of two main representants of Zingiberaceae family against B164A5 murine melanoma cells

Corina Danciu[1†], Lavinia Vlaia[2†], Florinela Fetea[3], Monica Hancianu[4], Dorina E Coricovac[5*], Sorina A Ciurlea[5], Codruţa M Şoica[6], Iosif Marincu[7*], Vicentiu Vlaia[8*], Cristina A Dehelean[4] and Cristina Trandafirescu[6]

Abstract

Background: *Curcuma longa* Linnaeus and *Zingiber officinale* Roscoe are two main representatives of *Zingiberaceae* family studied for a wide range of therapeutic properties, including: antioxidant, anti-inflammatory, anti-angiogenic, antibacterial, analgesic, immunomodulatory, proapoptotic, anti-human immunodeficiency virus properties and anticancer effects. This study was aimed to analyse the ethanolic extracts of *Curcuma rhizome* (*Curcuma longa* Linnaeus) and *Zingiber rhizome* (*Zingiber officinale* Roscoe) in terms of polyphenols, antioxidant activity and anti-melanoma potential employing the B164A5 murine melanoma cell line.

Results: In order to evaluate the total content of polyphenols we used Folin-Ciocâlteu method. The antioxidant activity of the two ethanolic extracts was determined by DPPH assay, and for the control of antiproliferative effect it was used MTT proliferation assay, DAPI staining and Annexin-FITC-7AAD double staining test. Results showed increased polyphenols amount and antioxidant activity for *Curcuma rhizome* ethanolic extract. Moreover, 100 μg/ml of ethanolic plant extract from both vegetal products presented in a different manner an antiproliferative, respectively a proapoptotic effect on the selected cell line.

Conclusions: The study concludes that *Curcuma rhizome* may be a promising natural source for active compounds against malignant melanoma.

Keywords: *Curcuma longa* Linnaeus, *Zingiber officinale* Roscoe, Polyphenols, Antioxidant, Melanoma

Background

For centuries, plants, plant products or pure active phytocompounds have been successfully used for the benefits of human health. *Zingiberaceae* family also known as ginger family comprises a number of approximately 52 genera and over 1300 species of aromatic plants [1]. Among this high number of representatives some species have been reported for their therapeutic properties both in classical and ethno medicine [2]. *Curcuma longa* Linnaeus and *Zingiber officinale* Roscoe are two main representatives of *Zingiberaceae* family studied for a wide range of therapeutic properties. *Curcuma longa* Linnaeus, popular name-turmeric, is an aromatic, nutraceutical plant. The vegetal product of this plant, the root, have been intensively used, under different pharmaceutical formulations in Indian traditional medicine (Ayurveda) for different ailments, namely for wounds, acne, parasitic infection (local administration) and common cold, urinary tract disease and liver disease (systemic administration) [3]. Numerous experimental studies regarding the therapeutic activity of turmeric reported a plethora of pharmacological

* Correspondence: dorinacoricovac@umft.ro; imarincu@umft.ro; vlaiav@umft.ro
†Equal contributors
[5]Department of Toxicology, Faculty of Pharmacy, University of Medicine and Pharmacy "Victor Babes", Eftimie Murgu Square, No. 2, Timisoara 300041, Romania
[7]Faculty of Medicine, University of Medicine and Pharmacy "Victor Babes", Eftimie Murgu Square, No. 2, Timisoara 300041, Romania
[8]Department of Organic Chemistry, Faculty of Pharmacy, University of Medicine and Pharmacy "Victor Babes", Eftimie Murgu Square, No. 2, Timisoara 300041, Romania
Full list of author information is available at the end of the article

properties of this vegetal extract, including: antioxidant, anti-inflammatory, anti-angiogenic, antibacterial, analgesic, immunomodulatory, proapoptotic, anti-human immunodeficiency virus properties, being also studied in arthritis, diabetes, Alzheimer's disease [4-8]. The major active compound responsible for the pharmacodynamic action is the polyphenol curcumin [9,10]. Additionally, this natural polyphenol has been described as an anticancer agent, both *in vitro* and *in vivo* on a wide range of cancer types, such as colon, pancreatic, liver, cervical, pulmonary, thymic, brain, breast and bone cancer [11-13]. Recent studies intensively support the role of polyphenols in the prevention of degenerative diseases, like cardiovascular affections and cancers. Different fruits, vegetables, cereals, olive oil, chocolate and beverages, such as green tea, and red wine represent main sources of natural polyphenols [14,15].

Together with turmeric, another exceedingly studied nutraceutical aromatic plant from *Zingiberaceae* family is *Zingiber officinale* Roscoe. Different types of extract from the root of ginger have been used in Ayurvedic and Chinese traditional herbal medicine in order to treat indigestion, vomiting, arthritis, rheumatism, pains, cramps, fever and infection [16]. The main pharmacological actions of active compounds extracted from ginger root reported by *in vitro* and *in vivo* test attributed to its active phytocompounds were: anti-inflammatory, antioxidant, antiemetic, anticancer, anticoagulant, immunomodulatory, antihyperglycemic, hypolipidemic, analgesic, and cardioprotective properties [6,16-19]. The main phytochemical constituents of the root, the vegetal product of this plant, responsible for the therapeutic action are gingerols, shogaols, paradols, gingerdiols, and zingerone [6,20]. Regarding its anticancer properties, recent studies have indicated a beneficial effect in case of liver, endometrial, ovarian and prostate cancer [21-24]. Furthermore, ginger was described as an anti-emetic agent in cancer chemotherapy [25]. Ginger was also reported to reduce the side effects of doxorubicin and cisplatin [26].

Skin cancers include basal cell carcinoma, squamous cell carcinoma and malignant melanoma. The first two types of skin cancer are the most frequent malignant neoplasms among fair-skinned population [27]. Recent studies report that in the last five decades the incidence of melanoma was also increasing especially in the case of white population [28]. Albeit malignant melanoma is less frequent than the other two types of skin cancer, is the most dangerous. It is responsible for most deaths due to its highly metastatic potential and resistance to chemotherapy [29]. One of the latest studies regarding the incidence in Europe shows that the highest rates are recorded in Nordic countries, especially in Switzerland, while Grece and other Mediterranean countries are at the opposite pole [30].

The aim of this study was to determine the total polyphenol content and the antioxidant activity of two ethanolic extracts obtained from the above mentioned species, and to perform preliminary *in vitro* tests regarding a possible antiproliferative and/or proapoptotic effect on murine melanoma cell line B164A5.

Results

Polyphenols content of the two ethanolic extracts

In the case of ethanolic extract obtained from *Curcuma rhizome*, the amount of polyphenols was 182 ± 0.6 mg GAE/g of dry plant material. A statistically significant decreased value was noticed in case of *Zingiber rhizome* ethanolic extract, namely 16 ± 0.15 mg GAE/g of dry plant material (Figure 1). Unpaired Student *t* test was used to determine the statistical difference between the two groups, the results were statistically significant (p value = 0.0097).

Antioxidant capacity

The investigation continued with the analysis of the antioxidant capacity of the two extracts. In this regard it was applied an assay based on the measurement of the reducing ability of antioxidants toward DPPH˙. Our results showed an increased antioxidant capacity for *Curcuma rhizome* ethanolic extract, 123.2 ± 5 μM T/100 g dry weight (dw) extract as compared to *Zingiber rhizome* ethanolic extract where a value of 17.7 ± 1.5 μM T/100 g dry weight (dw) (Figure 2). Unpaired Student *t* test was used to determine the statistical difference between the two groups, the results were not statistically significant (p value = 0.0249).

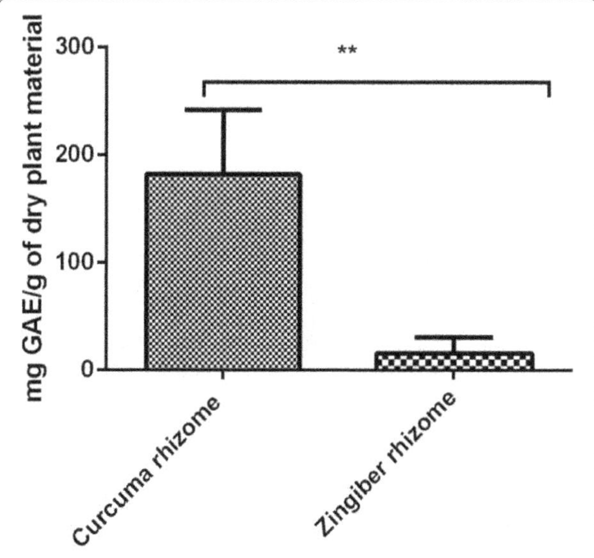

Figure 1 **Total pholiphenol content (mg P/1 ml extract) as revealed by Folin-Ciocalteu assay for *Curcuma rhizome* and *Zingiber rhizome*.**

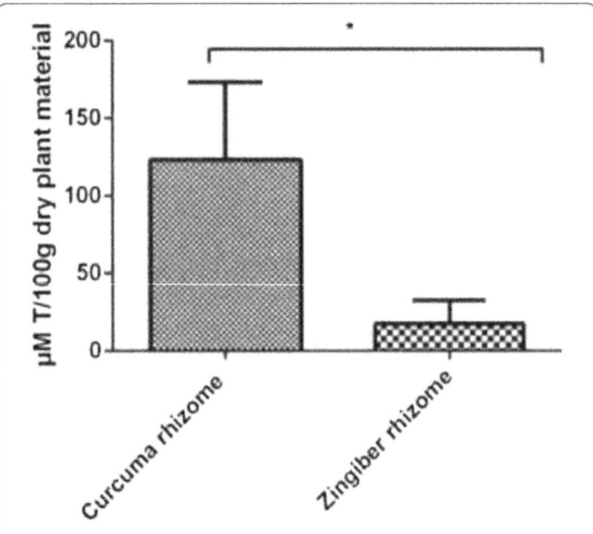

Figure 2 Antioxidant capacity (mM T/1 ml extract) as revealed by DPPH assay for *Curcuma rhizome* and *Zingiber rhizome.*

Antiproliferative activity of the two ethanolic extracts

In order to observe a possible antiproliferative or even cytotoxic effect of the two analyzed samples on murine melanoma B164A5 cell line, MTT proliferation assay was conducted as described in the Material and Methods section. 100 µg/ml of plant ethanol extracts showed after a period of incubation of 48 h an inhibition index of 38 ± 35 for *Curcuma rhizome* extract while in case of *Zingiber rhizome* extract the inhibition index was 17 ± 16 (Figure 3). Unpaired Student *t* test was used to determine the statistical difference between the two groups, the results were not statistically significant (p value = 0.4028). Preliminary studies employing lower concentrations of extracts were performed but the inhibition index was to low in order to be depicted.

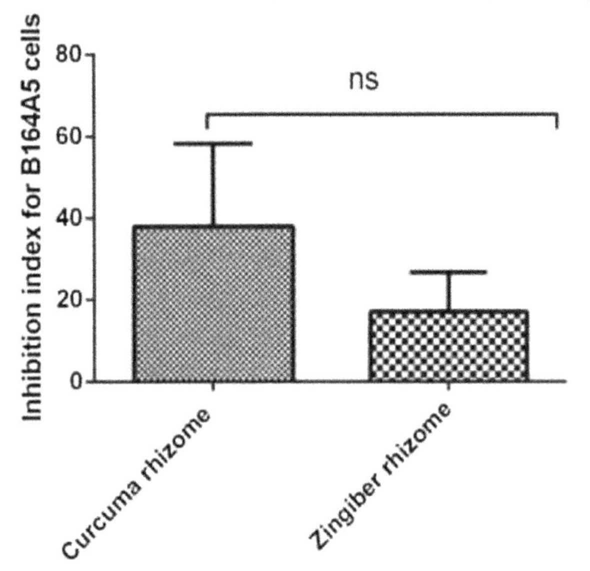

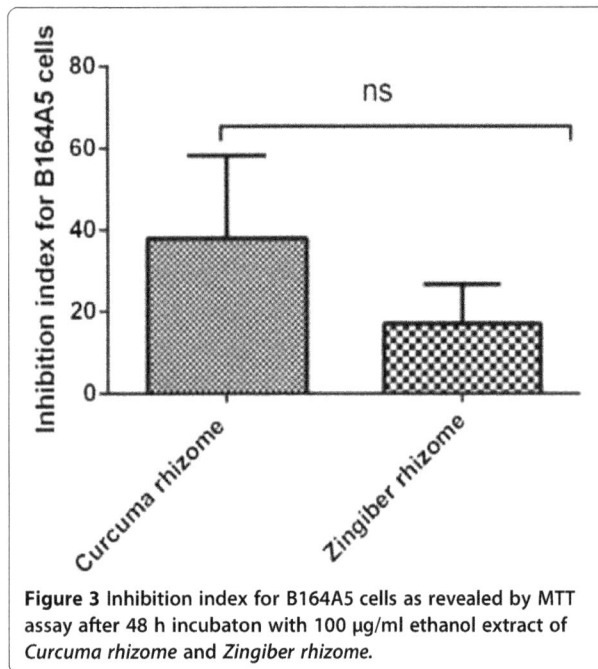

Figure 3 Inhibition index for B164A5 cells as revealed by MTT assay after 48 h incubaton with 100 µg/ml ethanol extract of *Curcuma rhizome* and *Zingiber rhizome.*

In addition, we decided to verify the apoptosis induced by the two ethanolic extracts on B164A5 cells using DAPI staining.After staining the cells with this reagent, the nucleus observed under fluorescent microscope was colored in blue. After 48 h incubation with the two extracts a number of cells presented nuclear fragmentation, condensed chromatin filaments or nuclear condensation as a sign of loss of cell membrane integrity. These characteristics were observed more frequently in case of the cells incubated with *Curcume rhizome* extract as compared to the ones incubated with *Zingiber rhizome* extract (Figure 4). Preliminary studies employing lower concentrations of extracts were performed but no or decreased signs of apoptosis could be detected.

In order to observe the apoptotic events (early apoptosis and late apoptosis) Annexin-FITC-7AAD double staining was performed. Results showed after an incubation period of 48 h a percentage of 5.935 ± 1.5 early apoptotic cells in case of *Zingiber rhizome* extract and 20.45 ± 0.77 early apoptotic cells in case of *Curcuma rhizome* extract. For late apoptotic cells the percentage was 10 ± 1.4 in case of *Zingiber rhizome* extract and 15.9 ± 1.55 in case of *Curcuma rhizome* extract (Figure 5). These results were statistically significant as revealed by Two-Way ANOVA followed by Bonferroni post test. Preliminary studies employing lower concentrations of extracts were performed but no or decreased signs of apoptosis could be detected.

Discussions

It is well known that natural polyphenols act as antioxidants therefore a diet rich in fruits, vegetables, cereals, extra virgin olive oil, red wine and tea (Mediterranean diet) would be of real practical interest in order to counteract some important pathologies like cardiovascular disease, some types of cancer, Alzheimer's disease, tooth decay or different infections [15,31,32]. Polyphenols are plant secondary metabolites and *Curcuma rhizome* and *Zingiber rhizome* have been reported as vegetal products that contain this type of phytochemicals [33-36]. It is well known that the amount of active agents is widely varying depending on the extraction conditions (solvent, temperature, extraction time) [37].

The aim of the present study was to analyze a possible anticancer potential of the two ethanolic extracts of curcuma and ginger root against B164A5 murine melanoma cell line. Extraction was performed by using ethanol as a solvent, solid: liquid ratio 1:50 at 70°C for 2 h. The extraction conditions were chosen based on a previous study conducted by the group of Surojanametakul *et al.*, who screened a wide range of parameters including solvent type, ratio and temperature, and have showed best results in terms of amount of extracted polyphenols [38].

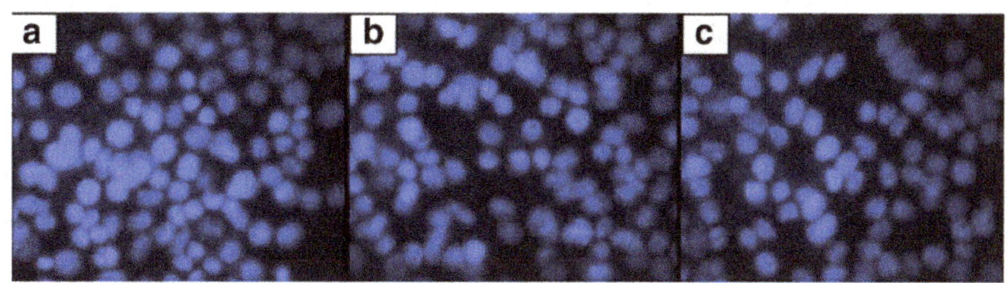

Figure 4 DAPI staining after 48 h incubaton of B164A5 cells with a) Medium; b) 100 µg/ml ethanol extract of *Curcuma rhizome*; c) 100 µg/ml ethanol extract of *Zingiber rhizome.*

In order to determine the amount of polyphenols we have chosen the Folin Ciocalteu assay. The procedure is convenient, simple, and reproducible [39]. Using other conditions not targeted to these two vegetal products, in a large screening study of 32 selected herbs, the group of Wojdyło *et al.*, detected 172 ± 0.12 mg GAE/g of dry plant material for *Curcuma rhizoma* [40]. In the same conditions as the ones we have used the group of Surojanametakul *et al.*, obtained for *Curcuma rhizoma* 146.65 mg GAE/g of dry plant material [38]. Other values for *Curcuma rhizoma* polyphenols were reported by the group of Mongkolsilp *et al.*, namely 122 mg GAE/g of dry plant material [41]. For two varieties of root of Malaysian *Zingiber officinale*, namely *Halia Bentong* and *Halia Bara*, the group of Ghasemzadeh *et al.*, recorded 10.22 ± 0.87 (*Halia Bentong*) and 13.5 ± 2.26 (*Halia Bara*) polyphenols expressed as mg gallic acid/g of dry plant material [42]. The group of Otunola *et al.*, reported a value of 22.09 total polyphenols expressed as mg gallic acid/g of dry plant material for ginger root [43].

Further investigations were conducted in order to determine the antioxidant capacity. We have measured the radical-scavenging activity of antioxidants against the free radical 1,1-diphenyl-2-picrylhydrazyl (DPPH).

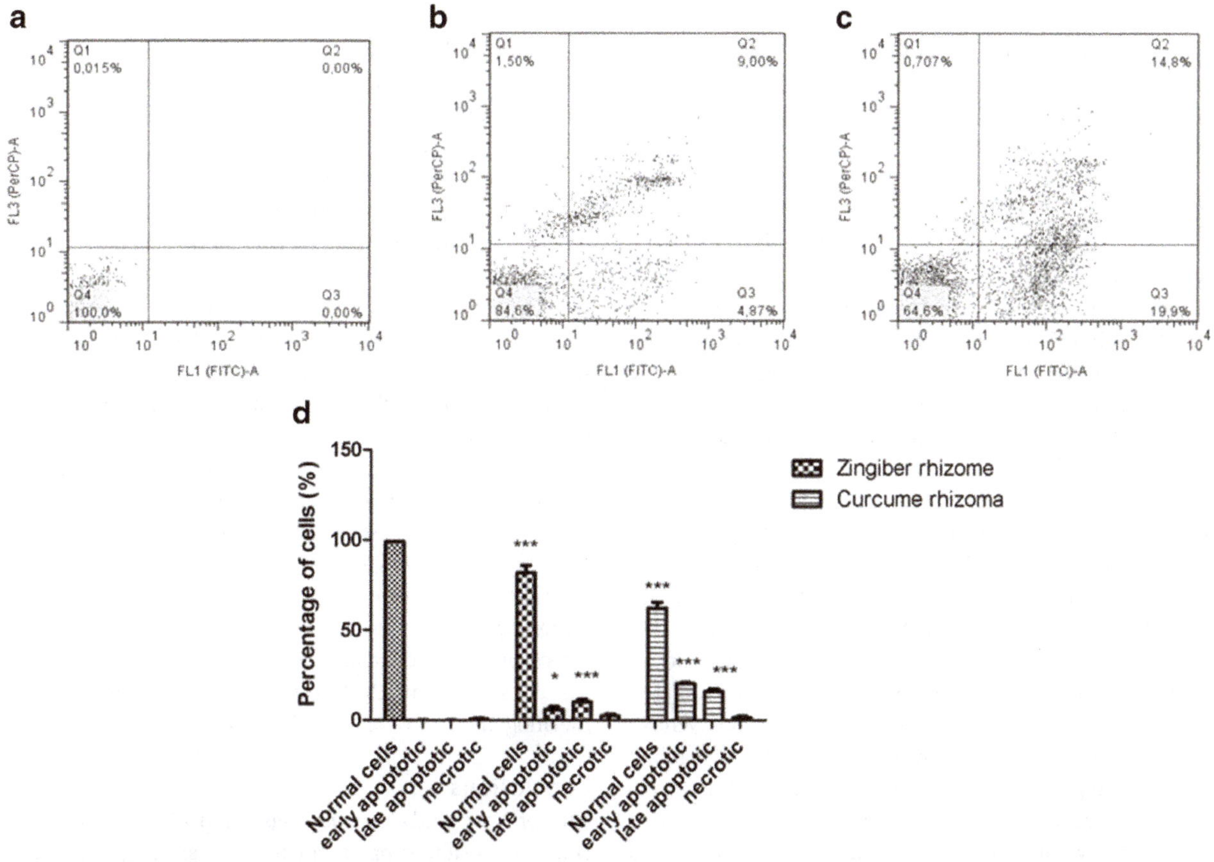

Figure 5 Annexin-FITC-7AAD double staining after 48 h incubaton of B164A5 cells with: a) Medium; b) 100 µg/ml ethanol extract of *Curcuma rhizome*; c) 100 µg/ml ethanol extract of *Zingiber rhizome*; d) The ensemble image together with statistical analyze of a,b,c.

DPPH is a very common assay used to test the antioxidant capacity of vegetal extracts [44]. Results showed an increased antioxidant capacity for *Curcuma rhizome* ethanolic extract, as compared to *Zingiber rhizome* ethanolic extract. It can be observed a direct correlation between the amount of polyphenols detected in the selected vegetal products and the antioxidant capacity. Polyphenols have been intensively investigated for their antioxidant properties. It seems that the mechanism involves the modulation of oxidative stress [14].

To reach the final aim of our study we have screen for an antiproliferative, respectively proapoptotic effect of the two ethanolic extracts on the murine melanoma B164A5 cell line. As explained in the results section MTT proliferation assay revealed that at a dose of 100 μg/ml the extracts present antiproliferative capacity. Several reports regarding the antiproliferative activity of the two ethanolic extracts have been described in the literature, but none targeted towards B164A5 murinic melanoma cells.

Aqueous turmenic extracts were found to inhibit the proliferation of cultured bovine smooth muscle cells [45]. *Curcuma longa* extract was also found to inhibit the proliferation of rat's hepatic stellate cells [46]. Additionally it was reported that the extract has cytotoxic effect with different IC_{50s} in breast and lung cancer cell lines [47,48]. Cytotoxicity was also observed in case of lymphocytes and Dalton's lymphoma [49]. Curcumin was found to sensitize pancreatic cancer cells to gemcitabine *in vitro* [50]. It was published that it inhibits proliferation and induce apoptosis in LNCaP prostate cancer cells *in vivo* [51]. Curcumin inhibits dose-dependent the SSC4-oral cancer cells, LoVo- human colorectal cancer cells, K1- papillary thyroid cancer cells, KKU100, KKU-M156, KKU-M213 - human biliary cancer cells [52-55]. Regarding melanoma, curcumin was reported as an antiproliferative agent on high metastatic B16F10 murinic melanoma cell line by targeting nucleotide phosphodiesterase 1A [56]. Furthermore curcumin was found to induce G2/M cell cycle arrest in case of human melanoma cells [57].

Ginger extract and 6-gingerol were found to inhibit the proliferation of rat colonic adenocarcinoma [58]. Moreover, it was proved that 10-shogaol, an important pharmacological compound from *Zingiber officinale* has the ability to promote growth of normal human skin cells [59]. Additionally literature reports G0/G1 arrest and apoptosis induced by ginger extract in case of HCT 116 and HT 29 colon cancer cell lines [60]. Chemopreventive efficacy of ginger extract was also described against hepatoma HepG2 and HLE cell lines [22]. Furthermore, aqueous extract of ginger was proclaimed to present antiproliferative activity in case of human non-small lung epithelium cancer (A549) cells and human cervical epithelial carcinoma (HeLa) [61].

The group of Lea *et al.*, published that ginger extract is active on some cancer cell lines like ovary -OVCAR and leukemia - K562, but not significantly active in case of other cancer cell lines like breast-MCF7, colon HT29, lung NCI460, prostate PCO3 and melanoma UACC62 [58,62].

Due to the fact that apoptosis is in balance with proliferation, we have also investigated a possible proapoptotic effect of selected extracts. DAPI staining was useful in order to observe first insights of apoptosis, but real quantification of both early and late apoptotic cells has been done by double staining Annexin-FITC-7AAD. As explained in the results section these assays revealed that at a dose of 100 μg/ml the extracts present proapoptotic capacity.

Curcuma aqueous extract was proclaimed with a proapoptotic activity against human colon carcinoma LS-174-T cells [63]. Turmerone, extracted by supercritical carbon dioxide was noted to induce apoptosis in hepatocellular carcinoma HepG2 cells [46]. The group of Ozaki *et al.*, demonstrated the role of curcumin in the induction of rabbit osteoclast apoptosis along with inhibition of bone resorption [64]. Curcumin was previously reported for its proapoptotic activity in human lung carcinoma A549 cells, NPC-TW 076, human nasopharyngeal carcinoma cells, human colon cancer colo 205 cells, murine myelomonocytic leukemia WEHI-3 cells, leukaemic Jurkat cells [63,65-67]. Curcumin was described as a proapoptotic agent against different human melanoma cells [57,68].

On the other hand, ginger extract was reported to trigger apoptosis in case of HCT 116 and HT 29 colon cancer cell lines [60]. Zerumbone, an active agent from *Zingiber aromaticum* was describd as a dose-dependent proapoptotic agent in case of HT-29, CaCo-2, and MCF-7 cancer cells [69]. Ethanolic extract of *Zingiber officinale* triggered apoptosis in case of $HepG_2$ - human hepatoma cell line [70]. [6] -paradol, a minor constituent of ginger and other related derivates were demonstrated to induce apoptosis in case of oral squamous carcinoma cell line KB and human promyelocytic leukemia (HL-60) cells [71,72]. In addition, [6]-gingerol, the main pharmacologically active principle from ginger root was reported to induce apoptosis in case of SCC-25 oral cavity cancer cell line, different human colorectal cancer cells, human cervical cancer HeLa cells and human promyelocytic leukemia (HL-60) cells [72-75]. Furthermore, terpenoids present in steam distilled extract of ginger had the capacity to induce apoptosis in endometrial cancer cells via p53 activation [22]. [6]-gingerol was reported to inhibit melanogenesis in B16F10 melanoma cells [76].

Conclusions

The present study can be considered of practical interest since our results supplement the data existing in the literature with new information regarding the antiproliferative, respectively proapoptotic effect of the selected extracts on murine melanoma B164A5 cell line. To the best of our knowledge this report was never done before. It is obvious that *Curcuma rhizome* ethanolic extract presented an increased effect in comparison with *Zingiber rhizome* ethanolic extract on B164A5 murine melanoma cell line regarding both proliferation and apoptosis. The increased anticancer activity may be correlated with the higher amount of polyphenols, respectively increased antioxidant capacity as detected by Folin Ciocalteu and DPPH assay. These preliminary findings are of great interest and further studies will be developed in order to characterize the extract and find pure active phytoconstituents that in a proper dose may exercise an increased antimelanoma activity. As a clear conclusion presented results indicate that *Curcuma rhizome,* a main representant of *Zingiberaceae* family may be a promising natural source for active compounds against malignant melanoma.

Methods

Vegetal extracts

Curcuma rhizome (*Curcuma longa* Linnaeus) and *Zingiber rhizome* (*Zingiber officinale* Roscoe) were achieved from University of Agricultural Sciences and Veterinary Medicine, Timisoara, Romania, Department of Plant Culture. Fresh rhizomes of both turmeric and ginger were cleaned, washed with deionised water, sliced and dried in the sun for one week, and dried again at 50°C in a drying stove for 6 hours. Dried rhizomes were cut in small pieces, and powdered by electronic mill. Extraction was performed by using ethanol as a solvent, solid: liquid ratio 1:50 at 70°C for 2 h [38].

The formal identification of the plant material was done by the specialist in the field Dr. Senior Lecturer Danciu Corina, Department of Pharmacognosy, Faculty of Pharmacy, University of Medicine and Pharmacy Victor Babes, Timisoara.A voucher specimen of this material has been deposited in a herbarium, available at UMFT Victor Babes, Timisoara-Department of Pharmacognosy, No. Ph-28.17.

Total polyphenolic content

Total polyphenolic content of selected vegetal products was determined by Folin Ciocalteu assay. The method is based on the measurement of optical density of a primer extract which by complexation with the Folin-Ciocalteu reagent absorbs in the visible domain at $\lambda = 720$ nm (multidetection Biotec spectofotometer, UV–VIS 190–900 nm). The necessary reagents are: bidistilled water, Folin Ciocalteu reagent (Merck, Romania), Na_2CO_3 (sodium bicarbonate) and ethanol (Sigma Aldrich, Romania). The assay was conducted on a microplate with 24 wells and the following quantities of reagents were used: 23 µl sample; 115 µl Folin Ciocalteu reagent; 345 µl Na_2CO_3 (7.5%) and 1.817 ml bidistilled water. Standard curve was done using different concentrations of gallic acid (mg/ml) as standard. Absorption at 765 nm was measured. Total phenol contents were expressed in gallic acid (Sigma Aldrich, Germany) equivalents (mg gallic acid/g dry weight - DW). All determinations were performed in triplicate.

Antioxidant capacity (DPPH assay)

This assay is based on the measurement of the reducing ability of antioxidants toward DPPH (2, 2′ diphenyl-1-picrylhydrazyl) radical. The DPPH radical-scavenging activity was determined using the method proposed by Brand-Williams *et al.* [77]. The assay was conducted on a microplate with 24 wells. DPPH (80 µM) was dissolved in pure ethanol (98%). The radical stock solution was prepared fresh. The mixture was shaken vigorously and allowed to stand at room temperature in the dark for 10 min. 200 µl of sample with 1.4 ml radical solution were added to each microplate well. The decrease in absorbance of the resulting solution was monitored at 515 nm for 30 min. The results were corrected for dilution and expressed in µM Trolox ((S)-(–)-6-hydroxy-2,5,7,8-tetramethylchroman-2- carboxylic acid) (Sigma Aldrich, Germany) per 100 g dry weight (DW). All determinations were performed in triplicate.

B164A5 melanoma cells

B164A5 cells were acquired from Sigma Aldrich (ECACC, origin Japan stored UK). The complete growth medium (culture medium) for this cells is DMEM (Dulbecco's Modified Eagle's Medium), supplemented with 10% FCS (Fetal Calf Serum), 1% Penicillin/Streptomycin mixture (Pen/Strep, 10.000 IU/ml) and 2% HEPES (4-(2-hydroxyethyl)-1-piperazineethanesulfonic acid). The cells were cultured by incubation at 37°C in 5% CO_2 atmosphere. At a confluence of 70-80% (every two or three days) the cells were passed using 0.25% Trypsin - 1 mM EDTA solution followed by centrifugation (5 minutes, 1200 rpm) and replated in T75 culture flasks at a subcultivation ratio of 1:10 to ensure optimal proliferation.

MTT proliferation assay

MTT kit was acquired from Roche, Germany. 100 µl cell suspension containing 6×10^3 B164A5 melanoma cells were seeded onto a 96-well microplate and attached to the bottom of the well overnight. Afterwards 100 µl of new medium containing 10% FCS and 100 µg/ml of plant ethanol extracts were added. After 48 h of incubation, 10 µl of MTT reagent from a stock solution of 5 mg/ml were added. The intact mitochondrial reductase converted and precipitated MTT as purple crystals during a 4 h contact period. After four hours the precipitated crystals were dissolved in 100 µl of solubilisation solution. Finally, the reduced MTT was spectrophotometrically analyzed at 570 nm, using a reference of 656 nm using an ELISA reader. Inhibition index was calculated as 1-absorbance sample X/absorbance sample blank.

DAPI (4, 6-Diamidino-2-phenylindole) staining

B16 cells were seeded at a concentration of 5×10^4 in a chamber slide system formed of 8 well glass slides in culture medium. After 24 h cells were incubated in medium containing 10%FCS and 100 µg/ml of plant ethanol extracts. The total volume added in chamber was 400 µl. Cells were incubated for 48 h and afterwards the medium was removed. Cells were washed with PBS and afterwards 400 µl of staining solution were added in each well. The staining solution consisted of a mixture of methanol and DAPI (Roche) as follows: 1 ml methanol: 2 µl DAPI (from a stock solution of 1 mg/ml). Cells were incubated for 5 minutes with this staining solution and afterwards washed with PBS and analyzed by fluorescence microscopy.

Annexin-FITC-7AAD double staining

Cells were cultivated in a 6 well plates at a density of 80% using normal medium. After 24 h medium containing 10% FCS and 100 µg/ml of plant ethanol extracts were added. After 48 h cells were detached using trypsin, washed with ice cold PBS and resuspended in 500 µl Annexin binding buffer (1.19 g HEPES NaOH pH =7.4; 4.09 g NaCl; 0.138 g $CaCl_2$ in 50 ml distillated water and diluted 1:10) at a concentration of 1×10^6 cells/ml. Cells were centrifugate 5 min at 1200 rpm, the supernatant was discarded and the cells were resuspended in 70 µl of Annexin binding buffer. 5 µl Annexin V-FITC (ImmunoTools) and 5 µl 7AAD (ImmunoTools) were added and cells were incubated 15 min on ice and in the dark. Samples were measured by FACS on FL1 and FL3 fluorescence channels using a BD Canto II FACS DIVA device. Untreated cells were used as negative control and cells treated with 500 nM staurosporine (LC laboratories) for 24 h were used for the compensations. Flow Jo soft (7.6.3) was used for data analysis.

Statistics

Unpaired Student t test or Two-Way ANOVA followed by Bonferroni post test were used to determine the statistical difference between various experimental and control groups.*, ** and *** indicate $p < 0.05$, $p < 0.01$ and $p < 0.001$ compared to control group. Results are presented as mean ± standard deviation (SD).

Abbreviations

DW: Dry weight; DPPH, 2: 2' diphenyl-1-picrylhydrazyl; DMEM: Dulbecco's Modified Eagle's Medium; FCS: Fetal Calf Serum; HEPES: 4-(2-hydroxyethyl)-1-piperazineethanesulfonic acid; DAPI: 4, 6-Diamidino-2-phenylindole; PBS: Phosphate buffered saline; GAE: Gallic acid equivalents; B164A5: Murine melanoma cells.

Competing interests

The authors declare that they have no competing interests.

Authors' contributions

CD, VV, MH and FF performed the experiments, analyzed/interpreted data and drafted the manuscript. DEC, SAC, CMS and IM analyzed/interpreted data. VV, CD and CT conceived the study, the experiments, also analyzed/interpreted data and finalized the manuscript. All co-authors reviewed and discussed the paper. All authors read and approved the final manuscript.

Acknowledgments

This study was published under the frame of European Social Found, Human Resources Development Operational Programme 2007–2013, project no. POSDRU/159/1.5/S/136893 obtained by postdoc. Danciu Corina.

Author details

[1]Department of Pharmacognosy, Faculty of Pharmacy, University of Medicine and Pharmacy Victor Babes", Eftimie Murgu Square, No. 2, Timisoara 300041, Romania. [2]Department of Pharmaceutical Technology, Faculty of Pharmacy, University of Medicine and Pharmacy Victor Babes", Eftimie Murgu Square, No. 2, Timisoara 300041, Romania. [3]Department of Chemistry and Biochemistry, University of Agricultural Sciences and Veterinary Medicine of Cluj-Napoca, Mănăştur Str.,No. 3-5, Cluj-Napoca 400372, Romania. [4]Department of Pharmacognosy, Faculty of Pharmacy, University of Medicine and Pharmacy "Gr.T.Popa", Iasi, Romania. [5]Department of Toxicology, Faculty of Pharmacy, University of Medicine and Pharmacy "Victor Babes", Eftimie Murgu Square, No. 2, Timisoara 300041, Romania. [6]Department of Pharmaceutical Chemistry, Faculty of Pharmacy, University of Medicine and Pharmacy "Victor Babes", Eftimie Murgu Square, No. 2, Timisoara 300041, Romania. [7]Faculty of Medicine, University of Medicine and Pharmacy "Victor Babes", Eftimie Murgu Square, No. 2, Timisoara 300041, Romania. [8]Department of Organic Chemistry, Faculty of Pharmacy, University of Medicine and Pharmacy "Victor Babes", Eftimie Murgu Square, No. 2, Timisoara 300041, Romania.

References

1. Kress WJ, Prince LM, Williams KJ. The phylogeny and a new classification of the gingers (Zingiberaceae): evidence from molecular data. Am J Bot. 2002;89:1682–96.
2. Tushar, Basak S, Sarma GC, Rangan L. Ethnomedical uses of Zingiberaceous plants of Northeast India. J Ethnopharmacol. 2010;132:286–96.
3. Chainani-Wu N. Safety and anti-inflammatory activity of curcumin: a component of tumeric (Curcuma longa). J Altern Complement Med. 2003;9:161–8.
4. Araújo CC, Leon LL. Biological activities of Curcuma longa L. Mem Inst Oswaldo Cruz. 2011;96:723–8.
5. Boaz M, Leibovitz E, Dayan YB, Wainstein J. Functional foods in the treatment of type 2 diabetes: olive leaf extract, turmeric and fenugreek, a qualitative review. Funct Foods Health Dis. 2011;1:472–81.
6. Mishra S, Palanivelu K. The effect of curcumin (turmeric) on Alzheimer's disease: an overview. Ann Indian Acad Neurol. 2008;11:13–9.
7. Ramadan G, Al-Kahtani MA, El-Sayed WM. Anti-inflammatory and anti-oxidant properties of Curcuma longa (turmeric) versus Zingiber officinale (ginger) rhizomes in rat adjuvant-induced arthritis. Inflammation. 2011;34:291–301.
8. Ringman JM, Frautschy SA, Cole GM, Masterman DL, Cummings JL. A potential role of the curry spice curcumin in Alzheimer's disease. Curr Alzheimer Res. 2005;2:131–6.
9. Sharma RA, Gescher AJ, Steward WP. Curcumin: the story so far. Eur J Cancer. 2005;41:1955–68.
10. Maheshwari RK, Singh AK, Gaddipati J, Srimal RC. Multiple biological activities of curcumin: a short review. Life Sci. 2006;78:2081–7.
11. Bar-Sela G, Epelbaum R, Schaffer M. Curcumin as an anti-cancer agent: review of the gap between basic and clinical applications. Curr Med Chem. 2010;17:190–7.
12. Darvesh AS, Aggarwal BB, Bishayee A. Curcumin and liver cancer: a review. Curr Pharm Biotechnol. 2012;13:218–28.
13. Shehzad A, Lee J, Lee YS. Curcumin in various cancers. Biofactors. 2013;39:56–68.

14. Scalbert A, Johnson IT, Saltmarsh M. Polyphenols: antioxidants and beyond. Am J Clin Nutr. 2005;81:215S–7.
15. D'Archivio M, Filesi C, Di Benedetto R, Gargiulo R, Giovannini C, Masella R. Polyphenols, dietary sources and bioavailability. Ann Ist Super Sanita. 2007;43:348–61.
16. Ali BH, Blunden G, Tanira MO, Nemmar A. Some phytochemical, pharmacological and toxicological properties of ginger (Zingiber officinale Roscoe): a review of recent research. Food Chem Toxicol. 2008;46:409–20.
17. Nicoll R, Henein MY. Ginger (Zingiber officinale Roscoe): a hot remedy for cardiovascular disease? Int J Cardiol. 2009;131:408–9.
18. Li Y, Tran VH, Duke CC, Roufogalis BD. Preventive and protective properties of *zingiber officinale* (ginger) in diabetes mellitus, diabetic complications, and associated lipid and other metabolic disorders: a brief review. Evid Based Complement Altern Med. 2012;2012:516870.
19. Siddaraju MN, Dharmesh SM. Inhibition of gastric H+, K+ −ATPase and Helicobacter pylori growth by phenolic antioxidants of Zingiber officinale. Mol Nutr Food Res. 2007;51:324–32.
20. Wilson R, Haniadka R, Sandhya P, Palatty PL, Baliga MS. Ginger (Zingiber officinale Roscoe) the Dietary Agent in Skin Care: A Review. In: Watson RR, Zibadi S, editors. Bioactive Dietary Factors and Plant Extracts in Dermatology. Karnataska: Humana Press; 2013. p. 103–11.
21. Habib SH, Makpol S, Hamid NAA, Das S, Ngah WZ, Yusof YA. Ginger extract (Zingiber Officinale) has anti-cancer and anti-inflammatory effects on ethionine-induced hepatoma rats. Clinics. 2008;63:807–13.
22. Liu Y, Whelan RJ, Pattnaik BR, Ludwig K, Subudhi E, Rowland H, et al. Terpenoids from Zingiber officinale (Ginger) induce apoptosis in endometrial cancer cells through the activation of p53. PLoS One. 2012;7: e53178. doi:10.1371/journal.pone.0053178.
23. Rhode J, Fogoros S, Zick S, Wahl H, Griffith KA, Huang J, et al. Ginger inhibits cell growth and modulates angiogenic factors in ovarian cancer cells. BMC Complement Altern Med. 2007;7:44.
24. Brahmbhatt M, Gundala SR, Asif G, Shamsi SA, Aneja R. Ginger phytochemicals exhibit synergy to inhibit prostate cancer cell proliferation. Nutr Cancer. 2013;65:263–372.
25. Haniadka R, Rajeev AG, Palatty PL, Arora R, Baliga MS. Zingiber officinale (ginger) as an anti-emetic in cancer chemotherapy: a review. J Altern Complement Med. 2012;18:440–4.
26. Pereira MM, Haniadka R, Chacko PP, Palatty PL, Baliga MS. Zingiber officinale Roscoe (ginger) as an adjuvant in cancer treatment: a review. J BUON. 2011;16:414–24.
27. Saladi RN, Persaud AN. The causes of skin cancer: a comprehensive review. Drugs Today. 2005;41:37–53.
28. Erdmann F, Lortet-Tieulent J, Schüz J, Zeeb H, Greinert R, Breitbart EW, et al. International trends in the incidence of malignant melanoma 1953–2008–are recent generations at higher or lower risk? Int J Cancer. 2013;132:385–400.
29. Wang L, Shi Y, Ju P, Liu R, Yeo SP, Xia Y, et al. Silencing of diphthamide synthesis 3 (Dph3) reduces metastasis of murine melanoma. PLoS One. 2012;7:e49988.
30. Forsea AM, Del Marmol V, de Vries E, Bailey EE, Geller AC. Melanoma incidence and mortality in Europe: new estimates, persistent disparities. Br J Dermatol. 2012;167:1124–30.
31. Santangelo C, Varì R, Scazzocchio B, Di Benedetto R, Filesi C, Masella R. Polyphenols, intracellular signalling and inflammation. Ann Ist Super Sanita. 2007;43:394–405.
32. Ferrazzano G, Amato I, Ingenito A, Zarrelli A, Pinto G, Pollio A. Plant polyphenols and their anti-cariogenic properties: a review. Molecules. 2011;16:1486–507.
33. Avwioro OG, Onwuka SK, Moody JO, Agbedahunsi JM, Oduola T, Ekpo OE, et al. Curcuma longa extract as a histological dye for collagen fibres and red blood cells. J Anat. 2007;210:600–3.
34. Prathapan A, Lukhman M, Arumughan C, Sundaresan A, Raghu KG. Effect of heat treatment on curcuminoid, colour value and total polyphenols of fresh turmeric rhizome. Int J Food Sci Technol. 2009;44:1438–44.
35. Miquel J, Bernd A, Sempere JM, Díaz-Alperi J, Ramírez A. The curcuma antioxidants: pharmacological effects and prospects for future clinical use. A review. Arch Gerontol Geriatr. 2002;34:37–46.
36. Ancy J, Thayumanavan B, Pournami PR. Characterization of polyphenol oxidase in ginger (Zingiber officinale R.). J Spices and Aromatic Crops. 2012;21:33–41.
37. Sasidharan S, Chen Y, Saravanan D, Sundram KM, Yoga Latha L. Extraction, isolation and characterization of bioactive compounds from plants' extracts. Afr J Tradit Complement Altern Med. 2010;8:1–10.
38. Surojanametakul V, Satmalee P, Saengprakai J, Siliwan D, Wattanasiritham L. Preparation of curcuminoid powder from turmeric root *(Curcuma Longa* Linn) for food ingredient Use. Kasetsart J (Nat Sci). 2010;44:123–30.
39. Cirilo G, Iemma F. Antioxidant Polymers: Synthesis, Properties, and Applications. Calabria: John Wiley & Sons; 2012.
40. Wojdyło A, Oszmiański J, Czemerys R. Antioxidant activity and phenolic compounds in 32 selected herbs. Food Chem. 2007;105:940–9.
41. Mongkolsilp S, Pongbupakit I, Sae-Lee N, Sitthihawom W. Radical scavenging activity and total phenolic content of medicinal plants used in primary health care. SWU J Pharm Sci. 2004;9:32–5.
42. Ghasemzadeh A, Jaafar HZE, Rahmat A. Antioxidant activities, total phenolics and flavonoids content in two varieties of Malaysia young ginger (Zingiber officinale Roscoe). Molecules. 2010;15:4324–33.
43. Otunola GA, Afolayan AJ. Evaluation of the polyphenolic contents and antioxidant properties of aqueous extracts of garlic, ginger, cayenne pepper and their mixture. J Applied Bot Food Qual. 2013;86:66–70.
44. Pyrzynska K, Pękal A. Application of free radical diphenylpicrylhydrazyl (DPPH) to estimate the antioxidant capacity of food samples. Anal Methods. 2013;5:4288–95.
45. Zhang W, Liu D, Wo X, Zhang Y, Jin M, Ding Z. Effects of Curcuma Longa on proliferation of cultured bovine smooth muscle cells and on expression of low density lipoprotein receptor in cells. Chin Med J. 1999;112:308–11.
46. Cheng SB, Wu LC, Hsieh YC, Wu CH, Chan YJ, Chang LH, et al. Supercritical carbon dioxide extraction of aromatic turmerone from Curcuma longa Linn. induces apoptosis through reactive oxygen species-triggered intrinsic and extrinsic pathways in human hepatocellular carcinoma HepG2 Cells. J Agric Food Chem. 2012;60:9620–30.
47. Ranjbari J, Alibakhshi A, Arezumand R, Pourhassan-Moghaddam M, Rahmati M, Zarghami N, et al. Effects of curcuma Longa extract on telomerase activity in lung and breast cancer cells. Zahedan J Res Med Sci. 2013;16:29–34.
48. Mohammad P, Nosratollah Z, Mohammad R, Abbas A, Javad R. The inhibitory effect of Curcuma longa extract on telomerase activity in A549 lung cancer cell line. Afr J Biotechnol. 2010;9:912–9.
49. Kuttan R, Bhanumathy P, Nirmala K, George MC. Potential anticancer activity of turmeric (Curcuma longa). Cancer Lett. 1985;29:197–202.
50. Kunnumakkara AB, Guha S, Krishnan S, Diagaradjane P, Gelovani J, Aggarwal BB. Curcumin potentiates antitumor activity of gemcitabine in an orthotopic model of pancreatic cancer through suppression of proliferation, angiogenesis, and inhibition of nuclear factor-κappaB-regulated gene products. Cancer Res. 2007;67:3853–61.
51. Dorai T, Cao YC, Dorai B, Buttyan R, Katz AE. Therapeutic potential of curcumin in human prostate cancer. III. Curcumin inhibits proliferation, induces apoptosis, and inhibits angiogenesis of LNCaP prostate cancer cells in vivo. Prostate. 2001;47:293–303.
52. Chen JW, Tang YL, Liu H, Zhu ZY, Lü D, Geng N, et al. Anti-proliferative and anti-metastatic effects of curcumin on oral cancer cells. Hua Xi Kou Qiang Yi Xue Za Zhi. 2011;29:83–6.
53. Guo L, Chen XJ, Hu YH, Yu ZJ, Wang D, Liu JZ. Curcumin inhibits proliferation and induces apoptosis of human colorectal cancer cells by activating the mitochondria apoptotic pathway. Phytother Res. 2013;27:422–30.
54. Song F, Zhang L, Yu HX, Lu RR, Bao JD, Tan C, et al. The mechanism underlying proliferation-inhibitory and apoptosis-inducing effects of curcumin on papillary thyroid cancer cells. Food Chem. 2012;132:43–50.
55. Prakobwong S, Gupta SC, Kim JH, Sung B, Pinlaor P, Hiraku Y, et al. Curcumin suppresses proliferation and induces apoptosis in human biliary cancer cells through modulation of multiple cell signaling pathways. Carcinogenesis. 2011;32:1372–80.
56. Abusnina A, Keravis T, Yougbaré I, Bronner C, Lugnier C. Anti-proliferative effect of curcumin on melanoma cells is mediated by PDE1A inhibition that regulates the epigenetic integrator UHRF1. Mol Nutr Food Res. 2011;55:1677–89.
57. Zheng M, Ekmekcioglu S, Walch ET, Tang CH, Grimm EA. Inhibition of nuclear factor-kappaB and nitric oxide by curcumin induces G2/M cell cycle arrest and apoptosis in human melanoma cells. Melanoma Res. 2004;14:165–71.
58. Brown AC, Shah C, Liu J, Pham JT, Zhang JG, Jadus MR. Ginger's (Zingiber officinale Roscoe) inhibition of rat colonic adenocarcinoma cells proliferation and angiogenesis in vitro. Phytother Res. 2009;23:640–5.
59. Chen CY, Cheng KC, Chang AY, Lin YT, Hseu YC, Wang HM. 10-shogaol, an antioxidant from zingiber officinale for skin cell proliferation and migration enhancer. Int J Mol Sci. 2012;13:1762–77.

60. Abdullah S, Abidin SAZ, Murad NA, Makpol S, Ngah WZW, Yusof YAM. Ginger extract (Zingiber officinale) triggers apoptosis and G0/G1 cells arrest in HCT 116 and HT 29 colon cancer cell lines. Afr J Biochem Res. 2010;4:134–42.
61. Choudhury D, Das A, Bhattacharya A, Chakrabarti G. Aqueous extract of ginger shows antiproliferative activity through disruption of microtubule network of cancer cells. Food Chem Toxicol. 2010;48:2872–80.
62. Leal PF, Braga ME, Sato DN, Carvalho JE, Marques MO, Meireles MA. Functional properties of spice extracts obtained via supercritical fluid extraction. J Agric Food Chem. 2003;51:2520–5.
63. Su CC, Lin JG, Li TM, Chung JC, Yang JS, Ip SW, et al. Curcumin-induced apoptosis of human colon cancer colo 205 cells through the production of ROS, Ca2+ and the activation of caspase-3. Anticancer Res. 2006;26:4379–89.
64. Ozaki K, Kawata Y, Amano S, Hanazawa S. Stimulatory effect of curcumin on osteoclast apoptosis. Biochem Pharmacol. 2000;59:1577–81.
65. Lin SS, Huang HP, Yang JS, Wu JY, Hsia TC, Lin CC, et al. DNA damage and endoplasmic reticulum stress mediated curcumin-induced cell cycle arrest and apoptosis in human lung carcinoma A-549 cells through the activation caspases cascade- and mitochondrial-dependent pathway. Cancer Lett. 2008;272:77–90.
66. Huang AC, Chang CL, Yu CS, Chen PY, Yang JS, Ji BC, et al. Induction of apoptosis by curcumin in murine myelomonocytic leukemia WEHI-3 cells is mediated via endoplasmic reticulum stress and mitochondria-dependent pathways. Environ Toxicol. 2013;28:255–66.
67. Korwek Z, Bielak-Zmijewska A, Mosieniak G, Alster O, Moreno-Villanueva M, Burkle A, et al. DNA damage-independent apoptosis induced by curcumin in normal resting human T cells and leukaemic Jurkat cells. Mutagenesis. 2013;28:411–6.
68. Bush JA, Cheung Jr KJ, Li G. Curcumin induces apoptosis in human melanoma cells through a Fas receptor/caspase-8 pathway independent of p53. Exp Cell Res. 2001;271:305–14.
69. Kirana C, McIntosh GH, Record IR, Jones GP. Antitumor activity of extract of Zingiber aromaticum and its bioactive sesquiterpenoid zerumbone. Nutr Cancer. 2003;45:218–25.
70. Harliansyah H, Murad NA, Ngah WZW, Yusof YAM. Antiproliferative, antioxidant and apoptosis effects of Zingiber officinale and 6-gingerol on HepG2 cells. Asian J Biochem. 2007;2:421–6.
71. Keum YS, Kim J, Lee KH, Park KK, Surh YJ, Lee JM, et al. Induction of apoptosis and caspase-3 activation by chemopreventive [6]-paradol and structurally related compounds in KB cells. Cancer Lett. 2002;177:41–7.
72. Lee E, Surh YJ. Induction of apoptosis in HL-60 cells by pungent vanilloids, [6]-gingerol and [6]-paradol. Cancer Lett. 1998;134:163–8.
73. Kim J. [6]-Gingerol induces apoptosis in oral cavity cancer cells. Otolaryngol Head Neck Surg. 2010;143:155–5.
74. Lee SH, Cekanova M, Baek SJ. Multiple mechanisms are involved in 6-gingerol-induced cell growth arrest and apoptosis in human colorectal cancer cells. Mol Carcinog. 2008;47:197–208.
75. Chakraborty D, Bishayee K, Ghosh S, Biswas R, Mandal SK, Khuda-Bukhsh AR. [6]-Gingerol induces caspase 3 dependent apoptosis and autophagy in cancer cells: drug-DNA interaction and expression of certain signal genes in HeLa cells. Eur J Pharmacol. 2012;694:20–9.
76. Huang HC, Chiu SH, Chang TM. Inhibitory effect of [6]-gingerol on melanogenesis in B16F10 melanoma cells and a possible mechanism of action. Biosci Biotechnol Biochem. 2011;75:1067–72.
77. Brand-Williams W, Cuvelier ME, Berset C. Use of free radical method to evaluate antioxidant activity. Lebensm Wiss Technol. 1995;28:25–30.

15

Effect of chlorocholine chlorid on phenolic acids accumulation and polyphenols formation of buckwheat plants

Oksana Sytar[1*†], Asel Borankulova[2†], Irene Hemmerich[3], Cornelia Rauh[3] and Iryna Smetanska[3,4]

Abstract

Background: Effect of chlorocholine chloride (CCC) on phenolic acids composition and polyphenols accumulation in various anatomical parts (stems, leaves and inflorescences) of common buckwheat (*Fagopyrum esculentum* Moench) in the early stages of vegetation period were surveyed.

Results: Treatment of buckwheat seeds with 2% of CCC has been increased content of total phenolics in the stems, leaves and inflorescences. On analyzing the different parts of buckwheat plants, 9 different phenolic acids – vanilic acid, ferulic acid, trans-ferulic acid, chlorogenic acid, salycilic acid, cinamic acid, *p*-coumaric acid, *p*-anisic acid, methoxycinamic acid and catechins were identified. The levels of identified phenolic acids varied not only significantly among the plant organs but also between early stages of vegetation period. Same changes as in contents of chlorogenic acid, ferulic acid, *trans*-ferulic acid were found for content of salycilic acid. The content of these phenolic acids has been significant increased under effect of 2% CCC treatment at the phase I (formation of buds) in the stems and at the phase II (beginning of flowering) in the leaves and then inflorescences respectively. The content of catechins as potential buckwheat antioxidants has been increased at the early stages of vegetation period after treatment with 2% CCC.

Conclusions: The obtained results suggest that influence of CCC on the phenolics composition can be a result of various mechanisms of CCC uptake, transforming and/or its translocation in the buckwheat seedlings.

Keywords: Chlorocholine chloride, Phenolic acids, Catechins, Buckwheat

Background

Buckwheat achene contain mostly carbohydrate, especially starch, 55,8% [1]. The starch content in buckwheat grains is 55,8%, in bran 40,7% and in the flour 78,4%. The protein content 11,7% in buckwheat grains, in bran 21,6% and in the flour 10,6% [1] Buckwheat protein content for different buckwheat species near 11-15% which is similar to the protein content in cereal grains. However, in cereals, 10-20% of the protein lies in the embryo, while 80-90% is found in the endosperm. In buckwheat 55% of the protein is located in embryo, 35% in the endosperm, and the reminder found in the hull [2]. Buckwheat can be grown in a nutrient-poor soil and requires only a short period from seeding to harvesting [3] and can be used as source for bread processing, especially development technology of using buckwheat seedlings with high phenolics content and high antioxidative capacities for bread processing.

In recent years, dietary plants such as buckwheat have attracted attention because they contain antioxidants that protect the human body from oxidative damage caused by free radicals. Buckwheat grain has a higher antioxidative activity than other cereal grains [4]; its antioxidative compounds include vitamins such as vitamins B1, B2 and E, as well as several phenolic compounds, which are found in the organs of buckwheat (leaf, stem and inflorescence) such as rutin, quercetin and proanthocyanidines (condensed tannins) [5,6].

Tissues of buckwheat seedlings accumulate large concentration of various phenolic compounds [7]. In buckwheat

* Correspondence: s-pi-r@hotmail.com
†Equal contributors
[1]Plant Physiology and Ecology Department, Taras Shevchenko National University of Kyiv, Institute of Biology, Volodymyrskya str., 64, Kyiv 01033, Ukraine
Full list of author information is available at the end of the article

seedlings has been found high content of chlorogenic acid [8] which is the most important cinnamic acid derivative [9]. In same time chlorogenic acid is the most potent functional inhibitor of the microsomal glucose-6-phosphate translocase (G6PT), is thought to possess cancer chemopreventive properties. It is also a promising precursor compound for the development of medicine that can resist AIDS virus HIV.

The phenolic compounds are important for plant due to their various biological functions including UV protection, pollen tube growth, antimicrobial activity, and insect resistance [10]. Simple phenolic acids such as *trans*-cinnamic and *p*-coumaric acids are precursors for more complex compounds including flavonoids, tannins, lignins and anthocyanins [10]. A series of naturally occurring phenolic acids with recognized anti-oxidant properties (derivatives of caffeic acid, rosmarinic acid, and trolox) have been conjugated with choline to account for the recognition by acetylcholinesterase. The synthesized hybrid compounds evidenced acetylcholinesterase inhibitory capacity of micromolar range (rationalized by molecular modeling studies) and good antioxidant properties and effects on human neuroblastoma cells for example [11].

CCC is an anti-gibberellin growth retardant. Treatment with 1.6 mM CCC resulted in the improved photosystem II (PSII) tolerance to UVB radiation, an increase in the contents of cytokinins, abscisic acid, and H_2O_2, which is one of molecule reactive oxygen species [12]. Exogenous CCC treatment has been found to improve crop performance under suboptimal growth conditions [13]; the high phenylalanine content in radish seedlings has been found under CCC treatment [14]. CCC controls the anthocyanin synthesis at the level of precursors so therefore it would be important for development of use CCC in the agriculture practice to study effect of CCC treatment on phenolic acids composition and their content in the plant crop such as buckwheat.

The changes of dynamics of total phenolics and phenolic acids formation in various anatomical parts (stems, leaves, inflorescences) of common buckwheat (*Fagopyrum esculentum* Moench.) in the early stage of vegetation period were surveyed what can be useful for development technology of growth buckwheat seedlings with high phenolics content and antioxidative capacities.

Results and discussion

Total phenolics

Phenolic compounds are plant metabolites characterized by the presence of several phenol groups. Some of them are very reactive in neutralizing free radicals by donating a hydrogen atom or an electron, chelating metal ions in aqueous solutions [15].

It was found that total phenolics content in leaves and stems of almond varieties changed according to season and plant organ [16]. In the variants with CCC treatment of buckwheat plants at phase I (formation of buds) it was visible tendency of increasing content of total phenolics in the leaves on 9% compared to the control, Figure 1.

In the phase II (at the beginning of flowering) in the stems of buckwheat plants has been observed tendency of increasing total phenolics content. Content of total phenolics in the leaves of buckwheat plants has been found tendency of increasing on 8% compared to the control. Content of total phenolics in the inflorescences in variant with CCC treatment was tendency of increasing on 8% compared to the control. Exogenous chlorocholine chloride (CCC) treatment has been found to improve crop performance under suboptimal growth conditions; however, the physiological mechanisms underlying the beneficial effects have not been fully understood. The treatment with certain concentration of CCC (e.g. 1.5–2.0 g L^{-1}) improves mineral nutrition and superoxide dismutases, peroxidase and catalase activities in potato leaves; which might have contributed to the higher tuber yield of the crop grown under suboptimal conditions [13]. The increasing of superoxide dismutases, peroxidase and catalase activities can be connected with developing oxidative stress and increasing of reactive oxygen species (ROS) in the plant tissues under CCC treatment. We suppose that in this case is possible to expect increasing of total phenolic content in variants with CCC treatment as stress response reaction for neutralization of ROS. It was found that CCC promotes anthocyanin synthesis in radish plants at the early stages of growth. A higher amount of total free amino acids, in particular phenylalanine, was present in CCC-treated seedlings compared to controls grown on distilled water [14]. The first reaction in the phenylpropanoid pathway is catalyzed by phenylalanine ammonia-lyase (PAL; EC 4.3.1.5.) converting L-phenylalanine to *trans*-cinnamic acid [17]. It has been shown that *de novo* synthesis of phenylalanine ammonia-lyase isoforms is induced by biotic and abiotic elicitors [18,19] and in a case with CCC it's just confirmed increasing of total phenolic content in the buckwheat experimental variant in the early stages of vegetation period.

Phenolic acids content and composition

On analyzing with HPLC the different parts of buckwheat plants, 9 different phenolic acids – vanilic acid, ferulic acid, *trans*-ferulic acid, chlorogenic acid, salycilic acid, cinamic acid, *p*-coumaric acid, *p*-anisic acid, methoxycinamic acid were identified (Tables 1 and 2).

The content of vanilic acid in the stems of buckwheat plants after treatment with 2% CCC has been increasing 25% and in the leaves at twice at the phase I (formation of buds). Then in the buckwheat leaves at phase II (at the beginning of flowering) has been shown decreasing of vanilic acid content on 16% compared to the control.

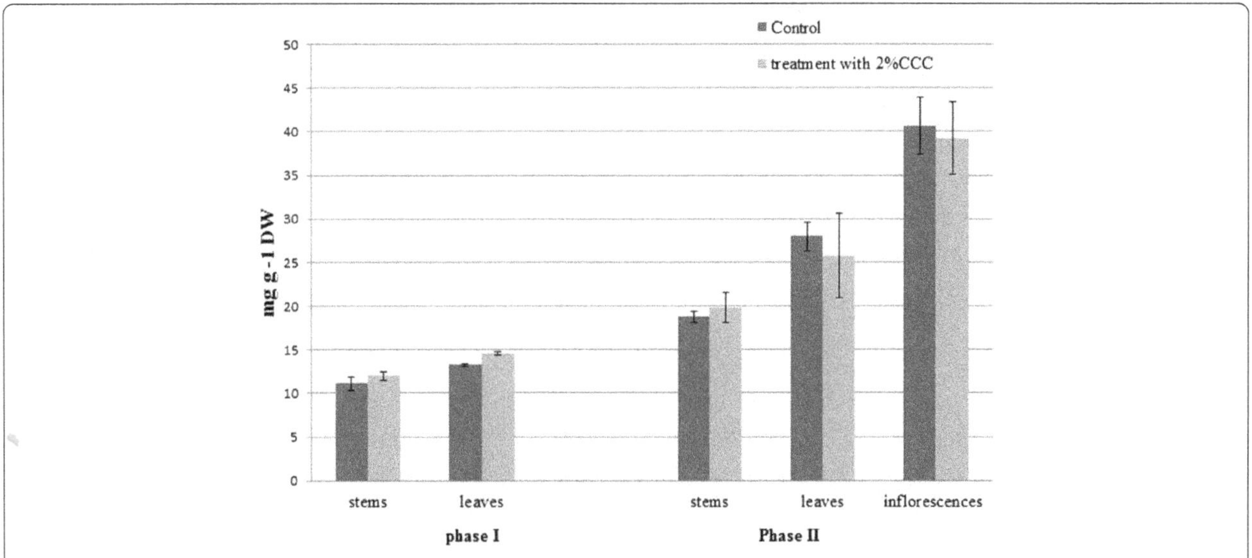

Figure 1 Total phenolics content in buckwheat plants at the phase I (formation of buds) and phase II (at the beginning of flowering) treated with different 2% CCC.

Such decreasing can be connected with redistribution and increasing of vanilic acid content in the inflorescences as estimated content of vanilic acid there was higher more than 2 times compared to the control.

The effect of CCC also was visible in the changes of chlorogenic acid content in buckwheat plants at the early stages of vegetation period. It can evidence about stress response of buckwheat plants on CCC treatment as chlorogenic acid is an important antioxidant in plants, which can protects against lipid peroxidation [20]. At the phase I (formation of buds) in the stems content of chlorogenic acid has been increased on 9% and at the phase II (at the beginning of flowering) it was higher twice compared to the control. At the phase I (formation of buds) content of chlorogenic acid was significant higher in the leaves on 8% and in the inflorescences at 2 times compared to the control.

Major wall-bound phenolics were 3,4-dihydroxybenzoic acid, p-coumaric acid and ferulic acid [21]. In the leaves at the phase I (formation of buds) content of *p*-coumaric acid has been shown tendency of increasing on 8% compared to the control. At the same time the content of *p*-coumaric acid in the inflorescences was higher on 42% compared to the control. At the phase II (beginning of flowering) the content of *p*-coumaric acid under the treatment with 2% CCC has been increasing in the stems on 54% compared to a control.

In the same time the increasing of *trans*-ferulic acid content in the leaves, stems and inflorescences of buckwheat plants at the phase I (formation of buds) and phase

Table 1 Content of phenolic acids in the stems of buckwheat cultivar Rubra after treatment with 2% CCC

	Stems			
	Phase I		Phase II	
	Control	2% CCC	Control	2% CCC
Vanilic acid	12,53 + 1,59	16,71 + 0,02*	15,21 + 0,75	38,21 + 8,96*
Chlorogenic acid	0,59 + 0,02	0,65 + 0,04*	1,53 + 0,14	3,25 + 0,35*
p-coumaric acid	8,79 + 1,67	9,55 + 0,96	9,73 + 0,15	21,10 + 5,60*
Ferulic acid	0,45 + 0,02	0,52 + 0,03*	0,68 + 0,05	7,83 + 1,20*
Trans-ferulic acid	19,62 + 3,45	21,89 + 2,35	49,01 + 7,87	59,82 + 2,29*
Salycilic acid	45,12 + 3,5	79,12 + 2,10*	59,78 + 2,16	360,39 + 49,22*
p-anisic acid	45,61 + 12,13	46,51 + 14,01	142,43 + 25,86	527,55 + 15,91*
Cinamic acid	4,02 + 1,23	5,10 + 1,32	4,34 + 0,23	8,71 + 1,55*
Methoxycinamic acid	5,36 + 0,81	5,74 + 0,96	6,76 + 0,63	43,16 + 4,46*

*Significant differences of these data were calculated using analysis of variance (ANOVA-Duncan's multiple test, SIGMASTAT 9.0).

Table 2 Content of phenolic acids in the leaves and inflorescences of buckwheat cultivar Rubra after treatment with 2% CCC

	Leaves				Inflorescences	
	Phase I		Phase II		Phase II	
	Control	2% CCC	Control	2% CCC	Control	2% CCC
Vanilic acid	16,54 + 1,30	18,57 + 2,72	44,61 + 3,26	37,20 + 2,18*	27,95 + 1,06	77,93 + 7,39*
Chlorogenic acid	1,25 + 0,07	1,35 + 0,04*	3,27 + 0,35	3,63 + 0,67	1,52 + 0,04	3,06 + 0,85*
p-coumaric acid	17,54 + 0,22	19,25 + 0,32*	46,05 + 3,66	41,24 + 2,78	4,53 + 0,06	7,82 + 1,70*
Ferulic acid	2,35 + 0,43	3,89 + 0,15*	7,35 + 0,53	8,37 + 0,38*	1,81 + 0,14	2,24 + 0,76*
Trans-ferulic acid	35,31 + 3,48	45,03 + 4,24*	29,03 + 4,18	53,13 + 4,89*	20,12 + 2,36	28,87 + 6,56*
Salycilic acid	125,36 + 4,51	134,01 + 1,21*	166,01 + 4,40	164,50 + 5,01	295,25 + 2,57	367,01 + 41,02*
p-anisic acid	48,71 + 2,63	49,81 + 2,31	599,04 + 52,06	459,54 + 83,56*	612,13 + 12,5	817,01 + 43,05*
Cinamic acid	8,20 + 1,54	9,25 + 1,45	9,56 + 2,39	13,51 + 2,92*	8,02 + 1,23	8,99 + 2,51
Methoxycinamic acid	9,26 + 0,02	9,83 + 0,03*	19,10 + 6,70	17,51 + 1,08	10,82 + 1,23	11,74 + 2,10

*Significant differences of these data were calculated using analysis of variance (ANOVA-Duncan's multiple test, SIGMASTAT 9.0).

II (beginning of flowering) has been estimated. The content of *trans*-ferulic acid has been increased under effect of 2% CCC in the phase I (formation of buds) in the stems 18% and in the leaves 22% compared to the control variant. Increasing of *trans*-ferulic acid content at the phase II (beginning of flowering) in the leaves (45%) and in the inflorescences (30%) compared to control has been estimated.

The content of ferulic acid has been significant increase in the leaves at the phase I (formation of buds). In the phase II (beginning of flowering) ferulic acid has been significant increase in the stems on 18%, in the leaves on 45% and in the inflorescences on 30% respectively.

Cinnamic acid is one of the basic phenylpropanoid with antioxidant activity, produced by plants in response to stressful conditions. Exogenous cinnamic acid increased growth characteristics in saline and non-saline conditions in maize plants. But effects of cinamic acid were more significant under saline conditions in comparison to non-saline conditions [22]. Cinamic acid relatively increased the leaf relative water content and the chlorophyll content, decreased plasma membrane permeability, mitigated membrane damage, inhibited the accumulation of malondialdehyde (product of membrane lipid peroxidation), and promoted the activity of membrane protective enzymes such as super oxide dismutase and peroxidase [23]. In same time cinnamic acid is a precursor in biosynthetic pathway of salicylic acid signaling molecule [24].

The content of cinamic acid during the phase I (formation of buds) in the stems, leaves was on control level. At the phase II (beginning of flowering) in the leaves has been found increasing of cinamic acid content on 29% and in the stems at twice compared to a control variant. In the inflorescences at the phase II (beginning of flowering) content of cinamic acid was on control level.

In higher plants, it is well established that salicylic acid derives from the shikimate-phenylpropanoid pathway [25]. Currently, it has been reported that this compound plays also a role in plants responses to abiotic stresses, such as drought, low and high temperatures, heavy metals, and osmotic stress [26-30]. Salycilic acid was also shown to influence a number of physiological processes, including seed germination, seedling growth, fruit ripening, flowering, ion uptake and transport, photosynthesis rate, stomata conductance, biogenesis of chloroplast [31-33].

Same changes as in content of other identified phenolic acids (chlorogenic acid, ferulic acid, *trans*-ferulic acid) were found for content of salycilic acid. The changes with increasing salycilic acid content has been observed in the buckwheat plants. The content of salycilic acid in phase I and phase II in the different part of buckwheat plants has been increased. The content of salycilic acid in the phase II (beginning of flowering) for inflorescences has been increased on 20%, for stems – on 80%.

It was estimated that metabolic pathway of salicylic acid rather than of chlorogenic acid is involved in the stress-induced flowering of *Pharbitis nil* (Japanese morning glory) plants [34]. The metabolic pathway from *t*-cinnamic acid to salycilic acid via benzoic acid is involved in the stress-induced flowering which can confirm also significant increasing of salicylic acid content in the stems of buckwheat plants in the phase II (beginning of flowering) compared to the phase I (formation of buds). At the same time significant increasing content of salycilic acid in the leaves and inflorescences in variant with CCC treatment in the phase II (beginning of flowering) is evidence about role of salycilic acid under plant stress conditions which could be occurred CCC treatment. Salycilic acid is an endogenous regulator of growth involved in a broad range of physiologic, metabolic and stress responses in plants [35].

At the phase II (beginning of flowering) has been observed increasing of *p*-anisic acid content in the stems on 73% and in the inflorescences 25% compared to the control. In the leaves at the phase II (beginning of flowering) has been shown significant decreasing of *p*-anisic acid content (23%). Same decreasing of vanilic acid content in the phase II (beginning of flowering) has been estimated. Such decreasing can be connected with redistribution and increasing of content these phenolic acids in the inflorescences compared to the control.

At phase I (formation of buds) content of methoxycinamic acid has been increased in the leaves of buckwheat on 6% compared to the control. During the phase II (beginning of flowering) has been shown significant increasing of methoxycinamic acid on 84% in the stems of buckwheat plants.

Catechins content

On analyzing with HPLC the different parts of buckwheat plants catechins has been estimated (Figure 2).

In the stems of buckwheat plants at phase II (beginning of flowering) in variant with CCC treatment has been shown catechins content higher 67% compared to the control. In the leaves have been estimated increasing of catechins content 32% compared to the control. In the inflorescences of buckwheat plants under CCC treatment content of catechins has been increased more than twice.

Catechins are a type of antioxidant found in the greatest abundance in the leaves of the tea plant *Camellia sinensis.* In smaller amounts, they are found in other foods such as wine, chocolate, berries, and apples. Their health benefits of have been under close examination since the 1990s, due to the strong association of tea with long life and health in many ancient cultures [36].

Watanabe has been identified 4 catechins in the buckwheat (*Fagopyrum esculentum* Moench) groats. The structures of these catechins were established as (-)-epicatechin, (+)-catechin 7-O-D-glucopyranoside, (-)-epicatechin 3-O-p-hydroxybenzoate, and (-)-epicatechin 3-O-(3,4-di-O-methyl) gallate on the basis of 1H, 13C. The antioxidant activity of the isolated compounds showed that the activity of catechins was superior to that of rutin, which is known as an antioxidant in buckwheat, at the same concentration [6].

The content of catechins as potential buckwheat antioxidants has been increased at the early stages of vegetation period after treatment with 2% CCC. It's known that catechins can be potential antioxidants among phenolic compounds.

Conclusions

It's known that vegetative mass of buckwheat plants is not used in food industry well and vegetative organs (leaves, stems and inflorescences) can contain higher antioxidants composition than buckwheat seeds. Therefore to find way of increasing content of buckwheat antioxidants is actual topic nowadays. In this research work we suggested simple idea to use chlorocholine chloride as factor which can increase content of phenolic compounds. The obtained results suggest that influence of CCC on the phenolics composition can be a result of various mechanisms of CCC uptake, transforming and/or its translocation in the buckwheat seedlings. The levels of identified phenolic acids varied not only significantly among the plant organs but also between early stages of vegetation period. Same changes as in contents of chlorogenic acid, ferulic acid, *trans*-ferulic acid were found for content of salycilic acid. The content of these phenolic acids has been significant

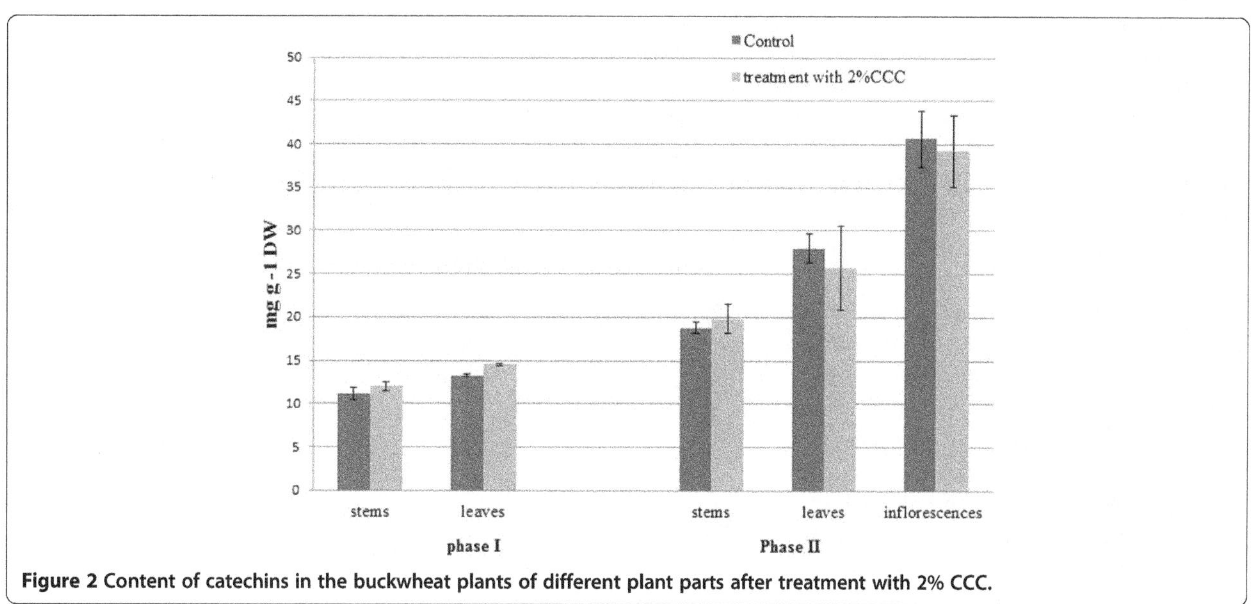

Figure 2 Content of catechins in the buckwheat plants of different plant parts after treatment with 2% CCC.

increased under effect of 2% CCC treatment at the phase I (formation of buds) in the stems and at the phase II (beginning of flowering) in the leaves and then inflorescences respectively. The enhanced accumulation of total phenolics, catechins and different changes for free phenolic acids (especially increasing content of chlorogenic acid and salycilic acid) can be explained probably by the synthesis of other unknown phenolic compounds or role some phenolic acids in the stress response reaction which could cause CCC treatment.

Methods

The common buckwheat (*Fagopyrum esculentum* Moench) cultivar Rubra has been used for this experimental work. Cultivar Rubra with high anthocyanins content 3.87 – 4.41 mg/100 g DW in the vegetative organ has been received by family selection method from chemo mutants from Taras Shevchenko National University of Kyiv.

Seeds were germinated between two layers of wet filter paper, which then were rolled and inserted in a 24 beaker containing 200 ml of tap water (Control) and solution of 2% CCC for 4 days (Experiment). Germination process was carried on in the darkness at 24 ± 1°C. After 4 days grown in such conditions the seedlings of buckwheat were taken to growth in the pots in 16/8 h night/day photoperiod and 65 ± 5% of relative humidity. Temperature in growth chamber was maintained at 24 ± 2°C for day and 18 ± 2°C during night period. The content of total phenols and phenolic acids has been evaluated in growth phase I (formation of buds), in phase II (at the beginning of flowering).

Determination of total phenolics

Total phenolics were determined by using Folin-Ciocalteu reagent [37]. 0.02 g powdered samples (freeze-dried) were extracted for 10 min with 500 mL of 70% methanol at 70°C. The mixtures were centrifuged at 3500 g for 10 min and the super-natants were collected in separate tubes. The pellets were re-extracted under identical conditions. Supernatants were combined and used for total phenolics assay and for HPLC analysis. For total phenolics assay 20 mL of extract was dissolved into 2 mL of distilled water. Two hundred microliters of dissolved extract were mixed with 1 mL of Folin-Ciocalteu reagent (previously diluted tenfold with distilled water) and kept at 25°C for 3–8 min; 0.8 mL of sodium bicarbonate (75 g L^{-1}) solution was added to the mixture. After 60 min at 25°C, absorbance was measured at 765 nm. The results were expressed as gallic acid equivalents.

HPLC analyses of flavanols and phenolics acids

The plant material was harvested and frozen in liquid nitrogen for the preventing of phenolic compound volatilization. Afterwards the samples were lyophilized. Further, finishing the freeze-drying process the material was grounded by flint mill (20000 g, 2 min). A total of 20 mg grounded samples from leaves suspension were extracted for 15 min using 0.75 mL 70% methanol (v/v, pH 4.0, phosphoric acid) in ultrasonic water bath on ice. Samples were centrifuged for 5 min at 6000 g. The supernatants were collected and the pellets were re-extracted twice more with 0.5 mL 70% methanol. Coumaric acid or cinnamic acid (40 mL of 3 mM solution) was added as internal standard to the first extraction. The combined supernatants from each sample were reduced to near dryness in a centrifugation evaporator (Speed Vac, SC 110) at 25°C.

Samples were added up to 1 mL with 40% acetonitrile. The samples were filtrated using 0.22 mm filters, and then analyzed with HPLC. The chromatography was performed using a Dionex UltiMate 3000 HPLC System with a diode array detector (DAD-3000) with a WPS-3000 SL auto sampler, LPG-3400SD pump and a TCC-3000RS Column Compartment (Dionex Corp., Sunnyvale, CA, USA).

Extracts (1 mL) were analyzed at a flow rate of 0.4 mL 1 min and a column temperature of 35°C. The column is Narrow-Bore Acclaim PA C16-column (3 mm, 120A, 2.1 × 150 mm, Dionex). A 49-min gradient program was used with 0,1% v/v phosphoric acid in ultrapure water (eluent A) and of 40% v/v acetonitrile in ultrapure water (eluent B) as follows: 0–5 min: 0.5% B, 1–9 min: 0–40% B, 9–12 min: 40% B, 12–17 min: 40–80% B, 17–20 min: 80% B, 20–24 min: 80–99% B, 24–32 min: 99–100% B, 32–36 min: 100–40% B, 36–49 min: 40–1% B. The gradient program was followed by a 4 min period to return to 0.5% B and a 5 min equilibration period resulting in a total duration of 39 min. The eluent was monitored at 290, 330, and 254 nm.

Statistical analysis

The means and standard deviations were calculated by the Microsoft Office Excel 2003. Significant differences of these data were calculated using analysis of variance (ANOVA-Duncan's multiple test, SIGMASTAT 9.0). All results were expressed as mean ± standard deviations from three and four replications.

Abbreviations

CCC: Chlorocholine chloride; ROS: Reactive oxygen species; HPLC: High-performance liquid chromatography.

Competing interests

The authors declare no financial conflict of interests.

Authors' contributions

SO carried out the all experiment, participated in the HPLC and drafted the manuscript. BA carried out the biochemical analysis and helped in writing the manuscript. HI participated in the HPLC analysis. CR participated in the design of the study and performed the statistical analysis. SI conceived of the study, and participated in its design and coordination and helped to draft the manuscript. All authors read and approved the final manuscript.

Acknowledgement
The authors thanks for the Grant for internship from Ministry of Education and Science of Ukraine (2013) for internship at the Institute of Food Technology and Food Chemistry Berlin University of Technology.

Author details
[1]Plant Physiology and Ecology Department, Taras Shevchenko National University of Kyiv, Institute of Biology, Volodymyrskya str., 64, Kyiv 01033, Ukraine. [2]Department of Technology of Food Products, Processing Industries and Biotechnology, Taraz State University named after MK Dulati, Suleimen Str., 7, Taraz 080012, Republic of Kazakhstan. [3]Department of Methods of Food Biotechnology, Berlin University of Technology, Institute of Food Technology and Food Chemistry, Koenigin Luise Str. 22, Berlin D-14195, Germany. [4]Agricultural Faculty, Department of Plant Food Processing, University of Applied Science Weihenstephan-Triesdorf, Steingruberstr. 2, Weidenbach 91746, Germany.

References

1. Bonafaccia G, Marocchini M, Kreft I: **Composition and technological properties of the flour and bran from common and tartary buckwheat.** *Food Chem* 2003, **80:**9–15.
2. Aufhammer W: **Pseudogetreidearten – Buchweizen, Reismelde und Amarant; Herkunft, Nutzung und Anbau.** *J Agr Crop Sci* 2003, **189**(3):197.
3. Ikeda K: **Buckwheat: composition, chemistry and processing.** *Adv Food Nutr Res* 2002, **44:**395–434.
4. Zielinski H, Kozlowska H: **Antioxidant activity and total phenolics in selected cereal grains and their different morphological fractions.** *J Agric Food Chem* 2000, **48**(6):2008–2016.
5. Watanabe M, Ohshi Y, Tsushida T: **Antioxidant compounds from buckwheat (*Fagopyrum esculentum* Moench) hulls.** *J Agric Food Chem* 1997, **45:**1039–1044.
6. Watanabe M: **Catechins as antioxidants from buckwheat (*Fagopyrum esculentum* Moench) groats.** *J Agric Food Chem* 1998, **46:**839–845.
7. Kim HJ, Park KJ, Lim JH: **Metabolomic analysis of phenolic compounds in buckwheat (*Fagopyrum esculentum* M.) sprouts treated with methyl jasmonate.** *J Agric Food Chem* 2011, **59**(10):5707–5713.
8. Sytar O, Zhenzhen C, Brestic M, Prasad MNV, Taran N, Smetanska I: **Foliar applied nickel on buckwheat (*Fagopyrum esculentum*) induced phenolic compounds as potential antioxidants.** *Clean - Soil, Air, Water* 2013, **41**(11):1129–1137.
9. Hahlbrock K, Scheel D: **Physiology and molecular biology of phenylpropanoid metabolism.** *Annu Rev Plant Physiol Plant Mol Biol* 1989, **40:**347–369.
10. Winkel-Shirley B: **Biosynthesis of flavonoids and effects of stress.** *Curr Opin Plant Biol* 2002, **5:**218–223.
11. Šebestík O, Marques SM, Falé PL, Santos S, Arduíno DM, Cardoso SM, Oliveira CR, Serralheiro MLM, Santos MA: **Bifunctional phenolic-choline conjugates as anti-oxidants and acetylcholinesterase inhibitors.** *J Enzyme Inhib Med Chem* 2011, **26**(4):485–497.
12. Kreslavskiia VD, Lubimova VY, Kotova LM, Kotov AA: **Effect of common bean seedling pretreatment with chlorocholine chloride on photosystem II tolerance to UVB radiation, phytohormone content, and hydrogen peroxide content.** *Russ J Plant Physiol* 2011, **58**(2):324–329.
13. Huiqun W, Langtao X, Jianhua T, Fulai L: **Foliar application of chlorocholine chloride improves leaf mineral nutrition, antioxidant enzyme activity, and tuber yield of potato (*Solanum tuberosum* L.).** *Scient Horticul* 2010, **125:**521–523.
14. Jain VK, Guruprasad KN: **Effect of chlorocholine chloride and gibberellic acid on the anthocyanin synthesis in radish seedlings.** *Physiol Plant* 1989, **75:**233–236.
15. Petti S, Scully C: **Polyphenols, oral health and disease: A review.** *J Dent* 2009, **37**(6):413–423.
16. Sivaci A, Duman S: **Evaluation of seasonal antioxidant activity and total phenolic compounds in stems and leaves of some almond (*Prunus amygdalus* L.) varieties.** *Biol Res* 2014, **47**. doi:10.1186/0717-6287-47-9.
17. Hanson KR, Havir EA: **Phenylalanine ammonia-lyase.** In *The biochemistry of plants.* Edited by Stumpf PK, Conn EE. New York: Academic; 1981:577–625.
18. Jahnen W, Hahlbrock K: **Differential regulation and tissue-specific distribution of enzymes of phenylpropanoid pathways in developing parsley seedlings.** *Planta* 1988, **173:**197–204.
19. Schmelzer E, Jahnen W, Hahlbrock K: **In situ localization of light-induced chalcone synthase mRNA, chalcone synthase, and flavonoid products in epidermal cells of parsley leaves.** *Proc Natl Acad Sci U S A* 1988, **85:**2989–2993.
20. Niggeweg R, Michael AJ, Martin C: **Engineering plants with increased levels of antioxidant chlorogenic acid.** *Nat Biotechnol* 2004, **22:**746–754.
21. Schützendübel A, Polle A: **Plant responses to abiotic stresses: heavy metal-induced oxidative stress and protection by mycorrhization.** *J Exp Bot* 2001, **53**(372):1351–1365.
22. Pramod KS, Ramendra S, Shivani S: **Cinnamic acid induced changes in reactive oxygen species scavenging enzymes and protein profile in maize (*Zea mays* L.) plants grown under salt stress.** *Annu Rev Plant Biol* 2013, **19**(1):53–59.
23. Xuezheng W, Hua W, Fengzhi W, Bo L: **Effects of cinnamic acid on the physiological characteristics of cucumber seedlings under salt stress.** *Front Agric China* 2007, **1**(1):58–61.
24. Hayat Q, Hayat S, Irfan M, Ahmad A: **Effect of exogenous salicylic acid under changing environment: A review.** *Environ Exp Bot* 2010, **8:**14–25.
25. Stitcher L, Mauch-Mani B, Metraux JP: **Systemic acquired resistance.** *Annu Rev Plant Pathol* 1997, **35:**235–270.
26. Molina A, Bueno P, Marín MC, Rodríguez-Rosales MP, Belver A, Venema K: **Involvement of endogenous salicylic acid content, lipoxygenase and antioxidant enzyme activities in the response of tomato cell suspension cultures to NaCl.** *New Phytol* 2002, **156:**409–415.
27. Nemeth M, Janda T, Horvath E, Paldi E, Szalai G: **Exogenous salicylic acid increases polyamine content but may decrease drought tolerance in maize.** *Plant Sci* 2002, **162:**569–574.
28. Munne-Bosch S, Peñuelas J: **Photo- and antioxidative protection, and a role for salicylic acid during drought and recovery in field-grown *Phillyrea angustifolia* plants.** *Planta* 2003, **217:**758–766.
29. Shi Q, Zhu Z: **Effects of exogenous salicylic acid on manganese toxicity, element contents and antioxidative system in cucumber.** *Environ Exper Bot* 2008, **63:**317–326.
30. Rivas-San Vicente M, Plasencia J: **Salicylic acid beyond defence: its role in plant growth and development.** *J Exp Bot* 2011, **62**(10):3321–3338.
31. Fariduddin Q, Hayat S, Ahmad A: **Salicylic acid influences net photosynthetic rate, carboxylation efficiency, nitrate reductase activity and seed yield in *Brassica juncea*.** *Photosynthetica* 2003, **41:**281–284.
32. Khodary SFA: **Effect of salicylic acid on the growth, photosynthesis and carbohydrate metabolism in salt stressed maize plants.** *Int J Agric Biol* 2004, **6:**5–8.
33. Hayat S, Fariduddin Q, Ali B, Ahmad A: **Effect of salicylic acid on growth and enzyme activities of wheat seedlings.** *Acta Agron Hung* 2005, **53:**433–437.
34. Hatayama T, Takeno K: **The metabolic pathway of salicylic acid rather than of chlorogenic acid is involved in the stress-induced flowering of *Pharbitis* nil.** *J Plant Physiol* 2003, **160**(5):461–467.
35. Hayata Q, Hayata S, Irfana M, Ahmad A: **Effect of exogenous salicylic acid under changing environment: a review.** *Environ Exp Bot* 2010, **68**(1):14–25.
36. Sytar O, Brestic M, Rai M, Shao HB: **Plant phenolic compounds for food, pharmaceutical and cosmetics production.** *J Med Plants Res* 2012, **6**(13):2526–2539.
37. Singleton VL, Rossi JA: **Colorimetry of total phenolics with phosphomolybdic-phosphotungstic acid reagents.** *Am J Enol Vitic* 1965, **16:**144–158.

16

In vitro antioxidant capacity and free radical scavenging evaluation of active metabolite constituents of *Newbouldia laevis* ethanolic leaf extract

Josiah Bitrus Habu[1] and Bartholomew Okechukwu Ibeh[2,3*]

Abstract

Background: The aim of the present study was to evaluate the *in vitro* antioxidant and free radical scavenging capacity of bioactive metabolites present in *Newbouldia laevis* leaf extract.

Results: Chromatographic and spectrophotometric methods were used in the study and modified where necessary in the study. Bioactivity of the extract was determined at 10 μg/ml, 50 μg/ml, 100 μg/ml, 200 μg/ml and 400 μg/ml concentrations expressed in % inhibition. The yield of the ethanolic leaf extract of *N.laevis* was 30.3 g (9.93%). Evaluation of bioactive metabolic constituents gave high levels of ascorbic acid (515.53 ± 12 IU/100 g [25.7 mg/100 g]), vitamin E (26.46 ± 1.08 IU/100 g), saponins (6.2 ± 0.10), alkaloids (2.20 ± 0.03), cardiac glycosides(1.48 ± 0.22), amino acids and steroids (8.01 ± 0.04) measured in mg/100 g dry weight; moderate levels of vitamin A (188.28 ± 6.19 IU/100 g), tannins (0.09 ± 0.30), terpenoids (3.42 ± 0.67); low level of flavonoids (1.01 ± 0.34 mg/100 g) and absence of cyanogenic glycosides, carboxylic acids and aldehydes/ketones. The extracts percentage inhibition of DPPH, hydroxyl radical ($OH^{\cdot}$), superoxide anion ($O_2^{\cdot-}$), iron chelating, nitric oxide radical (NO), peroxynitrite ($ONOO^-$), singlet oxygen ($^1O_2$), hypochlorous acid (HOCl), lipid peroxidation (LPO) and FRAP showed a concentration-dependent antioxidant activity with no significant difference with the controls. Though, IC_{50} of the extract showed significant difference only in singlet oxygen ($^1O_2$) and iron chelating activity when compared with the controls.

Conclusions: The extract is a potential source of antioxidants/free radical scavengers having important metabolites which maybe linked to its ethno-medicinal use.

Keywords: *Newbouldia laevis*, Phytochemicals, Ethanolic extraction, Antioxidants, Free radical scavengers, Bioactive constituents

Background

The African continent has one of the richest biodiversity in the world and abounds in plants of economic and medicinal importance which when developed would reduce expenditure on global drug development while meeting patient's health needs [1]. Current emphasis on healthy living based on antioxidant intake and the implication of oxidative stress molecules/free radicals on certain diseased condition [2] has generated renewed interest in screening for plants with high antioxidative properties. The identification and quantification of bioactive components that contribute to free radical scavenging activity and its consequent ethnopharmcological effect may provide link to specific drug discovery.

Newbouldia laevis is commonly known as African border tree. In Nigerian major languages it is called 'Aduruku' in Hausa, Ogirisi" in Igbo and Akoko in Yoruba [3]. *N. laevis* is a medium sized, sun loving, fast growing drought tolerant angiosperm which belongs to the Bignoniaceae family [4]. It grows up to a height of about 7–15 meters but is usually a shrub of 2–3 meters with many stemmed forming clumps of gnarled branches. In

* Correspondence: barthokeyibeh@yahoo.com
[2]Department of Biochemistry, College of Natural and Applied Sciences, Michael Okpara University of Agriculture Umudike, Umudike, Nigeria
[3]National Biotechnology Development Agency, Abuja, Nigeria
Full list of author information is available at the end of the article

sub-Saharan Africa, the plant is used in the management of a variety of ailments for example, the bark is chewed and swallowed for stomach pains and diarrhoea as well as toothache [5]. In Nigeria and Ivory Coast, the stem bark decoctions are used for treatment of epilepsy and convulsions in children [6]. Similarly, Senegalese use the stem bark for the treatment of rheumatism especially painful arthritis of the knee. The plant also has medicinal therapy against ear aches, sore feet and chest pain [7]. Currently, leaf and root extracts of *N. laevis* have been shown to possess antimalaria [8,9] and antimicrobial activities [10,11]. The leaves, stem and fruits have been used for febrifuge, wound dressing and stomach ache medication [12].

No extensive report on the presence, and free radical scavenging activity of basic metabolites from the leaves of *N. laevis* has been provided. Similarly, investigations of the plant have produced conflicting reports on the content of phytochemical compounds present in the plant leaf thus provide scarce and inaccurate information. Furthermore, the antioxidative potential of the plant leaf have not been critically evaluated. The study therefore, evaluated the principal metabolites present in the ethanolic leaf extract of the plant as well as the antioxidant potential and free radical scavenging activity of the leaf extract. The extract was examined for different reactive oxygen species (ROS) scavenging activities including hydroxyl, superoxide, nitric oxide, hydrogen peroxide, peroxynitrite, singlet oxygen and hypochlorous acid, iron chelating capacity, antioxidant activity and metabolic constituents.

Results

Extractive yield

The yield of the ethanolic leaf extract of *N. laevis* was 30.3 g (9.93%).

Phytochemical analysis

Preliminary phytochemical screening of *N.laevis* shows the presence of alkaloids, saponins, tannins, cardiac and steroidal glycosides, flavonoids, other metabolites were amino acids (Table 1) and vitamins A,C and E (Figure 1) while carboxylic acids, anthracene derivatives and aldehydes were absent. Evaluation of bioactive metabolic constituents gave high levels of saponins (6.2 ± 0.10), alkaloids (2.20 ± 0.03), cardiac glycosides (1.48 ± 0.22), amino acids, steroids (8.01 ± 0.04); moderate levels of tannins (0.09 ± 0.30), terpenoids (3.42 ± 0.67) and low levels of flavonoids (1.01 ± 0.34 mg/100 g) (Table 2).

Antioxidant vitamin composition found in the leaf extracts of *Newbouldia laevis*

The result shown in Figure 1 summarizes the composition of antioxidant vitamins present in the leaves of *N. laevis* grown in Nigeria a sub-Sahara African country.

Table 1 Phytochemical screening of basic metabolites of the leaf extracts of *Newbouldia laevis*

Plant metabolite	Extract content
Cyanogenic glycosides	+
Cardiac glycosides	++
Steroid glycoside	+
Saponins	+++
Tannins	++
Alkaloids	+++
Amino acids	+++
Terpenoids	++
Flavonoids	+
Carboxylic acids	-
Aldehyde/ketones	–
Ascorbic acid	+++
Anthracene derivatives	–

+ = Trace, ++ = high, +++ = Abundant, – = Absent.
Summary of TLC phytochemical identification of *N. leavis leaf extract.*

The concentration of vitamins measured in IU/100 g weight shows moderate levels of Vitamin A (188.28 ± 6.19) and high levels of vitamins C (515.53 ± 12 [25.7 mg/100 g]) and E (26.46 ± 1.08).

Antioxidant and free radical scavenging activity

The percentage inhibition of hydroxyl radical ($OH^{\cdot}$) and mannitol standard, superoxide anion ($O_2^{\cdot-}$)/quercetin, iron chelating /EDTA, and, nitric oxide radical (NO)/curcumin, peroxynitrite ($ONOO^-$)/gallic acid, singlet oxygen ($^1O_2$)/lipoic acid, hypochlorous acid (HOCl)/ascorbic acid, DPPH/ ascorbic acid, inhibition of lipid peroxidation (LPO) measured as TBARS and FRAP (Table 3) by *N. laevis* leaf extract showed a significant ($P < 0.05$) concentration-dependent antioxidant activity. The leaf extract had a comparable reduction capacity in all the concentrations measured when compared with the scavenging activity of known standards. The IC_{50} values of the extract showed significant difference only in singlet oxygen ($^1O_2$) (510.65 ± 9.54) vs lipoic acid standard (46.15 ± 1.16) and iron chelating (1225.05 ± 298.1) vs EDTA standard (1.27 ± 0.05) (Table 4). The FRAP of the extract at 400 μg/ml was 64 ± 2.52% (FRAP: 0.64) (Table 3) and that of the inhibition of lipid peroxidation (LPO) was 91.85 ± 0.34% (Table 3). The extract showed a good reducing power in a concentration dependent manner.

Discussion

The investigation reported here reveals the presence of secondary metabolites such as alkaloids, tannins, flavonoids and cardiac glycosides in the ethanolic leaf extract of *N. laevis* and the free radical scavenging activity

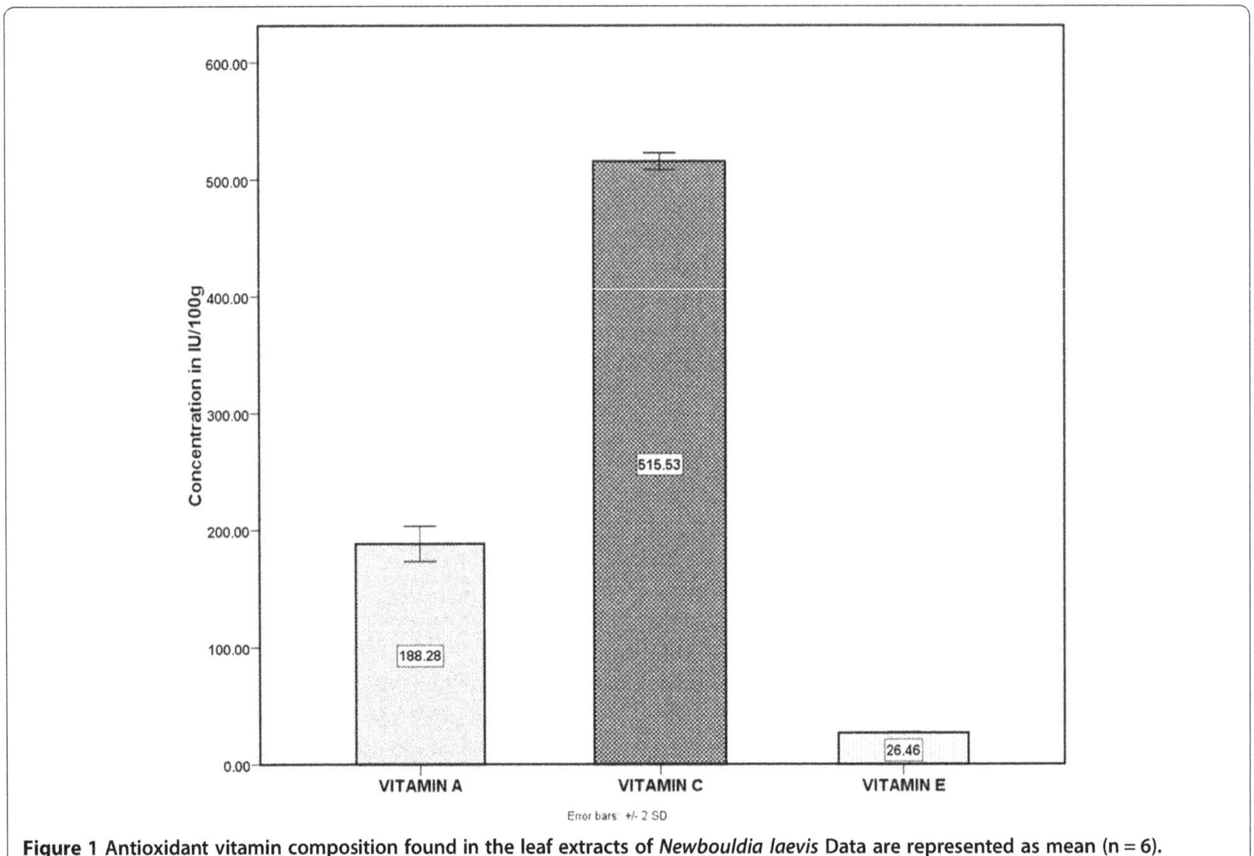

Figure 1 Antioxidant vitamin composition found in the leaf extracts of *Newbouldia laevis* Data are represented as mean (n = 6).

inherent in the plant species. The high antioxidant activity may relate to the plants' curative and/or management potential of many ailments claimed in its ethnomedicine. Earlier studies on the leaf and bark extracts of the Congolese *N. leavis* showed the absence of flavonoids, saponins, quinones, terpenes and steroids [13]. Although, recent phytochemical studies on the root, bark and stem of the plant have revealed the presence of alkaloid, quinoid and phenylpropanoid compounds [14].

Table 2 Phytochemical composition of metabolites found in the leave extracts of *Newbouldia laevis* (mg/100 g dry weight)

Plant metabolite	Composition
Cardiac glycosides	1.48 ± 0.22
Saponins	6.2 ± 0.10
Tannins	0.09 ± 0.30
Alkaloids	2.20 ± 0.03
Flavonoids	1.01 ± 0.34
Steroids	8.01 ± 0.04
Terpenoids	3.42 ± 0.67

Results are mean of sextuplicate determinations on a dry weight basis ± standard deviation.

Phytochemical results showed the absence of carboxylic acids, aldehyde/ketones and anthracene derivatives in the ethanolic leaf extract. However, the biological active components present in the extract were vitamins A, C and E, tannins, saponins, cardiac glycosides, flavonoids, alkaloids, steroids and terpenoids, this was corroborated by previous works on phytochemicals of *N. laevis* [15]. The discordant results from several other authors [13,16] on the bioactive metabolites (especially absence of saponins) present in *N. laevis* maybe as a result of the medium of extraction (i.e. solvent), storage and environmental factors. High levels of amino acids, saponins (6.2 ± 0.10), steroids (8.01 ± 0.04), alkaloids (2.20 ± 0.03) and terpenoids (3.42 ± 0.67) characterized *N. laevis* leaf extract. The phytochemicals identified have been shown to have curative effect on several disease pathogens, thus may relate to *N. laevis* widely ethno-medicinal use [9,11]. Saponins for instance have the ability to bind sterols of cell membrane and reduce choleasterol levels hence are widely used in conventional medicines exhibiting hypocholeasterolemic effects. Generally, it could be recalled that saponins form foams in aqueous solution which have haemolytic activity and choleasterol binding properties. They have natural tendency to ward-off microbes which makes them good candidates for treating fungal and yeast infections. These compounds served as

Table 3 Free radical scavenging potential of *Newbouldia laevis* measured as % inhibition

Antioxidant activity (% Inhibition)										
Conc (μg/ml)	Hydroxyl radical ($OH^{\cdot}$)	Superoxide anion ($O_2^{\cdot-}$)	Iron chelating	Nitric oxide radical (NO)	Peroxynitrite ($ONOO^-$)	Singlet oxygen ($^1O_2$)	Hypochlorous acid (HOCl)	DPPH	Lipid peroxidation (LPO)	FRAP (mM)
10	25.80 ± 0.03^a	30.60 ± 0.05^a	40.10 ± 0.02^a	41.06 ± 0.04^a	15.60 ± 0.01^a	53.64 ± 0.72^a	09.60 ± 0.12^a	42.64 ± 1.12^a	53.64 ± 0.82^a	08.07 ± 0.31^a
50	28.85 ± 0.01^a	40.65 ± 0.04^b	46.58 ± 0.04^a	49.75 ± 0.08^b	25.85 ± 0.04^b	59.10 ± 0.09^a	25.85 ± 0.31^b	45.85 ± 0.10^a	65.85 ± 0.09^a 1	18.10 ± 0.11^a
100	50.83 ± 0.11^b	62.74 ± 0.12^c	65.38 ± 0.07^b	55.13 ± 0.09^b	35.83 ± 0.09^c	71.62 ± 2.46^b	35.83 ± 0.05^c	65.85 ± 0.11^b	85.85 ± 0.14^b	30.15 ± 0.15^b
200	62.97 ± 0.04^c	69.79 ± 0.08^c	71.67 ± 0.09^c	61.86 ± 0.07^c	48.97 ± 0.09^d	86.16 ± 1.10^c	48.97 ± 0.09^d	79.85 ± 0.19^c	89.85 ± 0.16^b	50.09 ± 1.20^c
400	76.10 ± 0.02^d	81.11 ± 0.07^d	90.11 ± 0.08^d	80.08 ± 0.06^d	62.10 ± 0.11^e	96.00 ± 0.12^d	62.10 ± 0.11^e	85.85 ± 0.18^c	91.85 ± 0.34^c	64.01 ± 2.52^d

Data are expressed as mean ± standard deviation ($n = 6$); mean in the same column with different superscripts are significantly different using Duncan's multiple range test at $p < 0.05$.

Table 4 IC_{50} values of *Newbouldia laevis* scavenging activity and reference compounds

Activity	*N. Laevis* IC_{50}	Reference	IC_{50}
DPPH	51.4[#]	Ascorbic acid	55.4 ± 20.12**
Hydroxyl radical (OH)	497.21 ± 3.65[#]	Mannitol	571.45 ± 20.12**
Nitric oxide radical (NO)	92.42 ± 2.73[#]	Curcumin	90.82 ± 4.75 (6)**
Superoxide anion ($O_2^{\cdot-}$)	57.08 ± 1.22[#]	Quercetin	42.06 ± 1.35**
Peroxynitrite ($ONOO^-$)	1210.83 ± 23.85[#]	Gallic acid	876.24 ± 56.96 (6)**
Singlet oxygen ($^1O_2$)	510.65 ± 9.54	Lipoic acid	46.15 ± 1.16 (6) *
Hypochlorous acid (HOCl)	276.04 ± 12.01[#]	Ascorbic acid	235.95 ± 5.75 (6)**
Iron Chelating	1225.05 ± 298.1	EDTA	1.27 ± 0.05 (6)**

Units of IC_{50} for all activities are µg/ml. Data are expressed as mean ± S.D.
EDTA = Ethylenediamine tetraacetic acid. [#]indicates no significant difference where $^*p < 0.01$ and $^{**}p < 0.001$.

natural antibiotics that help the body to fight-off infections and microbial invasion and boost the effectiveness of certain vaccines. *N. laevis* inhibits *Staphylococcus aureus* and *Candida albicans* growth [16,17], recently the plant have been shown to also stimulate the activity of heapatic glucokinase, inhibiting glucose 6-phosphatase activity [18] thus serving as a good antidiabetic agent. The presence and concentration of these metabolites could explain the use of *N. laevis* in the treatment against various bacterial infections, sexually transmitted diseases and diabetes. The non-sugar part of saponins has a direct antioxidant activity which may contribute to the high free radical scavenging capacity of the plant leaf extract.

The trace level of cyanogenic glycoside could suggest the plant's very low toxicity when ingested in the form of traditional medicine (Table 1). Generally, flavonoids are widely distributed group of polyphenolic compounds, characterized by a common benzopyrone ring structure that has been reported to act as antioxidants in various biological systems. The biological function of flavonoids are extended to include protection against allergies, inflammation, free radicals, platelet aggregation, microbes, ulcers, heapatoxins, viruses and tumours [19]. Germann *et. al.,* [14] revealed the presence of newbouldioside A-C and phenylethanoid glycosides in the stem bark of *N. laevis.*

Quantitative analysis of vitamins A, C and E is indicative of an enhanced free radical scavenging capacity of the plant. The leaf extract could be said to have a moderate vitamin A (188.28 ± 6.19 IU/100 g), fairly high vitamins E (26.46 ± 1.08 IU/100 g) and C (515.53 ± 12 IU/100 g) content when compared with their respective standard references. However, comparison of the vitamins showed a higher vitamin C composition. The vitamin constituents of *N. laevis* may establish in part the efficient regulation of reactive oxygen species and scavenging activity observed in the plant extract investigated in addition to maintaining membrane fluidity and integrity. Vitamin C potentially regenerates vitamin E and renews its potency. A high vitamin E content of *N. laevis* thus suffices for its antioxidant activity which is responsible for stabilization of biomembrane structure. Vitamin A on the other hand, not only contributes to the plants free radical scavenging activity but also the immunostimulatory property of *N. laevis* [20].

Phenolic compounds are very important plant constituent with multiple biological functions including antioxidant activity much related to the radical scavenging ability of their OH groups. A number of studies have reported the relative correlation between phenol and antioxidant activity [21]. It could be seen that alternative solution to synthetic drugs resides in plant natural products mostly those with free radical scavenging property. DPPH has been widely used to evaluate the antioxidant activity of natural products from plant and microbial sources. The result of the present study showed that the *in vitro* free radical potential of the extract exhibited maximum free radical scavenging activity with a comparable IC_{50} value of the known standards, except in singlet oxygen quenching and iron chelating property. The antioxidant attributes of *N. laevis* leaf extract as affected by alkaline hydrolysis and the release of bound phenolics have limited experimental evidence with few investigators reporting on stem bark [22,23]. The investigated plant metabolites with redox properties plays an important role in absorbing and neutralizing free radicals, quenching singlet and triplet oxygen, or decomposing peroxides as reported in Tables 2, 3 and 4. A higher DPPH radical-scavenging activity is associated with a lower IC_{50} value thus the results presented here indicates a higher DPPH radical–scavenging activity of the extract though not significant when compared with ascorbic acid standard (Table 4). DPPH is a stable free radical at room temperature and accepts an electron or hydrogen radical to become a stable diamagnetic molecule which is generally regarded to be a model for lipophilic radical activity. The ferric reducing power of the extract at 400 µg/ml gave 64 ± 2.52% (FRAP: 0.64) and that of inhibition of lipid peroxidation (LPO) was 91.85 ± 0.34%. The inhibition of TBARS a measure of the oxidative stress was high suggesting that *N. laevis* is a good antioxidant source. As generally observed, the antioxidant reaction of *N. laevis is* concentration-dependent which means that an increase in antioxidant activity is linearly dependent on the ethanolic leaf extract concentration of the plant (Table 3).

Hydroxyl radicals are the major active oxygen species causing lipid peroxidation and various biological damage. *N. laevis* extract was able to remove the hydroxyl radicals from the sugar component of the MDA–like oxidant and prevented the oxidative reaction. The IC_{50} value indicates that the plant extract is a better hydroxyl radical scavenger than the standard mannitol. Similarly, superoxide anion a dangerous radical to cellular components can be removed by the efficient activity of flavonoids which scavenge superoxide anions [24]. As shown in Tables 3 and 4, the superoxide radical scavenging activities of the plant extract and the reference compound quercetin are increased markedly with increasing concentrations and are comparable (no significant difference).

Nitric oxide are important in inflammatory processes but at an increased level are directly toxic to tissues resulting in vascular damage and other ailments. This toxicity is heightened on reaction with superoxide radical to form a second reactive compound peroxynitrite anion ($ONOO^-$). *N. laevis* inhibits nitrite formation in the process of generating the radical (N) by direct competition with oxygen. Furthermore, the protonation of peroxynitrite ($ONOO^-$) forms a dangerous and highly reactive compound peroxynitrous acid (ONOOH) [25]. The plant extract inhibits the process by scavenging peroxynitrite. *N. laevis* exhibited comparable activity with the two standards curcumin (NO) and gallic acid ($ONOO^-$). HOCl inactivates catalase through breakdown of the heme prosthetic group. The plant extract inhibited catalase indicating its HOCl scavenging activity. Comparison with the ascorbic acid standard shows no significant difference. Conversely, singlet oxygen which induces hyperoxidation and oxygen cytotoxicity decreases antioxidative activity, also iron chelating effect which can stimulate lipid peroxidation are all reduced in a concentration dependent manner by the extract but not as efficient as the respective standards lipoic acid and EDTA.

Conclusions

All extracts at tested doses (10–400 µg mL-1) revealed good scavenging activity for DPPH, FRAP, hydroxyl radical ($OH^{\cdot}$), nitric oxide radical (NO), superoxide anion ($O_2^{\cdot -}$), peroxynitrite ($ONOO^-$), singlet oxygen ($^1O_2$), hypochlorous acid (HOCl), iron chelating and inhibition of TBARS in a dose-dependent manner. The activity maybe related to the presence and concentration of secondary metabolites present in *N.laevis* leaf extract.

Methods

Collection and identification of plant materials

Fresh matured leaves of *N. laevis* were harvested from farms in the Department of Forestry and Environmental Management, Michael Okpara University of Agriculture, Umudike Nigeria (Latitude 05^0 29^1 N to 05^0 42^1, Longitude 07^0 24^1 E to 07^0 33^1). The matured leaves were identified and confirmed by experts of the Department of Forestry, College of Natural Resources and Environmental Management, Michael Okpara, University of Agriculture Umudike, Nigeria. A voucher specimen with the number Ibeh 2011–23 was deposited in the University herbarium for future reference.

Sample preparation

The leaves of *N.laevis* were air-dried at room temperature and pulverized into a uniform material using a Thomas-Willey mini-milling machine (model 4, 3375-e25). Plant extraction (300 g of pulverized material) was done with 80% ethanol at 70°C by continuous percolation using Soxhlet extractor for 24 hours. The resulting extract was concentrated at 40°C in a rotary evaporator to yield a dark green mass of weight 30.3 g (9.93%). The obtained crude extract was packed ascetically in airtight plastic containers and stored at 4°C until required.

The percentage yield of the extract was calculated using the formula:

$$\%\,\text{Yield} = \frac{\text{weight of the extract}}{\text{weight of plant material}} \times \frac{100}{1}.$$

Phytochemical determination of the metabolites

For initial phytochemical detection of major metabolites of *N. laevis* thin-layer chromatography (TLC) on silica gel 60 F_{254} with layer thickness 0.25 mm (Merck, Darmstadt, Germany) was used after dissolving the extract (2 mg) in 2 ml ethanol. The plates were developed, then left to dry for about 10 min before they were viewed under UV fluorescence light at 254 and 366 nm. Spraying was done with the required detection reagent to determine the compounds present. For flavonoids, TLC was developed in n-butanol/acetic acid/water (4:1:5), then spots were visualized with 1% $AlCl_3$ solution in methanol under UV light (366 nm) (Ce 3041 Buck Scientific, UK). Alkaloids, saponins, tannins, anthraquinones, flavonoids, terpenoids, steroids and cardiac glycosides were all identified based on standard methods [26-28]. Quantitative determination was carried out by procedures previously described [29-31]. The concentration of vitamins A, E and C content of *N. laevis* was estimated using Barakat method [32] for vitamin C and Kirk and Sauya [33] for vitamins A and E.

Assessment of inhibition of lipid peroxidation

A modified version of the thiobarbituric acid reactive substances (TBARS) assay was used to assess the extent of lipid peroxides formed using egg yolk homogenate as lipid-rich media [34]. Egg homogenate (0.5 ml, 10% in distilled water v/v) was added to 0.1 ml of extract and the volume made up to 1 ml with distilled water. A volume of 0.05 ml of 0.07 M $FeSO_4$ was added to the above mixture and further incubated for 30 min, to induce lipid oxidation. Then 1.5 ml of 20% acetic acid (pH 3.5), 1.5 ml of 0.8% w/v TBA prepared in 1.1% w/v sodium duodecyl sulphate and 0.05 ml of 20% w/v TCA were sequentially added. The resulting mixture was vortexed and heated at 95°C for 60 min. After cooling, 5 ml of butan-1-ol was added and the mixture centrifuged at 3000 rpm for 10 min (Ultra-8 digital CR Scientific, Koningsweg, Netherlands). The absorbance of the organic upper layer was measured at 532 nm and converted to percentage inhibition using the formula: Varying concentrations (10 to 400 μg/ml) of the extract was used for all free radical scavenging (LPO,FRAP, $ONOO^-$,HOCl,$^1O_2$,NO, OH, Fe^{2+} chelation, DPPH, $O_2^{\cdot-}$) analysis. Free radical scavenging potential of *N. laevis* was measured as % Inhibition and the IC_{50} values determined in each parameter, comparison were made with corresponding reference compounds. It is imperative to note that the choice of assay standards were made to effectively evaluate the scavenging property of the extract using specific known and well characterized compounds.

$$\text{Inhibition of Lipid Peroxidation}(\%) = (1-E / C)X100 \tag{1}$$

Where C = absorbance of fully oxidized control and E = absorbance in the presence of extract.

Ferric reducing potential assay

The reductive potential (ferric reducing antioxidant power; FRAP) of *N. laevis* was determined based on the chemical reduction of Fe^{3+} to Fe^{2+} [35]. Briefly, 50 μl of the extract was added to 1.5 ml of freshly prepared and pre-warmed (37°C) FRAP reagent (300 mM acetate buffer, pH = 3.6, 10 mM tripyridyl-s-triazine (TPTZ) in 40 mM HCl and 20 mM FeCl3.6H2O in the ratio of 10:1:1) and incubated at 37°C for 10 min. The absorbance of the sample was read against reagent blank (1.5 ml FRAP reagent and 50 μl distilled water, [MI]) at 593 nm. Standard solutions of Fe2+ in the range of 100 to 1000 mM were prepared from ferrous sulphate (FeSO4.7H2O) using distilled water. Thus at low pH, the reduction of ferric tri (2-pyridyl)-1, 3, 5-triazine (Fe III TPTZ) complex to ferrous form (FRAP value) was measured by monitoring the change in absorption at 593 nm. Absorbance (A) readings were taken after 0.5 s and every 15 s thereafter during the monitoring period. The change in absorbance (ΔA_{593nm}) between the final reading selected and the M1 reading was calculated for each sample and related to (ΔA_{593nm}) of a Fe^{II} standard solution tested in parallel. The reaction was monitored for up to 8 min but the 4-min readings were selected for calculation of FRAP values. The final result was expressed as concentration of antioxidant having a ferric reducing ability equivalent to that of 1 mmol/L FeSO4. The calculation was done by:

$$\text{FRAP (mM)} = \frac{(\Delta A_{593nm}\ of\ sample from\ 0\text{-}4min)}{(\Delta A_{593nm} of\ standard\ from\ 0\ to\ 4\ min).} \times FRAP\,value\ of\ standard\ (1000\ mM) \tag{2}$$

DPPH based free radical scavenging activity

DPPH radical scavenging activity was detected for antioxidant activity by thin layer chromatography (TLC) screening through spotting a concentrated ethanolic solution of the extract on silica gel plates. The plates were developed in ethanol: ethyl acetate (2:1) then air-dried and sprayed with 0.2% w/v DPPH spray. The presence of yellow spots were detected. Radical scavenging activity of extracts was measured according to the DPPH spectrophotometric method [36] using vitamin C (Emzor Pharmaceutical Industries, Nigeria) as a reference antioxidant. Ethanol (1.0 ml) plus extract solution (2.5 ml) was used as blank while 1 ml of 0.3 mm DPPH plus ethanol (2.5 ml) was used as a negative control. The free radical scavenging properties of the extracts against 2, 2-diphenyl-1-picryl hydrazyl (DPPH) radical were measured at 518 nm, as an index of their antioxidant activity. IC_{50} values (the concentration of extracts required to scavenge 50% of DPPH free radicals) were also calculated. The absorbance (abs) of the resulting mixture measured at 518 nm was converted to percentage antioxidant activity (AA %) and thus calculated by the equation:

$$AA\% = \left[100-((\text{ABS sample} - \text{ABSblank}) \times 100)\right] / \text{ABScontrol} \tag{3}$$

Superoxide radical scavenging activity

Measurement of superoxide radical scavenging capacity of *N. laevis* extracts was done using a previously reported method [37] described by Fontana *et al.* The reaction mixture (1 ml) contained phosphate buffer (20 mM, pH 7.4), NADH (73 μM), nitroblue tetrazolium (NBT) solution (50 μM), Phenazine methosulphate (PMS) solution (15 μM) and various concentration of the plant extract as described elsewhere. The PMS/NADH system generates superoxide radicals, which reduce NBT to a purple formazan. This was incubated at 25°C for 5 mins and absorbance measured at 562 nm against the ethanol blank to determine the quantity of formazan. Thus the assay of SOD is based on the inhibition of the formation of

NADH-phenazine methosulphate-nitroblue tetrazolium formazan. Quercetin was used as a standard and the percentage inhibition of superoxide anion generation was calculated as previously described in equation 3.

Nitric oxide radical scavenging assay

Ebrahimzadeh *et al.* [38] procedure was adopted to determine the scavenging activity of the plant extracts against nitric oxide radical. Nitric oxide was generated from sodium nitroprusside and measured by the Greiss reaction. Curcumin was used as a standard. Curcumin inhibits induction of nitric oxide synthase and is a naturally occurring direct scavenger of nitric oxide. It reduces the amount of nitrite formed between oxygen and nitric oxide generated from sodium nitroprusside. The absorbance was measured at 596 nm and the percentage antioxidant activity calculated using the formula in equation 3.

Hydroxyl radical scavenging assay

The scavenging activity of the extract against hydroxyl radical was measured using the deoxyribose test-tube method [39] with minor changes. All solutions used was freshly prepared; 200 µL of 2.8 mM 2-deoxy-2-ribose, 5 µL of *N. laevis* leaf extract ,400 µL of 200 mM $FeCl_3$, 1.04 mM EDTA, 200 µL H_2O_2 (1.0 mM), 200 µL ascorbic acid (1.0 mM) and various concentrations (10–400 µg/ml) of the plant extract was mixed to form a reaction mixture. The mixture was incubated for 1 hour at 37°C. The extent of deoxyribose degradation was measured by TBA reaction. TCA (1.5 ml of 2.8% TCA) was added and kept for 20 mins. The solution was incubated at 90°C for 15 min to develop the colour. Afterwards, the solution was cooled and the absorbance measured at 532 nm against an appropriate blank solution Mannitol, a classical •OH scavenger was used as a positive control. The percentage antioxidant activity was calculated using the formula described in equation 3.

Peroxynitrite scavenging

Peroxynitrite ($ONOO^-$) was synthesized as described by previous methods [40]. An acidic solution (0.6 M HCl) of 5 ml H_2O_2 (0.7 M) was mixed with 5 ml 0.6 M KNO_2 on an ice bath for one second and 5 ml of ice-cold 1.2 M NaOH was added. Excess H_2O_2 was removed by treatment with granular MnO_2 prewashed with 1.2 M NaOH and the reaction mixture was left overnight at −20°C. Collection of peroxynitrite solution was achieved through the top of the frozen mixture and the concentration measured spectrophotometrically at 302 nm ($\varepsilon = 1670\ M^{-1}\ cm^{-1}$). The peroxynitrite scavenging activity was determined by Evans Blue bleaching assay [41] with slight modification. The reaction mixture contained 50 mM phosphate buffer (pH 7.4), 0.1 mM DTPA, 90 mM NaCl, 5 mM KCl, 12.5 µM Evans Blue, various concentrations of the plant extract (10–400 µg/ml) and 1 mM peroxynitrite in a final volume of 1 ml. The absorbance was measured at 611 nm after 30 min incubation at 25°C for. The percentage scavenging of $ONOO^-$ was calculated by comparing the results of the test and blank samples. Gallic acid was used as the standard.

Singlet oxygen scavenger

Production of singlet oxygen ($^1O_2$) was achieved by monitoring *N, N*-dimethyl-4-nitrosoaniline (RNO) bleaching, using a previously reported method [42,43]. Singlet oxygen was generated by a reaction between NaOCl and H_2O_2 and the bleaching of RNO monitored at 440 nm. The reaction mixture contained 45 mM phosphate buffer (pH 7.1), 50 mM NaOCl, 50 mM H_2O_2, 50 mM histidine, 10 µM RNO and various concentrations (10–400 µg/ml) of the plant extract in a final volume of 2 ml. It was incubated at 30°C for 40 min and the decrease in RNO absorbance was measured at 440 nm. The scavenging activity of sample was compared with that of lipoic acid, used as a standard compound.

Hypochlorous acid scavenging

Pedraza-Chaverrí *et al.* [44] description of hypochlorous acid scavenging activity was adopted with minor modification to determine the hypochlorous acid scavenging activity of *N.laevis.* Hypochlorous acid (HOCl) was prepared immediately before the experiment by adjusting the pH of a 10% (v/v) solution of NaOCl to 6.2 with 0.6 M H_2SO_4 and the concentration of HOCl was determined by measuring the absorbance at 235 nm using the molar extinction coefficient of 100 $M^{-1}\ cm^{-1}$. The scavenging activity was evaluated by determining the decrease in absorbance of catalase at 404 nm. The reaction mixture final volume (1 ml) contained 50 mM phosphate buffer (pH 6.8), catalase (7.2 µM), HOCl (8.4 mM) and increasing concentrations (10–400 µg/ml) of plant extract. The assay mixture was incubated at 25°C for 20 min and the absorbance measured against an appropriate blank. Ascorbic acid, a potent HOCl scavenger, was used as a standard.

Chelation power on ferrous (Fe^{2+}) ions

The ferrous ion chelating activity of the extract was evaluated *in vitro* as previously reported [45] with minor alterations. The reaction was carried out in HEPES buffer (20 mM, pH 7.2).Various concentrations (10–400 µg/ml) of the plant extract was added to a solution of 2 mM FeCl 2 (0.05 ml). The reaction was initiated by the addition of 5 mM ferrozine (0.2 ml) and the mixtures was then shaken vigorously and incubated at room temperature for 20 min. The absorbance of the solution was measured spectrophotometrically at 562 nm. The percentage inhibition of ferrozine-Fe^2+ complex formation (ferrous ion chelating ability) was calculated as [(A0 −A1/As)/A0] x100, where A0 is the absorbance of the control, and A1 is the absorbance of the plant extract and As the absorbance of a standard solution. EDTA was used as a standard.

Statistical analysis

The statistical analysis was done by one-way analysis of variance (ANOVA) using spss® version 18. The differences between the means were tested using posthoc LSD. A *p*-value of $p < 0.05$ was considered to be statistically significant and result presented as mean ± standard deviation. All assays were done in sextuplicate. The IC_{50} values were calculated by the formula $Y = 100^*A1/(X + A1)$, where $A1 = IC_{50}$, Y = response ($Y = 100\%$ when $X = 0$), X = inhibitory concentration. The IC_{50} values were compared by paired t tests and the antioxidant activity expressed in terms of IC_{50} (μg/ml concentration required to inhibit the radical formation by 50%).

Competing interests

The authors declare that they have no competing interests.

Authors' contributions

This work was carried out in collaboration between all authors. Author IBO conceptualized and designed the work, interpretation of results. Laboratory analysis and drafting of the original manuscripts and final approval of the version. HJB; involved in result interpretation and laboratory analysis. Critical revision of draft article for suitability and intellectual content and final approval of the version. All authors read and approved the final manuscript.

Author details

[1]Bioresources Development Centre Odi, Bayelsa, National Biotechnology Development Agency, Abuja, Nigeria. [2]Department of Biochemistry, College of Natural and Applied Sciences, Michael Okpara University of Agriculture Umudike, Umudike, Nigeria. [3]National Biotechnology Development Agency, Abuja, Nigeria.

References

1. Farombi EO. African indigenous plants with chemotherapeutic potentials and biotechnological plants with production of bioactive prophylactic agents. Afr J Biotechnol. 2003;2(12):662–7.
2. Bouayed J, Djilani A, Rammal H, Dicko A, Younos C, Soulimani R. Quantitative evaluation of the antioxidant properties of *Catha edulis*. J Life Sci. 2008;2:7–14.
3. Hutchinson J, Dalziel JM. Flora of West Tropical Africa, vol. II. London, S.W.I.: Crown Agents for Oversea Government and Adminstration 4, Millbank; 1963. p. 435–6.
4. Arbonnier M. Trees, Shrubs and Lianas of West African Dry Zones. Cote d'Ivorie: CIRAD, Margraf Publishers GMBH MNHN; 2004. p. 194.
5. Lewis WH, Manony PFE. Medical Botany: Plants Affecting Man's Health. New York, USA: John Wiley and Sons; 1977. p. 240.
6. Tor-anyin TA, Sha'ato R, Oluma HOA. Ethnobotanical Survey of antimalarial medicinal plants among the Tiv people of Nigeria. J Herbs Spices Med Plants. 2003;10(3):61–74.
7. Burkill HM. The useful Plants of West Tropical Africa, (Families A-D), vol. 1. 2nd ed. Kew, UK: Royal Botanic Gardens; 1985. p. 10. ISBN 094764301X.
8. Gbeassor M, Kedjagni AY, Koumagbo K, De Souza C, Agbo K, Aklikokou K, et al. In vitro antimalaria activity of six medicinal plants. Phytother Res. 2006;4(3):115–7.
9. Eyong KO, Folefoc GN, Kuete V, Beng VP, Krohn K, Hussain H, et al. Newbouldia quinine A. A napthoquinone-anthraquinone ether coupled pigment, as a potential antimicrobial and antimalaria agent from *Newbouldia laevis*. Phtochemistry. 2006;67(6):605–9.
10. Ogunlana EO, Ramstad E. Investigations into the antibacterial activities of local plants. Planta Med. 1975;27:534–60.
11. Ejele AE, Duru IA, Ogukwe CE, Iwu IC. Phytochemistry and antimicrobial potential of basic metabolites of piper umbellatum, piper guineense, Ocimum gratissimium and *newbouldia laevis* extracts. J Emerg Trends Eng Appl Sci (JETEAS). 2012;3(2):309–14.
12. Iwu MM. Handbook of African Medicinal Plants. London: CRC Press, Inc; 2000. p. 19.
13. Oliver-Bever B. Medicinal plants in Tropical West Africa. London: Cambridge University Press; 1986. p. 117–8. 168.
14. Germann K, Kaloga M, Ferreira D, Marais JP, Kolodziej H. Newbouldioside A–C Phenylethananoid Glycosides from the Stembark of *Newbouldia leavis*. Phytochemistry. 2006;67(8):805–11.
15. Anaduaka EG, Ogugua VN, Egba SI, Apeh VO. Investigation of some important phytochemical, nutritional properties and toxicological potentials of ethanol extracts of *Newbouldia laevis* leaf and stem. Afr J Biotechnol. 2013;12(40):5941–9.
16. Akerele JO, Ayinde BA, Ngiagah J. Comparative phytochemical and antimicrobial activities of the leaf and root bark of *Newbouldia laevis* seem (bignoniaceae) on some clinically isolated bacterial organisms. Niger J Pharm Sci. 2011;10(2):8–14.
17. Usman H, Osuji JC. Phytochemical and in-vitro antimicrobial assay of the leaf extract of *Newbouldia laevis*. Afr J Trad CAM. 2007;4(4):476–80.
18. Kolawole OT, Akanji MA. Effects of extracts of leaves *of Newbouldia laevis* on the activities of some enzymes of hepatic glucose metabolism in diabetic rats. Afr J Biotechnol. 2014;13(22):2273–81.
19. Miller A. Antioxidant flavonoids: structure, function and clinical usage. Altern Med Rev. 1996;1(2):103–11.
20. Niki E, Noguchi N, Tsuchihashi H, Naohiro G. Interaction among vitamin C, vitamin E, and 13-carotene13. Am J Clin Nutr. 1995;62(suppl):I322S–6.
21. Mayakrishnan V, Veluswamy S, Sundaram KS, Kannappan P, Abdullah N. Free radical scavenging potential of *Lagenaria siceraria* (Molina) Standl fruits extract. Asian Pac J Trop Med. 2013;6(1):20–6.
22. Ogulana OE, Ogunlana OO. Invitro assessment of antioxidant activity of *Newbouldia laevis*. J Med Plant Res. 2008;2(8):176–9.
23. Ogunlana OE, Ogunlana OO, Farombi OE. Assessment of the scavenging activity of crude methanolic stem bark extract of *Newbouldia Laevis* on selected free radicals. Adv Nat Appl Sci. 2008;2(3):249–54.
24. Robak J, Gryglewski IR. Flavonoids are scavengers of superoxide anions. Biochem Pharmacol. 1988;37:837–41.
25. Balavoine GG, Geletti YV. Peroxynitrite scavenging by different antioxidants. Part 1: convenient study. Nitric Oxide. 1999;3:40–54.
26. Harborne JB. Phytochemical Methods; A Guide to Modern Techniques of Plant Analysis. 2nd ed. London: Chapman and Hall; 1973. p. 49–279.
27. Harborne JB. Phytochemical Methods; A Guide to Modern Techniques of Plant Analysis. 2nd ed. London: Chapman and Hall; 1984. p. 4–16.
28. Trease GE, Evans WC. A Text Book of Pharmacognosy. Oxoford, UK: Elsb/Bailliere Tindal; 1987. p. 1055.
29. Trease GE, Evans WC. Pharmacognosy. 4th ed. USA: WB.Sounders; 1996. p. 243–83.
30. Sofowara A. Medical Plants and Traditional Medicine in Africa. Rep. Ibadan: Spectrum books LTD; 2006. p. 150.
31. Harbone JB. Methods of Extraction and Isolation. In: Phytochemical Methods. 3rd ed. London: Chapman and Hall; 1998. p. 42–98.
32. Barakat MZ, Shahab SK, Darwin N, Zahemy EI. Determination of ascorbic acid from plants. Anal Biochem. 1993;53:225–45.
33. Kirk RS, Sawyer R. Pearson's Chemical Analysis of Foods. 9th ed. Harlow, UK: Longman Scientific and Technical; 1991. p. 25.
34. Roberto G, Baratta MT. Antioxidant activity of selected essential oil components in two lipid model system. Food Chem. 2000;69(2):167–74.
35. Benzie FF, Strain JJ. Ferric reducing/antioxidant power assay: direct measure of total antioxidant activity of biological fluids and modified version for simultaneous measurement of total antioxidant power and ascorbic acid concentration. Methods Enzymol. 1999;299:15–23.
36. Mensor LI, Menezes FS, Leitao GG, Reis AS, Santos TC, Coube CS, et al. Screening of Brazilian plant extracts for antioxidant activity by the use of DPPH free radical method. Phytother Res. 2001;15:127–30.
37. Fontana M, Mosca L, Rosei MA. Interaction of enkephalines with oxyradicals. Biochem Pharmacol. 2001;61:1253–7.
38. Ebrahimzadeh MA, Pourmorad F, Hafezi S. Antioxidant Activities of Iranian Corn Silk. Turkish J Biol. 2008;32:43–9.
39. Halliwell B, Gutteridge J, Aruoma OL. The deoxyribose method: a simple

test-tube assay for determination of rate constants for reactions of hydroxyl radicals. Anal Biochem. 1987;165(1):215–9.

40. Beckman JS, Chen H, Ischiropulos H, Crow JP. Oxidative chemistry of peroxynitrite. Methods Enzymol. 1994;233:229–40.
41. Bailly F, Zoete V, Vamecq J, Catteu JP, Bernier JL. Antioxidant actions of ovothiol-derived 4-mercaptoimidazoles: glutathione peroxidase activity and protection against peroxynitrite-induced damage. FEBS Lett. 2000;486:19–22.
42. Chakraborty N, Tripathy BC. Involvement of singlet oxygen in 5-aminolevulinic acid-induced photodynamic damage of cucumber (*Cucumbis sativus* L.) *chloroplasts*. Plant Physiol. 1992;98:7–11.
43. Pedraza-Chaverrí J, Barrera D, Maldonado PD, Chirino Y, Macías-Ruvalcaba NA, Medina-Campos ON, et al. S-allylmercaptocysteine scavenges hydroxyl radical and singlet oxygen in vitro and attenuates gentamicininduced oxidative and nitrosative stress and renal damage *in vivo*. BMC Clin Pharmacol. 2004;4:5.
44. Pedraza-Chaverrí J, Arriaga-Noblecía G, Medina-Campos ON. Hypochlorous acid scavenging capacity of garlic. Phytother Res. 2007;21:884–8.
45. Haro-Vicente JF, Martinez-Gracia C, Ros G. Optimization of *in vitro* measurement of available iron from different fortificants in citric fruit juices. Food Chem. 2006;98:639–48.

17

Salinity-induced changes in the morphology and major mineral nutrient composition of purslane (*Portulaca oleracea* L.) accessions

Md. Amirul Alam[1*], Abdul Shukor Juraimi[2], M. Y. Rafii[2,3], Azizah Abdul Hamid[4], Farzad Aslani[2] and M. A. Hakim[3]

Abstract

This study was undertaken to determine the effects of varied salinity regimes on the morphological traits (plant height, number of leaves, number of flowers, fresh and dry weight) and major mineral composition of 13 selected purslane accessions. Most of the morphological traits measured were reduced at varied salinity levels (0.0, 8, 16, 24 and 32 dS m^{-1}), but plant height was found to increase in Ac1 at 16 dS m^{-1} salinity, and Ac13 was the most affected accession. The highest reductions in the number of leaves and number of flowers were recorded in Ac13 at 32 dS m^{-1} salinity compared to the control. The highest fresh and dry weight reductions were noted in Ac8 and Ac6, respectively, at 32 dS m^{-1} salinity, whereas the highest increase in both fresh and dry weight was recorded in Ac9 at 24 dS m^{-1} salinity compared to the control. In contrast, at lower salinity levels, all of the measured mineral levels were found to increase and later decrease with increasing salinity, but the performance of different accessions was different depending on the salinity level. A dendrogram was also constructed by UPGMA based on the morphological traits and mineral compositions, in which the 13 accessions were grouped into 5 clusters, indicating greater diversity among them. A three-dimensional principal component analysis also confirmed the output of grouping from cluster analysis.

Keywords: Purslane (*Portulaca oleracea* L.), NaCl, Salinity, Morphology, Mineral compositions

Background

Purslane (*Portulaca oleracea* L.) is the eighth most common plant distributed throughout the world, because it is an important heat- and drought-tolerant vegetable crop [9]. It is eaten fresh, cooked or dried, and cultivation has gained popularity across the world in recent years because the plant has been identified as a rich source of ω3 polyunsaturated fatty acids and antioxidants [3, 49]. Moreover, purslane is promising for providing both novel biologically active substances and essential compounds for human nutrition [15]. Purslane has proven to be more salt-tolerant than any other vegetable crop [4, 58] and can produce sufficient biomass under moderate salinity stress, which other vegetable crops cannot [32]. Salinity is possibly the most significant ecological factor that causes extensive crop yield losses globally, and its threat is escalating daily [48]. Increasing salinity reduces the average yield of major crops by more than 50 % [14], and these losses are of great concern, mainly in countries with agriculture-based economies. High concentrations of salt impose both osmotic and ionic stresses on plants, which lead to several morphological and physiological changes [30]. A clear stunting of plants has been observed to result from salinity stress [51]. Parida and Das [42] reported that the detrimental effects of high salinity in plants can result in plant death and/or decreased productivity. The earliest response is a reduction in the rate of leaf surface expansion, followed by a cessation of expansion as the stress intensifies [42]. Salinity stress causes an

*Correspondence: amirulalam@unisza.edu.my
[1] School of Agriculture Science and Biotechnology, Faculty of Bioresources and Food Industry, Universiti Sultan Zainal Abidin, Tembila Campus, 22200 Besut, Terengganu, Malaysia
Full list of author information is available at the end of the article

imbalance in the uptake of mineral nutrients and their distribution within the plants [23]. Furthermore, many nutrient interactions in salt-stressed plants can occur, which may have important consequences for growth [43]. Internal concentrations of major nutrients and their uptake have been frequently studied [17], but the relationship between micro-nutrient concentrations and soil salinity is rather complex and remains poorly understood [53]. Munns and Tester [41] stated that salt-tolerant species are able to grow and reproduce even in oceanic-level salinities. The only way to control the salinization process and to maintain the sustainability of landscapes and agricultural fields is to combat the salinization problems using environmentally safe and clean techniques and by using salt-tolerant species [13, 26]. Salt tolerant crop varieties are becoming essential in many areas of the world, including Malaysia, because of salt accumulation in soil, restrictions on groundwater use and saltwater intrusion into groundwater [29, 56]. Under the prevailing conditions of increasing salinity, it is necessary to incorporate salt-tolerant plants, which can withstand the increasing stress of salinity and can economically substitute existing crops. Therefore, this research was undertaken to study the effect of salinity on the morphological traits and mineral composition of purslane.

Results

Purslane morphological traits analysis

Plant height

The plant height of untreated control 13 purslane accessions differed very significantly ($P < 0.0001$) and ranged from 33.4 to 70 cm, with the highest plant height occurring in Ac9 and the lowest in Ac13 (Table 1). At the end of the salinity treatment, the plant height was highly reduced at 32 dS m^{-1} salinity followed by 24, 16 and 8 dS m^{-1} compared to the control plants (Table 1). However, some exceptions were also observed in the case of accession numbers Ac1, Ac2 and Ac8. Among all the 13 purslane accessions, the highest plant height reduction (>33 %) was recorded in Ac13 at 32 dS m^{-1} salinity, whereas the lowest reduction (3.28 %) was found in Ac5 at 8 dS m^{-1} salinity; both samples were ornamental purslane (Table 1). Interestingly, a slight increase (2.09 %) in plant height was also observed in Ac1 at 16 dS m^{-1} salinity stress compared to the control. Less than a 5 % reduction was observed in the case of Ac5, Ac6 and Ac9 at 8 dS m^{-1} salinity stress, while the same was observed in Ac5 at 16 dS m^{-1} salinity. Furthermore, at 24 dS m^{-1} salinity, less than a 10 % plant height reduction was recorded in Ac2 and Ac5, and even at the highest salinity stress (32 dS m^{-1}), the same reduction was noted in Ac1 and Ac2 (Table 1). On average across all accessions, a total of 6.99, 10.76, 16.23 and 20.18 % reductions in plant height were recorded, respectively, at 8, 16, 24 and 32 dS m^{-1} salinity, which were statistically significant values ($P < 0.05$; Table 1).

Number of leaves

Highly significant ($P < 0.001$) variation was observed in the number of leaves in the untreated control and 13 purslane accessions. The largest number of leaves (555)

Table 1 Effect of salinity on plant height of 13 purslane accessions

Purslane accessions	Plant height (cm)				
	Salinity level (dS m^{-1})				
	0	8	16	24	32
Ac1	43.10h	38.30h (11.14)	44.0e (+2.09)	36.7f (14.85)	39.7d (7.89)
Ac2	42.80h	39.20h (8.41)	37.4g (12.62)	40.5e (5.37)	39.6d (7.48)
Ac3	45.80j	42.10g (8.08)	41.7f (8.95)	40.8e (10.92)	37.7e (17.68)
Ac4	48.10e	45.50f (5.41)	42.7ef (11.23)	41.4e (13.93)	40.5d (15.8)
Ac5	42.70h	41.30g (3.28)	41.0f (3.98)	39.8e (6.79)	37.4e (12.41)
Ac6	59.61b	56.70b (4.88)	53.4b (10.42)	50.2b (15.78)	46.4b (22.16)
Ac7	56.40c	52.40d (7.09)	49.8c (11.7)	46.7c (17.19)	43.2c (23.4)
Ac8	52.90d	28.20e (8.88)	47.5c (10.21)	45.8cd (13.42)	44.1c (16.64)
Ac9	70.0a	67.40a (3.71)	63.8a (8.86)	57.6a (17.71)	53.3a (23.86)
Ac10	59.20b	54.40c (8.11)	51.8b (12.5)	44.7d (24.49)	42.8c (27.7)
Ac11	43.80gh	41.60g (5.02)	38.4g (12.33)	36.9f (15.75)	33.7f (23.06)
Ac12	45.30fg	41.42g (8.57)	36.7f (18.98)	33.4g (26.27)	32.8f (27.59)
Ac13	33.40i	29.60i (11.38)	25.7h (23.05)	24.23h (27.46)	22.1g (33.83)
Mean	49.47a	46.011b (6.99)	44.15c (10.76)	41.44d (16.23)	39.48e (20.18)

Mean values with different lower case letters in a row are significantly different at $P < 0.05$. Values in the parentheses indicate percent compared to the untreated control (0 dS m^{-1}) plants

'+' symbol denotes increase in plant height under salinity stress compared to control

was recorded in Ac13, which was a common purslane, and the lowest (351) was found in Ac3, which was an ornamental purslane (Table 2). The number of leaves in the salt-treated purslane accessions was substantially reduced with increasing salinity levels (Table 2). The highest reduction (43.6 %) was observed in Ac13 (common purslane) at the highest 32 dS m^{-1} salinity, whereas the lowest reduction (1.74 %) was noted in Ac11 (ornamental purslane) at 8 dS m^{-1} salinity compared to the control (Table 2). At 8 dS m^{-1} of salinity reduction, the number of leaves varied from 1.74 to 17.81 %, which increased to 4.28 to 27.69 % at 16 dS m^{-1} salinity. In contrast, less than a 10 % reduction was observed in Ac10 and Ac6 at 24 and 32 dS m^{-1} salinity, respectively (Table 2). Interestingly, a consequent and significant ($P < 0.05$) increase in number of leaves was also found in Ac5 and Ac9 with increasing of salinity levels compared to the control accessions (Table 2). The mean values of all of the accessions revealed a total of 8.31, 13.73 and 20.82 % reduction and 24.77 % increase in the number of main branches, respectively, at 8, 16, 24 and 32 dS m^{-1} salinity levels, which were statistically significant increases ($P < 0.05$; Table 2).

Flowering

The numbers of flowers in the untreated control compared to 13 purslane accessions differed very significantly ($P < 0.0001$) and ranged between 6.63 and 63.47, with the highest flower numbers occurring in Ac12, which was a common purslane, and the lowest values were in Ac8, which was an ornamental purslane (Table 3). Highly significant reductions in the number of flowers were observed at the highest, 32 dS m^{-1}, salinity compared to the control as well as at other salinity levels (Table 3). The highest reduction (96.48 %) in the number of flowers was recorded in Ac13 at the highest, 32 dS m^{-1}, salinity, which was a common purslane, whereas the lowest reduction in the number of flowers (3.86 %) was observed in Ac5 at the lowest, 8 dS m^{-1}, salinity compared to the control, which was an ornamental purslane (Table 3). All 13 purslane accessions and 4 salinity levels (except the control) had less than a 5 % reduction in the number of flowers recorded in Ac5 and Ac7 at 8 dS m^{-1} salinity, whereas a 15–56 % reduction occurred in the number of flowers that were observed at 16 dS m^{-1} salinity. Further augmented salinity levels at 24 and 32 dS m^{-1} salinity reductions in the number of flowers varied from 31–72 to 44–97 %, respectively, compared to the control (Table 3). The mean values of all of the accessions revealed 17.74, 37.79, 51.36 and 70.78 % reductions in the number of flowers at 8, 16, 24 and 32 dS m^{-1} salinities, respectively, which were statistically significant reductions ($P < 0.0001$; Table 3).

Fresh weight

Highly significant ($P < 0.0001$) variation was observed in the fresh weights in the untreated control and the 13 purslane accessions. The highest fresh weight (341.03 g)

Table 2 Effect of salinity on number of leaves in 13 purslane accessions

Purslane accessions	Number of leaves				
	Salinity level (dS m^{-1})				
	0	8	16	24	32
Ac1	525.30ab	431.7bc (17.81)	381.4c (27.39)	350.7b (33.24)	333.61cd (36.49)
Ac2	501.20ab	413.8cd (17.44)	362.4c (27.69)	351.8b (29.81)	333.3cd (33.49)
Ac3	350.80de	314.2ef (10.43)	260.8d (25.66)	220.37c (37.18)	241.3ef (31.21)
Ac4	489.60ab	417.5cd (14.73)	403.3bc (17.63)	349.9b (28.53)	321.7cd (34.29)
Ac5	405.80cd	417.5cd (+2.88)	411.7bc (+1.45)	420.6ab (+3.65)	428.7a (+5.64)
Ac6	456.80bc	427.7bc (6.37)	411.8bc (9.85)	409.7ab (10.31)	411.2ab (9.98)
Ac7	490.40ab	444.2bc (9.42)	413.3bc (15.72)	388.7ab (20.74)	349.7b–d (28.29)
Ac8	527.20ab	511.1a (3.05)	489.7a (7.11)	449.3a (14.78)	431.5a (18.15)
Ac9	353.60de	361.2de (+2.15)	383.7c (+8.51)	359.8b (+1.75)	380.3a–c (+7.55)
Ac10	372.60d	363.4de (2.47)	356.7c (4.28)	348.3b (6.52)	333.4cd (10.52)
Ac11	487.80ab	479.3ab (1.74)	453.3ab (7.07)	389.9ab (20.07)	288.4de (40.88)
Ac12	282.60e	273.9f (3.08)	255.7d (9.52)	201.5c (28.69)	196.4f (30.5)
Ac13	555.40a	461.5a–c (16.91)	419.13bc (24.54)	351.4b (36.73)	313.4cd (43.57)
Mean	446.08a	409.0b (8.31)	384.84c (13.73)	353.23d (20.82)	335.61e (24.77)

Mean values with different lower case letters in a row are significantly different at $P < 0.05$. Values in the parentheses indicate percent compared to the untreated control (0 dS m^{-1}) plants

'+' symbol denotes increase in number of leaves under salinity stress compared to control

Table 3 Effect of salinity on number of flowers in 13 purslane accessions

Purslane accessions	Number of flowers				
	Salinity level (dS m^{-1})				
	0	8	16	24	32
Ac1	34.65g	26.53e (23.43)	17.23g (50.27)	13.27h (61.7)	8.92h (74.26)
Ac2	46.33c	39.63bc (14.46)	27.89e (39.8)	23.22d (49.88)	11.18f (75.87)
Ac3	33.77g	30.32g (10.22)	15.39h (54.43)	17.88f (47.05)	12.38e (63.34)
Ac4	43.28e	37.13cd (14.21)	28.6e (33.92)	19.8e (54.25)	10.58g (75.56)
Ac5	38.3f	36.82cd (3.86)	31.28c (18.33)	26.37b (31.14)	21.14a (44.8)
Ac6	44.57d	39.57bc (11.21)	34.53b (22.53)	25.55c (42.67)	17.11c (61.61)
Ac7	44.37d	42.58ab (4.03)	37.39a (15.73)	33.31a (24.93)	19.41b (56.25)
Ac8	6.63k	5.13h (22.62)	3.47j (47.66)	3.08f (53.54)	2.43j (63.35)
Ac9	18.78j	15.12g (19.48)	12.39i (34.03)	11.12i (40.78)	7.44i (60.38)
Ac10	25.53h	22.13f (13.32)	17.81g (30.24)	13.24h (48.14)	7.11i (72.15)
Ac11	23.34i	21.58f (7.54)	17.69g (24.21)	11.28i (51.67)	7.23i (69.02)
Ac12	63.47a	33.53d (47.17)	29.68d (53.24)	19.42e (69.4)	13.47c (78.78)
Ac13	57.61b	45.32a (21.33)	25.61f (55.55)	16.23g (71.83)	2.03j (96.48)
Mean	36.97a	29.39b (17.74)	23.0c (37.79)	17.98d (51.36)	10.8e (70.78)

Mean values and ± SE with different lower case letters in a row are significantly different at $P < 0.05$. Values in the parentheses indicate percent compared to the untreated control (0 dS m^{-1}) plants

was recorded in Ac8, which was an ornamental purslane, and the lowest (103.67 g) was found in Ac13, which is a common purslane (Table 4). The fresh weights of the salinity stressed purslane accessions were also significantly affected with the highest levels (378.15 g) occurring in Ac9 at 24 dS m^{-1} salinity and the lowest (86.98 g) in Ac12 at 32 dS m^{-1} salinity compared to the control (Table 4). Increases in fresh weights with increasing salinity were recorded in Ac1 at 8 dS m^{-1} salinity, in Ac9 at 16, 24 and 32 dS m^{-1} salinity and in Ac13 at 8 and 16 dS m^{-1} salinity levels compared to the control (Table 4). At 8 dS m^{-1} salinity levels, the fresh weight reductions varied between 1 and 37 %, with the lowest reduction (0.89 %) in Ac3 and the highest (36.42 %) reduction in Ac8. In contrast, 3–43, 2–48 and 4–55 % fresh weight reductions were recorded in 16, 24 and 32 dS m^{-1} salinity, respectively. On average over all of the accessions, 14.36, 18.88, 21.02 and 26.09 % reductions in fresh weight were observed at 8, 16, 24 and 32 dS m^{-1} salinity, respectively, which were statistically significant ($P < 0.05$; Table 4).

Dry weight

The dry matter (DM) content in the untreated control plants was significantly different ($P < 0.0001$) from the 13 purslane accessions and ranged from 7.94 to 20.67 g pot^{-1}, with the highest DM content occurring in Ac6 and the lowest in Ac5 (Table 5). The dry matter content was also significantly reduced by NaCl-induced salinity stress in all 13 purslane accessions, with increasing of salinity levels occurring, except in Ac1 at 8 dS m^{-1} salinity, in Ac9 at 16, 24 and 32 dS m^{-1}, in Ac12 and in Ac13 at 8 dS m^{-1} salinities, where significant increases in the dry matter content were recorded (Table 5). In contrast, the highest dry matter reduction (63.47 %) was found in Ac6 at 32 dS m^{-1} salinity, and the lowest reduction (1.64 %) was noted in Ac5 at 24 dS m^{-1} salinity, whereas the highest increase (54.19 %) in dry matter content was recorded in Ac9 at 24 dS m^{-1} salinity, following the lowest increase (1.83 %) in Ac13 at 8 dS m^{-1} salinity (Table 5). The mean values of all the accessions revealed 11.24, 20.91, 23.05 and 32.88 % reductions in the dry matter content at 8, 16, 24 and 32 dS m^{-1} salinity, respectively, which were statistically significant ($P < 0.0001$; Table 5).

Micro and macro mineral elements

Phosphorus (P) content in purslane

Significant ($P < 0.0001$) variations were also observed in the P content of the untreated control and 13 purslane accessions. The phosphorus content differed from 0.25 to 0.71 %, with the highest value observed in Ac13 and the lowest in Ac4 (Table 6). Both the negative and positive effects of different salinity levels were noted in the phosphorus content in all 13 purslane accessions. In most of the accessions, the phosphorus content was found to increase at the initial (8 dS m^{-1}) augmented salinity stress, with some exceptions in Ac3, Ac4 and Ac13 compared to the control (Table 6). Further salinity increases reduced the P content in all of the purslane accessions up to the highest salinity levels,

Table 4 Effect of salinity on fresh weight of 13 purslane accessions

Purslane accessions	Fresh weight (g)				
	Salinity level (dS m^{-1})				
	0	8	16	24	32
Ac1	225.0e	231.33b (+2.81)	203.78b (9.43)	192.49b (14.45)	187.37b (16.72)
Ac2	213.58f	203.14e (4.88)	193.58c (9.36)	188.93bc (11.54)	171.58d (19.66)
Ac3	190.5g	188.79f (0.89)	177.93d (6.59)	181.37bc (4.79)	177.68c (6.73)
Ac4	187.0g	149.16g (20.24)	112.32g (39.94)	117.3g (37.27)	106.31h (43.15)
Ac5	134.16i	121.31h (9.58)	129.48e (3.39)	131.28f (2.15)	113.58g (15.34)
Ac6	279.0c	229.0b (17.92)	174.72d (37.38)	154.68e (44.56)	159.87e (42.69)
Ac7	230.0e	223.51c (2.82)	174.97d (23.93)	168.94d (26.55)	151.6f (34.09)
Ac8	341.03a	216.82d (36.42)	197.4bc (42.11)	180.29cd (47.13)	156.61ef (54.08)
Ac9	305.17b	248.61a (18.53)	346.97a (+13.69)	378.15a (+23.91)	355.68a (+16.55)
Ac10	149.17h	114.53i (23.22)	103.43h (30.66)	98.26h (34.13)	89.30j (40.14)
Ac11	242.0d	185.0f (23.55)	174.83d (27.76)	169.56d (29.93)	161.79e (33.14)
Ac12	129.48i	112.94i (12.77)	105.14h (18.79)	93.4h (27.27)	86.98j (32.82)
Ac13	103.66j	113.52i (+9.51)	119.81f (+15.58)	101.3h (2.78)	99.11i (4.39)
Mean	209.98a	179.82b (14.36)	170.34c (18.88)	165.84d (21.02)	155.19e (26.09)

Values with different lower case letters in a row are significantly different at $P < 0.05$. Values in the parentheses indicate percent compared to the untreated control (0 dS m^{-1}) plants

'+' symbol indicates % increase in fresh weight compared to control

Table 5 Effect of salinity on dry weight of 13 purslane accessions

Purslane accessions	Dry weight (g)				
	Salinity level (dS m^{-1})				
	0	8	16	24	32
Ac1	16.27bc	17.13bc (+5.29)	13.1bc (13.1)	9.23e–g (9.23)	8.33ef (8.33)
Ac2	15.39bc	13.58d (11.76)	10.34d (10.34)	9.18e–g (9.18)	8.42d–f (8.42)
Ac3	23.55a	21.39a (21.39)	19.29a (19.29)	20.16b (20.16)	18.51b (18.51)
Ac4	15.55bc	13.97cd (13.97)	10.7d (10.7)	8.95e–g (8.95)	8.66d–f (8.66)
Ac5	7.94d	6.45f (6.45)	6.74f (6.74)	7.81fg (7.81)	5.39g (5.39)
Ac6	20.67ab	17.77ab (17.77)	10.02d (10.02)	14.38c (14.38)	7.55e–g (7.55)
Ac7	15.91bc	13.24d (13.24)	14.17b (14.17)	12.77cd (12.77)	12.04c (12.04)
Ac8	16.63bc	11.57de (11.57)	11.16cd (11.16)	10.92de (10.92)	9.98c–e (9.98)
Ac9	15.5bc	13.48d (13.48)	21.3a (21.3)	23.9a (23.9)	23.4a (23.4)
Ac10	10.10cd	8.60ef (8.6)	7.44ef (7.44)	7.13g (7.13)	6.89fg (6.89)
Ac11	18.84ab	14.0cd (14.0)	13.43bc (13.43)	12.47cd (12.47)	11.08cd (11.08)
Ac12	11.99cd	14.41b–d (14.41)	9.46de (9.46)	7.72fg (7.72)	6.33fg (6.33)
Ac13	12.05cd	12.27d (12.27)	11.34cd (11.34)	9.58ef (9.58)	7.93e–g (7.93)
Mean	15.41a	13.68b (11.24)	12.19c (20.91)	11.86d (23.05)	10.35e (32.88)

Mean values with different lower case letters in a row are significantly different at $P < 0.0001$. Values in the parentheses indicate percent compared to the untreated control (0 dS m^{-1}) plants

'+' symbol indicate % increase in dry weight compared to control

whereas a complete reduction in the P content at all 4 salinity levels was noted in Ac5 and Ac7 compared to the control (Table 6). Consequent reductions in the P contents were found to increase with increasing salinity stress, and the highest reduction (69.43 %) was seen in Ac5 at the highest salinity levels at 32 dS m^{-1}, whereas the highest increase (183.07 %) was noted in Ac4 at the lowest salinity levels (8 dS m^{-1}) compared to the

Table 6 Effect of salinity on P content in 13 purslane accessions

Purslane accessions	P content (%, DW basis)				
	Salinity level (dS m^{-1})				
	0	8	16	24	32
Ac1	0.32i	0.43h (+36.39)	0.14k (56.33)	0.11j (63.92)	0.11j (64.87)
Ac2	0.37g	0.42i (+14.17)	0.15j (58.86)	0.11j (69.21)	0.13h (63.49)
Ac3	0.37g	0.52f (+37.97)	0.49c (+31.02)	0.35e (7.49)	0.34c (8.56)
Ac4	0.25j	0.72a (+183.07)	0.61b (+139.76)	0.38d (+47.64)	0.23g (8.66)
Ac5	0.42d	0.20k (53.32)	0.20i (53.32)	0.15h (65.64)	0.13hi (69.43)
Ac6	0.41e	0.42i (+0.97)	0.39f (8.03)	0.32f (22.63)	0.32d (22.63)
Ac7	0.34h	0.14l (58.63)	0.13k (60.42)	0.13i (62.80)	0.12ij (63.99)
Ac8	0.54c	0.66c (+22.35)	0.40e (25.88)	0.37d (30.73)	0.25f (52.70)
Ac9	0.57b	0.59d (+3.16)	0.49c (14.91)	0.46b (19.47)	0.45a (21.40)
Ac10	0.32i	0.34j (+6.25)	0.28h (14.06)	0.24g (26.25)	0.24g (26.25)
Ac11	0.39f	0.46g (+19.74)	0.35g (9.87)	0.32f (17.14)	0.29e (25.71)
Ac12	0.42de	0.57e (+38.31)	0.41d (0.96)	0.41c (1.45)	0.37b (11.08)
Ac13	0.71a	0.70b (1.27)	0.78a (+9.31)	0.49a (31.31)	0.34c (51.76)
Mean	0.42b	0.47a (+13.66)	0.37c (11.61)	0.29d (29.51)	0.26e (38.66)

Mean values with different lower case letters in a row are significantly different at $P < 0.05$. Values in the parentheses indicate percent compared to the untreated control (0 dS m^{-1}) plants

'+' symbol indicates % increase of P content

control (Table 6). On average, over all of the accessions, 13.66 % increase, 11.61, 29.51 and 38.66 % reductions in P content were recorded, respectively, at 8, 16, 24 and 32 dS m^{-1} salinities and were statistically significant ($P < 0.05$; Table 6).

Sodium (Na) content in purslane

The accession differences in sodium concentrations in purslane were highly pronounced ($P < 0.0001$), ranging from 0.26 to 0.77 % under control conditions, with the highest in Ac11 and the lowest in Ac1 (Table 7). Sodium concentrations were observed to increase progressively with increasing salinity in most of the purslane accessions, with the exception of Ac11, where a significant decrease in the Na concentration was found at all 4 salinity levels compared to the control (Table 7). In salinity-stressed purslane, the highest increase (257.6 %) in Na concentration was observed in Ac1 at 32 dS m^{-1} salinity, whereas zero effect from salinity stress was recorded in Ac6 at 24 dS m^{-1} salinity compared to the control (Table 7). On the contrary, at the beginning in Ac13, a significant increase in Na concentration was observed; however, further increased salinity resulted in Na concentrations that declined significantly compared to the control (Table 7). On average over all of the accessions, 34.8, 54.5, 56.1 and 68.5 % increases in Na concentrations were recorded, respectively, at 8, 16, 24 and 32 dS m^{-1} salinities and were statistically significant ($P < 0.05$; Table 7).

Potassium (K) content in purslane

The potassium content varied greatly ($P < 0.0001$) among all 13 untreated purslane accessions, with the highest content (8.20 %) observed in Ac11, and the lowest content (3.30 %) in Ac1. Interestingly, the K content in Ac10 (an ornamental purslane) and in Ac12 (a common purslane) was found to be similar (5.98 %), which was also statistically non-significant. Augmented salinity stresses also significantly ($P < 0.05$) reduced the K content in all 13 purslane accessions, except in Ac1 and Ac10 at 8 dS m^{-1} salinity, in which a slight increase (4.51 and 8.79 %, respectively) in the K content was recorded compared to the control (Table 8). Throughout the salinity treatments, the K contents were increasingly reduced with increasing salinity levels, and the highest reduction (60.6 %) was observed in Ac5 at 32 dS m^{-1} salinity, whereas the lowest (0.84 %) was seen in Ac12 at the lowest (8 dS m^{-1}) salinity stress compared to the control (Table 8). On average, over all of the accessions, 13.08, 25.18, 31.93 and 37.40 % reductions in K content were recorded, respectively, at 8, 16, 24 and 32 dS m^{-1} salinity and were statistically significant values ($P < 0.05$; Table 8).

Calcium (Ca) content in purslane

Calcium concentrations in purslane accessions observed in the range of 4.17–1.40 % in untreated control plants with the highest levels in Ac6 and the lowest in Ac10 (Table 9). The calcium content was heavily affected by salinity, with clear differences among accessions where

Table 7 Effect of salinity on Na content in 13 purslane accessions

Purslane accessions	Na content (%, DW basis)				
	Salinity level (dS m^{-1})				
	0	8	16	24	32
Ac1	0.26l	0.46i (+72.12)	0.63f (+141.22)	0.85c (+225.57)	0.94a (+257.63)
Ac2	0.46d	0.61d (+31.61)	0.72d (+54.69)	0.86b (+85.18)	0.90b (+95.33)
Ac3	0.32j	0.34k (+7.19)	0.63f (+97.50)	0.72g (+124.69)	0.79e (+146.56)
Ac4	0.29k	0.55g (+91.29)	0.50i (+73.52)	0.73f (+153.31)	0.75g (+160.63)
Ac5	0.43f	0.59e (+36.47)	0.71e (+63.28)	0.76e (+75.98)	0.71i (+63.05)
Ac6	0.43g	0.71b (67.61)	0.62g (+44.37)	0.43l (0.0)	0.71h (+67.61)
Ac7	0.62b	0.72a (+17.40)	0.76b (+24.23)	0.86a (+40.16)	0.76f (+23.41)
Ac8	0.35h	0.36j (+3.15)	0.97a (+178.80)	0.61j (+73.35)	0.67k (+41.40)
Ac9	0.32j	0.58f (+80.56)	0.72d (+124.14)	0.83d (+160.19)	0.88c (+174.95)
Ac10	0.33i	0.54h (62.19)	0.44k (+32.04)	0.65i (+95.51)	69.2d (+107.19)
Ac11	0.77a	0.71b (7.64)	0.71c (5.96)	0.20m (74.74)	0.43l (44.43)
Ac12	0.49c	0.68c (+38.89)	0.49j (0.00)	0.68h (+38.89)	0.70j (+43.0)
Ac13	0.44e	0.57f (+29.38)	0.61h (+36.97)	0.44k (1.10)	0.37m (17.73)
Mean	0.42d	0.57c (+34.75)	0.65b (+54.51)	0.66b (+56.14)	0.71a (+68.49)

Mean values with different lower case letters in each column are significantly different at $P < 0.05$. Values in the parentheses indicate percent compared to the untreated control (0 dS m^{-1}) plants

'+' symbol indicates % increase of Na content

Table 8 Effect of salinity on K content in 13 purslane accessions

Purslane accessions	K content (%, DW basis)				
	Salinity level (dS m^{-1})				
	0	8	16	24	32
Ac1	3.33h	3.48f (+4.51)	3.18h (4.51)	2.80d (15.79)	2.55h (23.31)
Ac2	3.8g	3.48f (8.55)	3.13h (17.76)	2.83d (25.66)	2.65h (30.26)
Ac3	5.78de	5.35d (7.36)	4.13f (28.57)	4.03c (30.30)	3.75de (35.06)
Ac4	7.18b	5.05e (29.62)	4.70d (34.49)	4.05c (43.55)	3.40f (52.61)
Ac5	6.30c	3.70f (41.27)	3.05h (51.59)	2.90d (53.97)	2.48h (60.60)
Ac6	5.15f	4.88e (5.34)	4.48e (13.11)	4.00c (22.33)	3.10g (39.81)
Ac7	5.60e	3.78f (32.59)	3.60g (35.71)	3.20d (42.86)	2.98g (46.88)
Ac8	5.18f	5.08d (1.93)	4.95c (4.35)	4.50bc (13.04)	4.13bc (20.29)
Ac9	6.28c	5.88c (6.37)	5.40a (13.94)	5.10a (18.73)	6.50a (+3.59)
Ac10	5.98d	6.50b (+8.79)	5.15bc (13.81)	4.34bc (27.33)	4.30b (28.03)
Ac11	8.20a	7.70a (6.10)	4.98c (39.33)	4.80ab (41.46)	3.55ef (56.71)
Ac12	5.98d	5.93c (0.84)	5.23ab (12.55)	4.48bc (25.10)	4.00cd (33.05)
Ac13	7.33b	5.33d (27.30)	4.95c (32.42)	4.75ab (35.15)	4.23bc (42.32)
Mean	5.85a	5.08b (13.08)	4.38c (25.18)	3.98d (31.93)	3.66e (37.40)

Mean values with different lower case letters in a row are significantly different at $P < 0.05$. Values in the parentheses indicate percent compared to the untreated control (0 dS m^{-1}) plants

'+' symbol indicates % increase of K content

both an increase and decrease in Ca concentrations were observed (Table 9). An increase in Ca concentrations throughout the 4 salinity levels was found in Ac4, Ac9, Ac10, Ac12 and Ac13. However, in Ac7, the only increase was seen at 24 dS m^{-1} salinity and in Ac8 at 8 and 16 dS m^{-1} salinity, compared to the control. However, the highest increase (145 %) in Ca concentration due to salinity stress was observed in Ac10 at 8 dS m^{-1} salinity, followed by 123 % increase in Ac13 and 109 % increase in Ac10 at 32 dS m^{-1} salinity compared to the

Table 9 Effect of salinity on Ca content in 13 purslane accessions

Purslane accessions	Ca content (%, DW basis)				
	Salinity level (dS m^{-1})				
	0	8	16	24	32
Ac1	2.74f	1.65i (39.80)	1.13k (58.77)	1.06l (61.11)	1.33l (51.46)
Ac2	1.94i	1.72i (11.33)	1.60j (17.59)	1.60k (17.59)	1.25m (35.72)
Ac3	3.28b	2.13g (34.93)	3.06b (6.83)	2.49g (24.15)	2.38g (27.32)
Ac4	2.08h	3.72b (+78.85)	2.68d (+28.85)	3.09e (+48.86)	3.17d (+52.31)
Ac5	2.72f	1.22j (55.32)	2.01h (26.22)	1.67j (38.57)	1.92j (29.45)
Ac6	4.17a	2.70e (35.12)	2.38f (42.99)	3.35c (19.58)	3.88b (6.91)
Ac7	2.66g	1.84h (30.72)	2.18g (18.07)	3.69b (+38.86)	2.12h (20.18)
Ac8	2.82e	3.62c (+28.41)	3.07b (+9.09)	2.50g (11.36)	1.65k (41.48)
Ac9	3.02d	4.20a (+39.26)	3.46a (+14.85)	5.26a (+74.27)	5.14a (+70.29)
Ac10	1.40l	3.43d (+145.14)	1.60j (+14.29)	2.36h (+68.57)	2.93e (+109.14)
Ac11	3.07c	2.55f (16.83)	1.90i (38.0)	1.93i (37.22)	1.98i (35.66)
Ac12	1.73j	3.49d (+101.85)	2.83c (+63.89)	3.22d (+86.11)	2.79f (+61.57)
Ac13	1.48k	2.06g (+38.86)	2.42e (+63.78)	2.79f (+88.65)	3.312c (+123.35)
Mean	2.55c	2.64ab (+3.73)	2.33e (8.38)	2.69a (+5.76)	2.60c (+2.23)

Mean values with different lower case letters in a row are significantly different at $P < 0.05$. Values in the parentheses indicate percent compared to the untreated control (0 dS m^{-1}) plants

'+' symbol indicates % increase of Ca content

control (Table 9). However, the highest decrease (61 %) in Ca concentration was recorded in Ac1 at 24 dS m^{-1} salinity followed by 58.8 % decrease at 16 dS m^{-1} salinity in the same accession and a 55 % decrease in Ac5 at 8 dS m^{-1} salinity compared to the control (Table 9). On average over all of the accessions, 3.73 % increase, 8.38 % decrease, 5.76 % and 2.23 % increase in Ca concentration were recorded, respectively, at 8, 16, 24 and 32 dS m^{-1} salinity, which were statistically significant values ($P < 0.05$; Table 9).

Magnesium (Mg) content in purslane

The Mg content in 13 untreated purslane accessions also significantly ($P < 0.0001$) varied, with the highest concentration (2.03 %) observed in Ac1 and the lowest concentration (0.82 %) in Ac2 (Table 10). The Mg concentration in the purslane accessions was also significantly ($P < 0.05$) affected by augmented salinity stress. The highest salinity stress increase (83 %) in Mg concentration was observed in Ac13 at 16 dS m^{-1} salinity followed by a 64.8 % increase in Ac12 at the same salinity and a 48.4 % increase in Ac13 at 24 dS m^{-1} salinity compared to the control (Table 10). In contrast, the highest reduction (61 %) in Mg content due to salinity stress was observed in Ac1 at 24 dS m^{-1} salinity followed by 60.5 % at 32 dS m^{-1} salinity and 56 % at 16 dS m^{-1} salinity in the same accessions, respectively, compared to the control (Table 10). On average, over all of the accessions, 8.92, 1.83, 5.38 and 10.35 % reductions in Mg concentration were recorded, respectively, at 8, 16, 24 and 32 dS m^{-1} salinity and were statistically significant ($P < 0.05$; Table 10).

Iron (Fe) content in purslane

Thirteen untreated control purslane accessions greatly varied in Fe concentration and ranged between 9.30 and 56.0 ppm, with the highest concentration observed in Ac6 and the lowest in Ac7 (Table 11). Varied levels of salinity also significantly ($P < 0.05$) affected the Fe concentration. At 8 dS m^{-1} salinity, the Fe content was found to increase in all purslane accessions, except Ac1, where a decrease in Fe content was recorded at all salinity levels. At this lower salinity level, the highest increase (344.8 %) in Fe content was seen in Ac5, followed by 278 % in Ac4, respectively, compared to the control (Table 11). However, a further increase in salinity also continued to increase the Fe content in Ac4, Ac5, Ac7, Ac8, Ac9, Ac10 and Ac13 but at a decreasing rate. However, Ac2, Ac3, Ac6, Ac11 and Ac12 exhibited reductions in Fe content when the salinity levels changed to 16 dS m^{-1} from 8 dS m^{-1} (Table 11). Furthermore, NaCl-induced the highest reduction (74.9 %) in Fe content in Ac6 at 32 dS m^{-1} salinity, followed by a 64 % reduction in Ac3 at the same salinity levels compared to the control (Table 11). On average, over all of the accessions, 66.7 and 10.5 % increases at 8 and 16 dS m^{-1} salinity, and 21 and 35.7 % reductions in Fe concentrations were recorded at 24 and 32 dS m^{-1} salinity, respectively, which were statistically significant ($P < 0.05$; Table 11).

Table 10 Effect of salinity on Mg content in 13 purslane accessions

Purslane accessions	Mg content (%, DW basis)				
	Salinity level (dS m^{-1})				
	0	8	16	24	32
Ac1	2.03a	1.29g (36.41)	0.89j (56.21)	0.79l (61.14)	0.80k (60.55)
Ac2	0.82k	0.93j (+13.58)	1.04h (+27.45)	1.06k (+29.90)	0.85j (+3.92)
Ac3	1.83c	1.19h (34.79)	0.90i (50.77)	1.43g (21.66)	1.56c (14.44)
Ac4	1.92b	1.89a (1.25)	1.37f (28.60)	1.69a (11.90)	1.41g (26.30)
Ac5	1.44f	1.29g (10.03)	1.36f (5.01)	1.29j (10.31)	1.15i (20.6)
Ac6	1.59d	1.36cd (14.32)	1.89a (+18.59)	1.52e (4.77)	1.61b (+1.01)
Ac7	1.38g	1.32f (4.07)	1.51c (+9.88)	1.58d (+15.12)	1.43f (+3.78)
Ac8	1.36g	1.64b (+19.94)	1.41e (+3.52)	1.32i (2.93)	1.38h (+1.47)
Ac9	1.48e	1.38c (7.01)	1.48d (0.00)	1.61c (+8.36)	1.48d (0.00)
Ac10	1.57d	1.36de (13.59)	1.51c (3.65)	1.65b (+4.95)	1.74a (+11.14)
Ac11	1.03h	1.15i (+10.93)	1.32g (+27.47)	1.40h (+35.61)	1.46e (+41.81)
Ac12	0.92j	0.58k (36.52)	1.52c (+64.78)	0.56 m (39.13)	0.80k (13.48)
Ac13	0.99i	1.34ef (34.88)	1.82b (+83.06)	1.47f (+48.39)	0.77l (221.90)
Mean	1.41a	1.29d (8.92)	1.39b (1.84)	1.34c (5.38)	1.27e (10.35)

Mean values with different lower case letters in a row are significantly different at $P < 0.05$. Values in the parentheses indicate percent compared to control (0 dS m^{-1}) plants

'+' symbol indicates % increase of Mg content

Table 11 Effect of salinity on Fe content in 13 purslane accessions

Purslane accessions	Fe content (ppm)				
	Salinity level (dS m^{-1})				
	0	8	16	24	32
Ac1	25.70e	21.90i (14.79)	19.80h (22.92)	14.60g (43.19)	12.20gh (52.53)
Ac2	26.30e	42.40f (+61.22)	19.70h (25.10)	21.80cd (17.11)	29.80b (+13.31)
Ac3	28.60d	45.60e (+59.44)	19.50h (31.82)	18.30f (36.01)	10.20hi (64.34)
Ac4	16.90g	63.90b (+278.11)	28.10f (+66.27)	19.80ef (+17.16)	14.60ef (13.61)
Ac5	14.50h	64.50b (+344.83)	42.7b (+194.48)	20.9de (+44.14)	16.20de (+11.72)
Ac6	55.50a	91.40a (+64.68)	39.50c (28.83)	25.80b (53.51)	13.90fg (74.95)
Ac7	9.30i	16.70j (+79.57)	15.70j (+68.82)	11.10i (+19.35)	23.40c (+151.61)
Ac8	30.70d	33.80h (+10.10)	30.90e (+0.65)	33.80c (+10.10)	22.60d (26.38)
Ac9	21.40f	36.0g (+68.22)	23.10g (+7.94)	12.90h (39.72)	10.40hi (51.40)
Ac10	19.80f	44.0ef (+122.22)	22.70g (+14.65)	20.30de (+2.53)	16.00de (19.19)
Ac11	29.30d	45.80e (+56.31)	26.80f (8.53)	13.10gh (55.29)	8.40i (71.33)
Ac12	43.30c	52.00d (+20.09)	33.20d (25.33)	22.60c (47.81)	22.50c (48.04)
Ac13	50.30b	61.30c (+21.87)	88.90a (+76.74)	58.30a (+15.90)	38.70a (23.06)
Mean	28.60c	47.60a (+66.66)	31.60b (+10.50)	22.60cd (21.07)	18.40d (35.71)

Mean values with different lower case letters in a row are significantly different at $P < 0.05$. Values in the parentheses indicate percent compared to the untreated control (0 dS m^{-1}) plants

'+' symbol indicates % increase of Fe content

Zink (Zn) content in purslane

The zinc content also varied greatly among all 13 untreated control purslane accessions, with the highest Zn content (0.74 ppm) in Ac12 and the lowest (0.31 ppm) in Ac9 (Table 12). Aggravated salinity stress caused significant changes in the Zn content among the purslane accessions. At the lowest salinity levels (8 dS m^{-1}), an increase in Zn concentration was seen in all 13 purslane accessions compared to the control, with the highest increase (182.6 %) in Ac6 followed by a 48.6 % increase in

Table 12 Effect of salinity on Zn content in 13 purslane accessions

Purslane accessions	Zn content (mg L^{-1})				
	Salinity level (dS m^{-1})				
	0	8	16	24	32
Ac1	0.43cd	0.51e–g (+18.60)	0.53bc (+23.26)	0.46b (+6.98)	0.4cd (6.98)
Ac2	0.41ed	0.49f–h (+19.51)	0.45cd (+9.76)	0.37cd (9.76)	0.33ef (19.51)
Ac3	0.4d–f	0.47gh (+17.51)	0.5bc (+25.0)	0.39bc (2.50)	0.46bc (+15.0)
Ac4	0.35e–g	0.62c (+77.14)	0.4d (+14.29)	0.38cd (+8.57)	0.4cd (+14.29)
Ac5	0.49bc	0.54de (+10.20)	0.56b (+14.29)	0.44bc (10.20)	0.52b (+6.12)
Ac6	0.46b–d	1.3a (+182.61)	0.46cd (0.00)	0.42bc (8.70)	0.3f (34.78)
Ac7	0.33fg	0.49f–h (+48.48)	0.48b–d (+45.45)	0.37cd (+12.12)	0.36de (+9.09)
Ac8	0.42c–e	0.52de (+23.81)	0.28e (33.33)	0.31d (26.19)	0.3f (28.57)
Ac9	0.31g	0.23i (25.81)	0.28e (9.68)	0.38cd (+22.58)	0.38de (+22.58)
Ac10	0.44cd	0.45h (+2.27)	0.41d (6.82)	0.42bc (4.55)	0.32ef (27.27)
Ac11	0.39d–f	0.56d (+43.59)	0.48b–d (+23.08)	0.37cd (5.13)	0.35de (10.26)
Ac12	0.74a	1.1b (+48.68)	0.93a (+25.68)	0.84a (+13.51)	0.85a (+14.86)
Ac13	0.53b	0.63c (+18.87)	0.45cd (15.09)	0.37cd (30.19)	0.23g (56.60)
Mean	0.44c	0.61a (+38.77)	0.48b (8.95)	0.42d (3.16)	0.40d (8.77)

Mean values with different lower case letters in a row are significantly different at $P < 0.05$. Values in the parentheses indicate percent compared to the untreated control (0 dS m^{-1}) plants

'+' symbol indicates % increase of Zn content

Ac12 and 48.5 % increase in Ac7, respectively (Table 12). The Zn concentration continued to increase with further increases in salinity levels at 16 dS m^{-1} salinity in Ac1, Ac2, Ac3, Ac4, Ac5, Ac7, Ac11 and Ac12, but the Zn concentration decreased in percentage compared to the control (Table 12). Meanwhile, the highest reduction (56.6 %) in Zn content due to salinity stress was found in Ac13 at 32 dS m^{-1} salinity, followed by a 34.7 % reduction in Ac6 at 32 dS m^{-1} salinity and 33.3 % reduction in Ac8 at 16 dS m^{-1} salinity, respectively, compared to the control (Table 12). On average over all of the accessions, a 38.8 % increase, 8.95, 3.2 and 8.8 % reduction in the Zn concentration were recorded at 8, 16, 24 and 32 dS m^{-1} salinity levels, respectively, which were statistically significant ($P < 0.05$; Table 12).

Salt salinity relationships

The sodium–calcium ratio was found to increase with lower levels of salinity but decreased polynomially ($R^2 = 0.956$) at the highest level of salinity (Fig. 1). The sodium–potassium ratio was influenced by the different levels of salinity in purslane and the ratios increased polynomially ($R^2 = 0.994$) with salinity (Fig. 1). The potassium–phosphorus ratio declined with lower levels of salinity stress but later tended to increase polynomially ($R^2 = 0.854$) with increased salinity levels (Fig. 1). The magnesium–calcium ratio decreased initially but later increased with increasing salinity levels ($R^2 = 0.909$) (Fig. 1). The zinc to iron ratio was also found to decrease at the beginning of salinity stress but to later increase polynomially ($R^2 = 0.935$) with increasing salinity levels (Fig. 1).

Correlation matrix

The correlation matrixes for seven mineral cations in purslane at different salinity levels are presented in Table 13. Phosphorus had a strong positive correlation ($P \leq 0.001$) with potassium and was negatively correlated ($P \leq 0.05$) with sodium and positively correlated ($P \leq 0.05$) with calcium and iron, whereas no statistically significant correlation was found with magnesium and zinc. Whereas potassium was significantly correlated ($P \leq 0.05$) with sodium and calcium, the positive correlations observed with magnesium and iron and the negative associations observed with zinc were not statistically significant. In contrast, sodium was negatively correlated with iron (Table 13).

Cluster and principal component analysis (PCA)

To assess the patterns of variation, a UPGMA cluster analysis and PCA were performed using the measured parameters. All 13 purslane accessions were grouped into five distinct clusters at a 1.19 similarity coefficient level (Fig. 2). Among the 5 clusters, Ac9 was separated from the others and formed cluster V, Ac12 solely constituted cluster IV, and Ac13 was alone in cluster III. Cluster II was the largest group, consisting of Ac3, Ac4, Ac8, Ac10, and Ac11. The cluster I was formed with Ac1, Ac2, Ac5

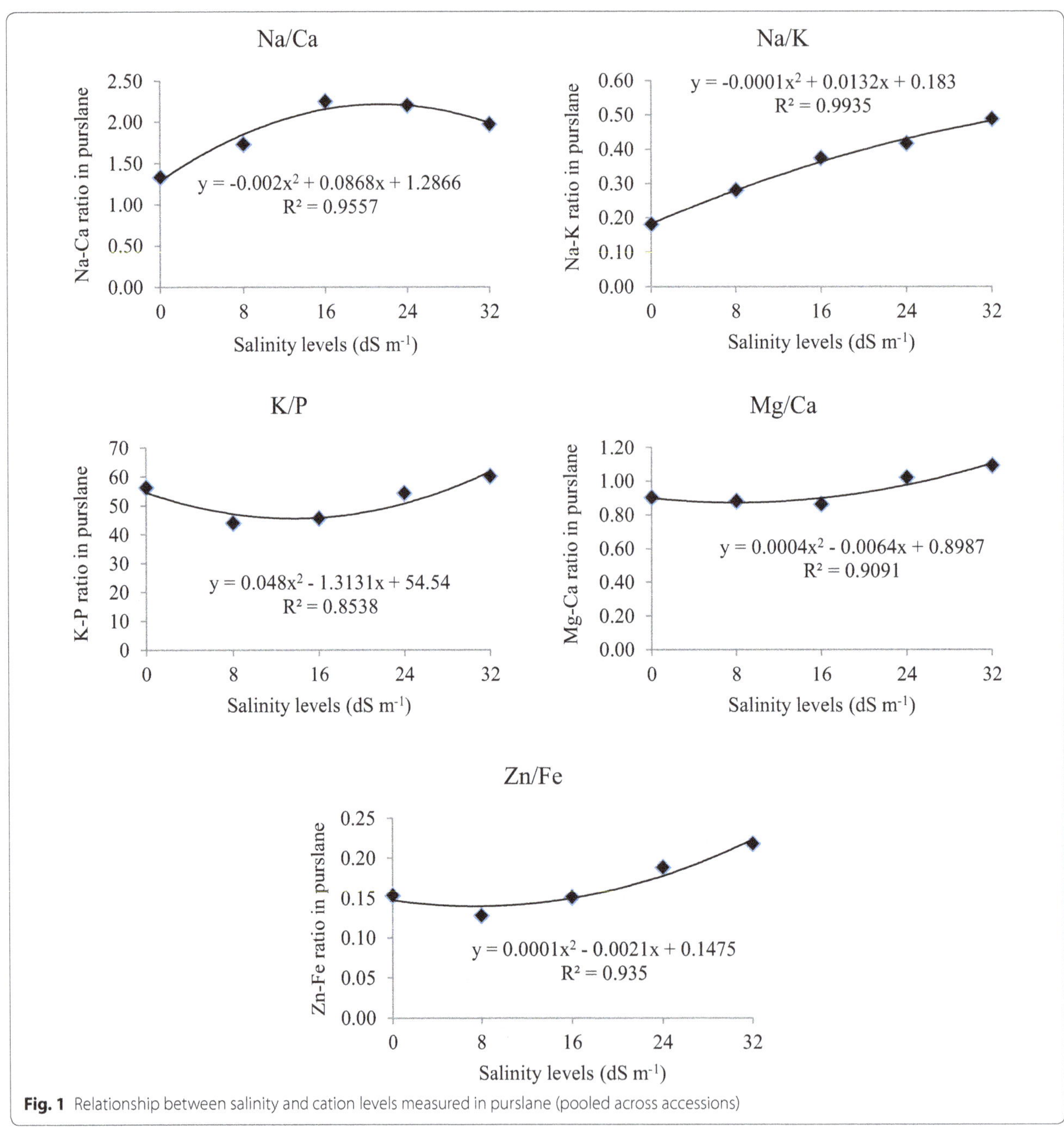

Fig. 1 Relationship between salinity and cation levels measured in purslane (pooled across accessions)

and Ac7. The biplot of the 13 salinity tolerant purslane accessions, representing the variations among the measured parameters, are shown in Fig. 3. The patterns of the cluster analysis were also confirmed with a PCA with a three-dimensional (3D, Fig. 4) plot, which also gave results similar to those of the dendrogram (Fig. 2). The principal components analysis (PCA) confirmed 82.9 % of the total variation among all of the accessions studied (Table 14).

Discussions

Important morphological traits, i.e., plant height, number of leaves, number of flowers, fresh weight and dry weight, and concentrations of major macro- and micro-minerals, i.e., Na, P, K, Ca, Mg, Fe and Zn, in 13 untreated and salt-treated purslane accessions were investigated in this study. The results indicated that the untreated control plants greatly varied in the above-mentioned parameters representing morphological traits and mineral contents.

Table 13 Pearson's correlation coefficients between micro and macro minerals

Factors	P	K	Na	Ca	Mg	Fe	Zn
P	1						
K	0.73**	1					
Na	−0.62*	−0.56*	1				
Ca	0.61*	0.64*	−0.09 ns	1			
Mg	0.14 ns	0.30 ns	−0.26 ns	0.50 ns	1		
Fe	0.58*	0.21 ns	−0.59*	0.07 ns	−0.05 ns	1	
Zn	0.04 ns	−0.03 ns	−0.09 ns	−0.02 ns	−0.50 ns	0.30 ns	1

ns non-significant

*, ** Significance at 5 and 1 % levels, respectively

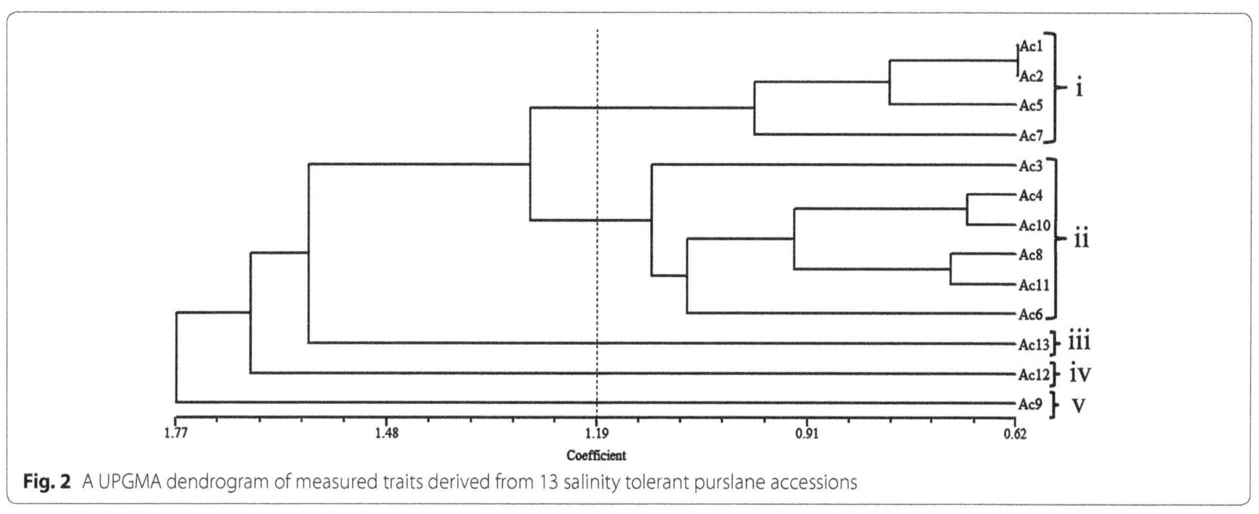

Fig. 2 A UPGMA dendrogram of measured traits derived from 13 salinity tolerant purslane accessions

Salt treatment also significantly influenced all of the investigated parameters in this study. The responses of the 13 purslane accessions to salt treatment were very different from each other and did not follow any particular trend, indicating vast diversity among the purslane accessions collected from different locations in western peninsular Malaysia.

Among the morphological traits, plant height varied greatly in the untreated control 13 purslane accessions. The plant heights ranged from 33 to 70 cm (an approximately twofold difference from lowest to highest, Table 1); the number of leaves ranged from 282 to 556 (an approximately twofold difference from lowest to highest, Table 2); the number of flowers ranged from 6 to 64 (an approximately tenfold difference from lowest to highest, Table 3). The fresh weight varied from 103 to 342 g (an approximately fourfold difference from lowest to highest, Table 4) and the dry weight ranged from 7 to 24 g (an approximately threefold difference from lowest to highest, Table 5).

NaCl-induced salinity had significant impacts on the plant height, number of leaves, numbers of flowers, fresh weights and dry weights of the 13 purslane accessions. However, the responses of the individual accessions were very different from each other. One general trend was that treatments with the highest 32 dS m^{-1} salinity caused significant reductions in all measured traits for most accessions compared to 24 dS m^{-1} salinity. The effects of 8, 16 and 24 dS m^{-1} salinity were variable; either increasing or declining (or remaining similar) in these parameters compared to the untreated control plants. An increase in plant height was recorded only in Ac1 at 16 dS m^{-1} salinity and was a very small increase (2 %) compared to the control. Consecutive and significant decreases in plant height were observed in the remaining 12 purslane accessions. At 8 and 16 dS m^{-1} salinity, the highest reduction (>46 and >48 %, respectively) was observed in Ac8 compared to the control and to all other accessions (Table 1). Ali et al. [7] and Kafi and Rahimi [32] reported significant plant height reductions in purslane at 24 mM of salinity stress. Salinity stress-induced reductions in plant height have also been observed in rice [24] and in turfgrass [55, 56]. In contrast, 13.17 % increases in plant height in *Pennisetum alopecuroides* grass at 100 mM salinity stress have been described by Mane et al. [34].

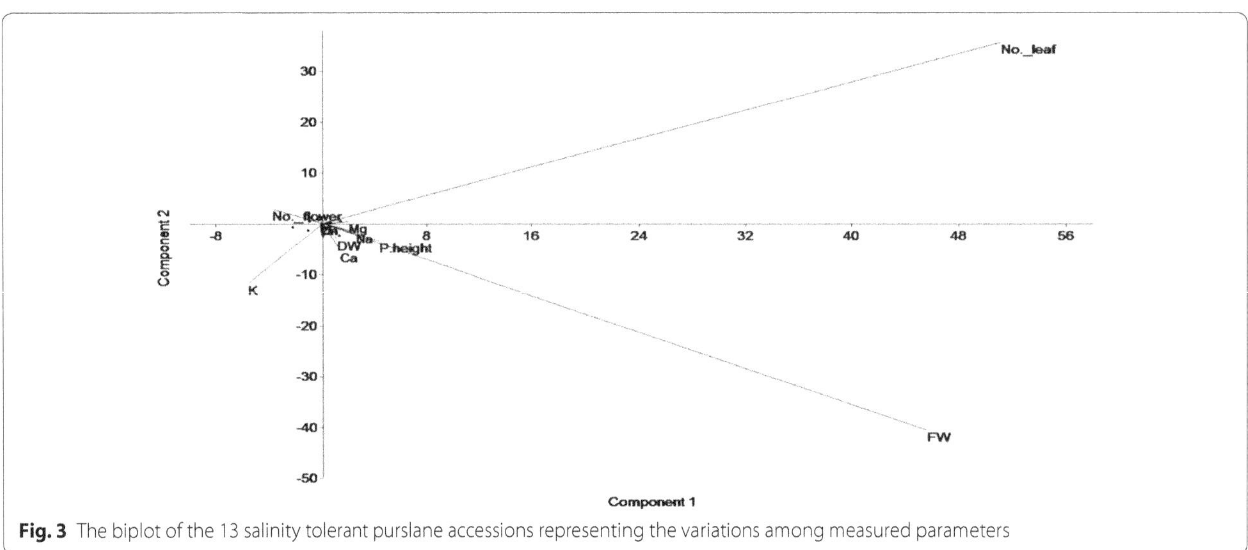

Fig. 3 The biplot of the 13 salinity tolerant purslane accessions representing the variations among measured parameters

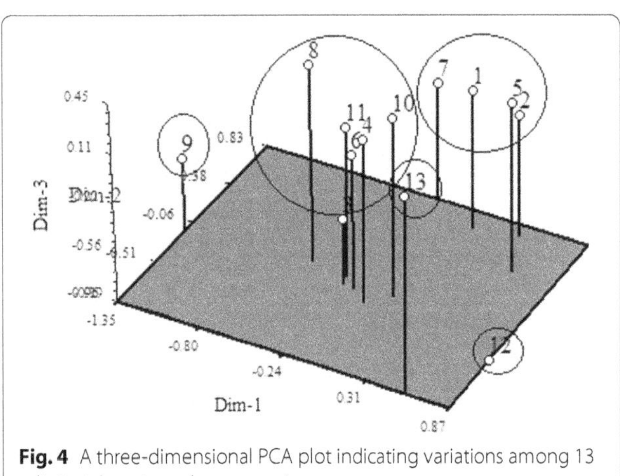

Fig. 4 A three-dimensional PCA plot indicating variations among 13 salinity tolerant purslane accessions

Table 14 Principal component analysis and percentage variation with first four principal components for 13 purslane accessions

Variable	Eigen vector			
	PC1	PC2	PC3	PC4
Eigen value	39.81	31.61	17.78	10.3
Percent	33.2	26.3	14.8	8.6
Cumulative	33.2	59.5	74.3	82.9
Ph	0.429	−0.156	0.058	−0.25
#Leaf	0.112	−0.137	−0.623	−0.153
#Flower	−0.314	−0.077	0.181	−0.632
FW	0.422	−0.164	0.134	0.086
DW	0.36	0.027	0.33	0.075
Na	0.068	−0.471	0.237	−0.161
P	0.114	0.505	0.047	0.102
K	0.196	0.439	0.057	0.103
Ca	0.34	0.275	0.247	−0.372
Mg	0.297	0.109	−0.33	−0.472
Fe	−0.208	0.386	−0.189	−0.234
Zn	−0.308	0.129	0.427	−0.195

Purslane is a succulent, leafy vegetable plant, and it produces an abundant number of leaves. Therefore, the shedding of leaves is the first symptom of salinity stress. Throughout the experiment, the shedding of leaves was observed to increase with increasing salinity levels from 8 up to 32 dS m^{-1} salinity. The highest reduction (43.57 %; approximately threefold higher from lowest to highest) in leaves was found in Ac13 compared to the control (Table 2). At 8 and 16 dS m^{-1} salinity levels, non-significant ($P > 0.05$) differences were observed in Ac1, Ac2, Ac4 and Ac5. However, at 24 and 32 dS m^{-1} salinity levels, Ac1, Ac2 and Ac4 varied non-significantly. Similarly, at 24 and 32 dS m^{-1} salinity, Ac10, Ac11 and Ac12 also varied non-significantly (Table 3). Ahmad et al. [1] reported a reduction in the number of leaves in *Rosa hybrida* L. due to slight increases in salinity. Augmented salinity induced a reduction in the leaf numbers in Jojoba plants following the application of a higher salinity treatment (120.7 mM NaCl) in Ali et al. [8].

The numbers of flowers also significantly varied among the 13 salinity-stressed purslane accessions throughout the experimental period. Accession-wide responses to different salinity levels were also very significant. The highest level of flower reduction was 96.48 % (approximately fivefold higher from lowest to highest) in Ac13 at 32 dS m^{-1} salinity, followed by 78.78 % (approximately twofold higher from lowest to highest) in Ac12, 75.87 %

(approximately fivefold higher from lowest to highest) in Ac2, 75.56 % (approximately fivefold higher from lowest to highest) in Ac4, and 74.26 % (approximately threefold higher from lowest to highest) in Ac1 at 32 dS m^{-1} salinity, compared to the control (Table 3). Both of the common purslane varieties were the most affected accessions compared to the ornamental accessions. Very highly significant reductions in the number of flowers in Cumin (*Cuminum cyminum* L.) have been reported by Hassanzadehdelouei et al., [27] at 11 dS m^{-1} salinity. In wheat (*Triticum aestivum*), high reductions in the numbers of spikelets have also been described by Ranjbar [44] at 20 dS m^{-1} salinity. The shedding or reduction of flower numbers from salt stressed plants may be due to a lack of optimum water uptake from roots. Abiotic stresses are known to affect meiosis during gamete production and male sterility appears to be more common than female gamete sterility [46].

Purslane is a succulent plant containing approximately 90 % water or more, in both the leaves and stems. Therefore, the fresh weight of purslane was comparatively higher than the dry weight. The elevated salinity stress caused a very high and significant reduction in the fresh weight of purslane and the fresh weight reduction increased with increasing salinity augmentation. In contrast, some of the accessions also achieved a significant increase in the fresh weight of purslane after salinity application. The highest reduction in fresh weight was observed in Ac8 at all 4 salinity levels of 8 dS m^{-1} (36.43 %), at 16 dS m^{-1} (42.1 %), at 24 dS m^{-1} (47.13 %) and at 32 dS m^{-1} (54.1 %), respectively, compared to the control as well as in other accessions (Table 4). Accession fresh weight at 8–32 dS m^{-1} salinity showed 42.69 and 32.82 % reductions (approximately threefold higher from lowest to highest) were found in Ac6 and Ac12, respectively, compared to the control (Table 4). In contrast, salinity induced the highest increase in the fresh weight in Ac9 (23.9 %; approximately twofold higher from lowest to highest) at 24 dS m^{-1} salinity, but with a further increase in salinity to 32 dS m^{-1}, a smaller increase of 16.5 % was seen compared to the control (Table 4). Salinity-induced fresh weight reduction is a common phenomenon that occurs in most cultivated crop plants and trees. The reductions in fresh weight due to salinity stress have been investigated by several scientists in several tomato crops [37] and in *Ocimum basilicum* [36]. The increase in fresh weight in *Pennisetum alopecuroides* at 100 mM salinity has also been reported by Mane et al. [34]. In this study, a reduction in biomass accelerated with increasing salinity, which is obvious because of the disturbances in physiological and biochemical activities under saline conditions [16], which may be due to the reduction in leaf area and number of leaves [19].

Plant dry matter content is a functional parameter that is used to assess plant strategy resource acquisition and use [20]. NaCl-induced salinity significantly affected the total dry matter production in all 13 purslane accessions. The dry matter production in purslane was very low compared to the fresh weight due to very high water content in the leaves and stems. The highest significant reduction in dry matter content (63.47 %; approximately a fivefold reduction from lowest to highest) was recorded in Ac6 at 32 dS m^{-1} salinity followed by 51.52 % (approximately a fourfold reduction from lowest to highest) in the same accessions at 16 dS m^{-1} salinity, respectively, compared to the control (Table 7). In contrast, in Ac9, a successive increase in dry matter content was noted from 16 to 32 dS m^{-1} salinity, where the highest increase (54.2 %) occurred at 24 dS m^{-1} salinity; however, the increasing rate of dry matter content later decreased with increasing salinity compared to the control (Table 5). The significant decrease in dry matter content of sugar beet cultivars was described by Dadkhah and Grrifiths [18] at 350 mM salinity stress. A level of 250 mM salinity resulted in the shoot and root dry matter contents exhibiting a marked decrease in hybrid maize varieties [21]. Worldwide, several authors have published reports on dry matter reductions in different crops under salinity stressed conditions in rice [24], turfgrass (Uddinn et al. 2012; [54]) and *Solanum quitoense* Lam. [22]. However, an increase in dry matter content in *Pennisetum alopecuroides* at 100 mM salinity has also been reported by Mane et al. [34]. This induced dry matter production under salinity conditions might be due to the accumulation of inorganic ions and organic solutes for osmotic adaptation, whereas a decrease in the dry matter content at the highest salinity levels might be due to the inhibition of hydrolysis in reserved nutrients and their translocation to the growing shoots [34].

The major micro- and macro-mineral contents of 13 untreated and salt-treated purslane accessions were also determined in our study. Clear and highly significant ($P \leq 0.001$) accession variations were observed across all measurements of Na^+, P, K^+, Ca^{2+}, Mg^{2+}, Fe and Zn content in the untreated 13 purslane accessions. Among the measured mineral contents in purslane, the potassium content was highest, followed by the sodium, magnesium, calcium, phosphorus, iron and zinc contents (Tables 6, 7, 8, 9, 10, 11, 12). Aggravated salinity stress also had a very significant impact on all of the measured micro and macro minerals of purslane. At lower salinity stress (8 dS m^{-1}), a common trend was identified: the mineral contents of P, Na^+, Fe and Zn increased compared to the control and reductions were observed of the remaining minerals (Table 6, 7, 8, 9, 10, 11, 12). However, after applying the next salinity level (16 dS m^{-1}), the mineral content of the majority of the purslane accessions reduced significantly, and only a few continued to increase, but with decreasing rates. The phosphorous content increased at 8 dS m^{-1} salinity in most of the accessions, but later increasing levels of salinity tended to significantly decrease and continued to decrease up to the highest salinity levels compared to the control (Table 6). In agreement with our findings, Zuazo et al. [59] opined that the increase in phosphorus content at lower salinity (2.5 dS m^{-1}) in mango stems and Hirpara et al. [28] showed a decrease in phosphorus content in *Butea monosperma* at the highest (13 dS m^{-1}) salinity stress. As in P, an increasing trend was also observed in the sodium content with increasing salinity in most of the purslane accessions, although there were also reductions in some of the accessions (Table 7). Several researchers have found that salinity stress increased the Na^+ content in *Butea monosperma* [28], *Salvadora persica* seedlings [43], *Andrographis paniculata* plants [52] and in common purslane [55]. In contrast, the K^+ content was very significantly reduced in most of the purslane accessions at most salinity levels, with some exceptions for certain accessions and salinity levels (Table 8). Similar results have also been described by Talei et al. [52] in *Andrographis paniculata* plants and in common purslane by Uddin et al. [55]. NaCl-induced salinity stress caused both an increase and decrease in Ca^{2+} content in this study and different accessions responded differently at various levels of salinity stress (Table 9). The augmented salinity stress increases in calcium content have been reported in *Salvadora persica* seedlings [43] and in *Andrographis paniculata* [52]. However, Uddin and Juraimi [54] found a reduction in calcium content in turfgrass species. Similar trends were also observed in the case of Mg^{2+} content in 13 salinity stressed purslane accessions (Table 10). Zuazo et al. [59] described an increase in magnesium content in mango stems but a decrease in roots in different salinity regimes. Talei et al. [52] also reported increased magnesium in *Andrographis paniculate,* and Uddin and Juraimi [54] showed a decrease in turfgrass species. The iron content significantly increased at lower salinity levels but later tended to decrease with increasing salinity levels (Table 11). The increase in Fe^{2+} concentration due to lower salinity stress in mango rootstocks has been reported by Zuazo et al. [59]. Salinity stress reductions in iron contents have also been found in prose millet in *Andrographis paniculata* plants [52]. A similar trend was also found for the zinc contents at the lowest salinity levels in all purslane accessions, except Ac9, where reductions were recorded at 8 and 16 dS m^{-1} salinity but at increased salinity levels at 24 and 32 dS m^{-1}, a significant but similar state of reduction was found (Table 12). Similar results have also been described by Talei et al. [52] in *Andrographis paniculata* plants.

There are three major constraints to plant growth in saline substrates: (a) a water deficit (drought stress) arising from low water potential of saline rooting media; (b) ion toxicity associated with the excessive uptake of mainly Na^+ and Cl^-; and (c) nutrient imbalances [35]. Salt-stressed plants mainly adopt three mechanisms to cope with the three constraints: (a) osmotic adjustment by inorganic and/or organic solutes; (b) salt inclusion/exclusion; and (c) ion discrimination [57]. From our previous findings [3–5] among the 13 accessions in our study, two accessions (Ac7 and Ac9) were found to be salt tolerant; six accessions (Ac3, Ac5, Ac6, Ac10, Ac11 and Ac12) were moderately tolerant; and the remaining five (Ac1, Ac2, Ac4, Ac8 and Ac13) accessions were identified as moderately susceptible to salinity stress on the basis of biomass production. Osmotic adjustment through increased Na influx (Table 6) and ion discrimination, Ca/Na, Na/K and Mg/Ca in particular (Fig. 1), seem to be the key factors in salt tolerance among these purslane accessions. Continued control over Na influx and osmotic adjustment through increased Na^+ uptake are probably both important facets of the physiology of purslane plant ability to cope with a saline environment. For instance, from among the two most salt tolerant accessions, Ac7 accumulated less Na compared to Ac9 (Table 6), which indicated the enhanced ability of Ac7 to restrict the entry of Na into the shoot, which is commonly termed "salt exclusion". However, Ac9 exhibited a better ability to adjust osmotic balance with greater inclusion of Na in the shoots, which is commonly termed "salt inclusion". Halophytic or salt tolerant species differ from salt-sensitive ones in having restricted uptake or the ability to transport Na^+ and Cl^- to the leaves despite an effective compartmentalization of these ions. This is critical for preventing the build-up of toxic ions in the cytoplasm [11, 38]. In salt excretory plants, salt is kept away from photosynthesizing or meristematic cells. In these plants, an osmotic balance is generally achieved via extensive accumulation of organic solutes and/or inorganic ions. However, in plants where salt inclusion is the prime mechanism, the accumulation of some inorganic ions (predominantly Na^+ and Cl^-) regulates the osmotic adjustment [31].

However, over all genotypes, salt tolerance was not correlated with shoot Na accumulation, suggesting considerable variation in the salinity tolerance among accessions and the possible existence of a range of salt tolerant mechanisms, both between and within purslane accessions [6]. Accession Ac9, in particular, maintained better vegetative growth despite accumulating higher Na. This might indicate salt tolerance in the discontinuous distribution of Na ions from leaf to leaf and cell to cell within the leaves, as has been explained by Ashraf et al. [12]. The shoot analyses reported here suggest that a nutritional disturbance of K and Ca has a role in shoot growth inhibition and may play a role in genotypic tolerance. This study indicated that the more tolerant accessions (Ac9) had higher K and Ca accumulation (though Ac7 only had greater Ca) in saline control conditions. Jones and Gorham [31] also reported that plants with greater salt tolerance were more efficient users of K and Ca under saline conditions.

Increased Na/Ca, Na/K and Mg/Ca ratios with increasing salinity (Fig. 1) indicated ion discrimination between Na, K, Ca and Mg. This suggested that Na, K, Ca and Mg also played a role in salt tolerance in purslane. Munns and James [39] claimed that all plants discriminate to some extent between Na and K. It is therefore possible that K/Na and Ca/Na discrimination is associated with salt tolerance. Ion imbalance, particularly when caused by Ca^{2+} and K^{+}, is the most important and widely studied phenomenon affected by salt stress, which is directly influenced by the uptake of Na^{+} and Cl^{-} ions [38, 40]. Ashraf et al. [12] reported that one of the most important physiological mechanisms of salt tolerance is the selective absorption of K^{+} by plants from the saline media and that the maintenance of better concentrations of K^{+} and Ca^{+} and limit on the Na^{+} uptake are vital for salt stress tolerance in plants, as has been seen in this study with purslane. Higher K^{+}/Na^{+} or Ca^{2+}/Na^{+} ratios are characteristic tissue salt tolerance traits and are often used as criteria for screening for salt tolerance [11, 39, 50].

Cluster analysis and PCA, as a multivariate technique, can group individuals or objects on the basis of their characteristics. Individuals with similar descriptions are mathematically congregated within the same cluster [2]. Distance, similarity and relatedness of varieties are the foundation of this method. The UPGMA constructed dendrogram revealed 5 clusters where Ac9, Ac12 and Ac13 were most different from all of the others, indicating the highest salt tolerance and the highest diversity compared to other accessions. To improve variety development, the most judicious combination can be made with Ac9, Ac12 and Ac13 with Ac1, Ac2 or Ac4 or Ac10 or Ac8 or Ac11, which would bring about the greater genetic diversity [10]. Whereas according to biplot analysis of all the measured parameters, number of leaves showed the highest correlation with fresh weight (FW) and positioned at the opposite direction of average line of the component 1 (Fig. 3). Among measured minerals K and Ca also showed highest correlation and positioned at the lower level of both component 1 and 2 (Fig. 3).

Conclusions

In conclusion, although there were significant variations among all 13 purslane accessions among the measured parameters, in general, this research indicated high salt tolerant crop plants that are capable of producing a satisfactory amount of dry matter content, which is a fundamental requirement of any salt tolerant plant species. Throughout the experiment, accession wise complex results were found among morphological traits. Different accessions exhibited different performances under exposure to different levels of salinity stress. However, one common trend was that all of the accessions were affected at the highest salinity level compared to the control, while some were also affected at moderate or lower salinity levels. Most of the measured morphological traits were reduced under varied salinity regimes, but plant height was found to increase in Ac1 at 16 dS m^{-1} salinity and Ac13 was the most affected accession. However, the highest reduction in the leaves and number of flowers was recorded in Ac13 at 32 dS m^{-1} salinity compared to the control. The highest fresh and dry weight reductions were noted in Ac8 and Ac6 at 32 dS m^{-1} salinity, respectively, whereas the highest increase in both fresh and dry weight was found in Ac9 at 24 dS m^{-1} salinity compared to the control. In contrast, at the lower salinity levels, all of the measured minerals were found to increase and later decrease with increasing salinity, but the performances of the accessions were different with regard to the salinity levels. Overall, among all 13 purslane accessions, considering morphological development and mineral contents, Ac9 was the most salt tolerant purslane accession that produced the highest amount of fresh and dry weight, and Ac13 was the most affected accession. It was also found that ornamental purslane showed more salt tolerance than common purslane. Therefore, we can suggest both types of purslane for consumer and commercial production as a fresh vegetable source in any type of soil, especially for saline agriculture.

Methods

Purslane accessions and study location

There are approximately 7 types of purslane available in Malaysia. In our study, 13 different purslane accessions were collected from varied locations in western peninsular Malaysia [3]. Among those, 11 were ornamental purslane (Ac1–Ac11) and two were common purslane (Ac12 and Ac13). The experiment was conducted in a Field-2 glasshouse at the Faculty of Agriculture, at the University of Putra Malaysia (UPM) from July to October, 2013, and all of the chemical analyses were performed at the Plant Physiology and Analytic Lab, at the Department of Crop Science, in the Faculty of Agriculture, UPM, Malaysia, and the histological study was performed at the Botany Laboratory in the same department.

Planting and cultural practices

Seedlings of the two common purslane varieties and cuttings of the 11 ornamental purslane accessions (ornamental purslane do not produce seeds) were first grown in plastic trays filled with rice field top soils (38.96 % sand, 11.05 % silt and 49.88 % clay) with pH 4.8, 2.64 % organic carbon, 1.25 g cc^{-1} bulk density and CEC of 7.06 meq 100 g^{-1} soil. The soil nutrient status was 0.17 % total N, 5.67 ppm available P, 15.6 ppm available K, 3357 ppm Ca and 319 ppm Mg. Soil water retention was 30.72 % (wet basis) and 46.17 % (dry basis) at field capacity. The soil belonged to the Serdang series.

Five 10-day-old seedlings or cuttings for each accession were transplanted into plastic pots (24 × 22 × 20 cm) filled with the same prepared soil mentioned above. The plants were allowed to recover from transplanting shock, and full establishment occurred over 29 days. During this time, the plants were irrigated with tap water as and when necessary. No fertilizer was used. Five levels of salinity (0, 8.0, 16.0, 24.0 and 32.0 dS m^{-1}) were used in this study, which were prepared using NaCl (Merck, Darmstadt, Germany) and distilled water. Salt treatment was initiated 30 days after transplanting (DAT) and continued until the end of the study. In each pot, 200 mL of saline water was applied on alternate days in the treatment. The control plants received 200 mL of distilled water. The experiment was organized in a two factorial (purslane accessions × salinity) randomized complete block designs with three replications. Whole plants were harvested from ground level, 60 days after transplanting. The plants were washed under tap water and kept in a cool dry place for 3 days and the fresh weights were recorded. After that, the samples were transferred into an oven and left for 3 days at 40 °C to avoid sudden heat burning. Finally, the oven temperature was balanced at 50 °C and left for complete drying. The dry weights of the whole plants in each treatment and replication were recorded before grinding.

Data collection and analysis

Morphological data collection

Plant height The average plant heights of the five plants in each pot were measured in cm from salt treated and untreated control plants. The percentages of increase and/or decrease in plant height due to salinity stress were calculated using the following formula:

$$\text{Percentage of plant height changes} = \frac{\text{Control treatment value} - \text{Salinized treatment value}}{\text{Control treatment value}} \times 100$$

Number of leaves The shedding of leaves is a prominent symptom of salinity stress in purslane. The percentage of shedding of leaves compared to untreated control plants were calculated using following formula:

$$\text{Percentage of sheeding of leaves} = \frac{\text{Control treatment value} - \text{Salinized treatment value}}{\text{Control treatment value}} \times 100$$

Number of flowers Purslane blooms daily, so the total numbers of flowers were counted every day and were recorded. The percentages of flower reductions were calculated using the following formula:

$$\text{Percentage of flower reduction} = \frac{\text{Control treamtnet value} - \text{Salinized treatment value}}{\text{Control treatment value}} \times 100$$

Fresh weight The 60-day-old harvested fresh and surface moisture-free purslane plants were weighed using an electric balance, and the mean fresh weight (FW) was calculated. The reduction in fresh biomass with the reduction percentage from salinity stress was also measured using the above formula.

Dry weight The mean dry weights (DW) were calculated from the oven-dried samples. The dry matter reduction with the percentages due to salinity stress over the control was measured using the following formula:

$$\text{Percentage of dry matter reduction} = \frac{\text{Control treamtnet value} - \text{Salinized treatment value}}{\text{Control treatment value}} \times 100$$

Micro- and macro-mineral analysis

The P, Na, K, Ca, Mg, Fe and Zn contents from the control and the salinity-stressed purslane dry samples were analysed using the digestion method [33] and were determined using an Atomic Absorption Spectrophotometer (AAS; Perkin Elmer, 5100, USA). For this purpose, the ground powder samples of 0.25 g were weighed and poured into a digestion tube. Then, 5 mL of concentrated sulphuric acid (H_2SO_4) were added and kept overnight or at least for 2 h until the plant materials properly moistened. Then, 2 mL of 50 % hydrogen peroxide (H_2O_2) was slowly added and the digestion tube was placed in a digestion block, where the digester block was set to heat for 45 min at 285 °C temperature. After 45 min, the tube was removed and allowed to cool before 2 mL of 50 % H_2O_2 was added again. After that, it was maintained for the heating as well as cooling process and repeated until the digested solution became colourless or clear. The cleared cool sample was then filtered and the final volume was made into 100 mL by adding distilled water for the analysis.

Multivariate analysis

A cluster analysis was performed to construct a dendrogram based on the similarity matrix data using the unweighted pair group method with arithmetic averages (UPGMA) and the *SHAN* clustering program. All of the analyses were performed with the *NTSYS-pc 2.10* software [45]. The binary data were also subjected to a PCA (Principal Component Analysis) to investigate the structure of our collection. The PCA of the 13 purslane accessions were calculated using the EIGEN module of *NTSYS-pc 2.10* software [45]. The biplot analysis was done using Past: Palaeontological Statistics software package [25].

Statistical analysis

All recorded data were subjected to analysis of variance using the SAS statistical software package version 9.3 [47]. Data were submitted to analysis of variance (ANOVA) and the means compared by Tukey's multiple range test ($P < 0.05$). Pearson's correlation coefficient analyses were done to assess the associations between different parameters.

Abbreviations

P: phosphorus; K: potassium; Ca: calcium; Mg: magnesium; Fe: iron; Zn: zinc; dS m^{-1}: deci Siemens per meter; Ac: accession; NaCl: sodium chloride; UPM: Universiti Putra Malaysia; cm: centimeter; g and g cc^{-1}: gram, grams per cubic centimeter; CEC: cation exchange capacity; ppm: parts per million; meq: mill equivalents; g and g^{-1}: gram and per gram; Mg: milligram; DAT: days after transplanting; mL: milliliter; FW and DW: fresh weight and dry weight; AAS: atomic absorption spectrophotometer; H_2SO_4 and H_2O_2: sulfuric acid and hydrogen peroxide; SAS: statistical analysis system; ANOVA: analysis of variance.

Authors' contributions

MAA was the main researcher/student of this study and prepared the manuscript. ASJ was the main supervisor of the student and help in manuscript writing. MYR was the co-supervisor, helped in draft preparation and statistical analysis. AAH was also the co-supervisor of the research, helped in nutritional analysis and draft preparation. FA and MAH helped in data analyzing, editing and finalizing the manuscript. All authors read and approved the final manuscript.

Author details

[1] School of Agriculture Science and Biotechnology, Faculty of Bioresources and Food Industry, Universiti Sultan Zainal Abidin, Tembila Campus, 22200 Besut, Terengganu, Malaysia. [2] Department of Crop Science, Faculty of Agriculture, Universiti Putra Malaysia, UPM Serdang, 43400 Serdang, Selangor, DE, Malaysia. [3] Institute of Tropical Agriculture, Universiti Putra Malaysia, UPM Serdang, 43400 Serdang, Selangor, DE, Malaysia. [4] Faculty of Food Science and Technology, Universiti Putra Malaysia, UPM Serdang, 43400 Serdang, Selangor, DE, Malaysia.

Acknowledgements

The authors sincerely acknowledge UPM Research University Grant (01-02-12-1695RU) for financial support of the project and IGRF (International Graduate Research Fellowship, UPM) for Ph.D. Fellowship.

Competing interests

The authors declare that they have no competing interests.

References

1. Ahmad I, Khan MA, Qasim M, Ahmad R, Tauseef-Ussamad. Growth, yield and quality of *Rosa hybrida* L. as influenced by NaCl salinity. J Ornam Plants. 2013;3(3):143–53.
2. Ahmadikhah A, Nasrollanejad S, Alisha O. Quantitative studies for investigating variation and its effect on heterosis of rice. Int J Plant Prod. 2008;2(4):297–308.
3. Alam MA, Juraimi AS, Rafii MY, Hamid AA, Aslani F. Collection and identification of different purslane (*Portulaca oleracea* L.) accessions available in Western Peninsular Malaysia. Life Sci J. 2014;11(6):431–7.
4. Alam MA, Juraimi AS, Rafii MY, Hamid AA, Aslani F. Screening of purslane (*Portulaca oleracea* L.) accessions for high salt tolerance. Sci World J. 2014;2014:1–12.
5. Alam MA, Juraimi AS, Rafii MY, Hamid AA, Aslani F, Hasan MM, Zainudin MM, Uddin MK. Evaluation of antioxidant compounds, antioxidantactivities and mineral composition of 13 collected purslane (*Portulaca oleracea* L.) accessions. Biomed Res Int. 2014;2014:1–10.
6. Alam MA, Juraimi AS, Rafii MY, Hamid AA, Aslani F, Alam MZ. Effects of salinity and salinity-induced augmented bioactive compounds in purslane (*Portulaca oleracea* L.) for possible economical use. Food Chem. 2015;169:439–47.
7. Ali AKS, Mohamed BF, Dreyling D. Salt tolerance and effects of salinity on some agricultural crops in the Sudan. J For Prod Ind. 2014;3(2):56–65.
8. Ali EF, Bazaid S, Hassan FAS. Salt effects on growth and leaf chemical constituents of *Simmondsia chinensis* (Link) schneider. J Med Plants Stud. 2013;1(3):22–4.
9. Anastacio A, Carvalho IS. Accumulation of fatty acids in purslane grown in hydroponic salt stress conditions. Int J Food Sci Nutr. 2013;64(2):235–42.
10. Arolu IW, Rafii MY, Hanafi MM, Mahmud TMM, Latif MA. Molecular characterizations of *Jatropha curcas* germplasm using inter simple sequence repeat (ISSR) markers in Peninsular Malaysia. Aust J Crop Sci. 2012;6(12):1666–73.
11. Ashraf M. Some important physiological selection criteria for salt tolerance in plants. Flora. 2004;199:361–76.
12. Ashraf M, Athar HR, Harris PJC, Kwon TR. Some prospective strategies for improving crop salt tolerance. Adv Agron. 2008;97:45–110.
13. Beltrão J, Brito J, Neves MA, Seita J. Salt removal potential of turfgrasses in golf courses in the Mediterranean Basin. WSEAS Trans Environ Dev. 2009;5(5):394–403.
14. Bray EA, Bailey-Serres J, Weretilnyk E. Responses to abiotic stresses. In: Gruissem W, Buchannan B, Jones R, editors. Biochemistry and molecular biology of plants. Rockville, MD: American Society of Plant Physiologists; 2000. p. 1158–249.
15. Carvalho IS, Mónica T, Maria B. Effect of salt stress on purslane and potential health benefits: oxalic acid and fatty acids profile. In: The proceedings of the international plant nutrition colloquium XVI. UC Davis: Department of Plant Sciences; 2008.
16. Craine JM. Reconciling plant strategy theories of Grime and Tilman. J Ecol. 2005;93:1041–52.
17. Cramer CL, Edwards K, Dron M, Liang X, Dildine SL, Bolwell P, Dixon RA, Lamb CJ, Schuch W. Phenylalanine ammonia-lyasegene organization and structure. Plant Mol Biol. 1989;12:367–83.
18. Dadkhah AR, Grrifiths H. The effect of salinity on growth, inorganic ions and dry matter partitioning in sugar beet cultivars. J Agric Sci Technol. 2006;8:199–210.
19. Dong Y, Ji T, Dong S. Stress responses to rapid temperature changes of the juvenile sea cucumber (*Apostichopus japonicus* Selenka). J Ocean Univ China. 2007;6:275–80.
20. Duru M, Khaled RAH, Ducourtieux C, Theau JP, Quadros FLF, Cruz P. Do plant functional types based on leaf dry matter content allow characterizing native grass species and grasslands for herbage growth pattern? J Plant Ecol. 2008;201(2):421–33.
21. Eker S, Cömertpay G, Konufikan O, Ülger AC, Öztürk L, Çakmak I. Effect of salinity stress on dry matter production and ion accumulation in hybrid maize varieties. Turk J Agric For. 2006;30:365–73.
22. Flórez SL, Lasprilla DM, Chaves B, Fischer G, Magnitskiy S. Growth of lulo (*Solanum quitoense* Lam.) plants affected by salinity and substrate1. Rev Bras Frutic Jaboticabal. 2008;30(2):402–8.

23. Glenn EP, Brown JJ, Blumwald E. Salt tolerance and crop potential of halophytes. Crit Rev Plant Sci. 1999;18:227–55.
24. Hakim MA, Juraimi AS, Musa MH, Ismail MR, Selamat A. Salinity effect on vegetative growth and chlorophyll contents of six dominant weed species in Malaysian coastal rice field. J Food Agric Environ. 2013;11(3&4):1479–84.
25. Hammer O, Harper Dat, Ryan PD. Past: palaeontological statistics software package for education and data analysis. 2009. http://folk.uio.no/ohammer/past.USA. Accessed 23 Feb 2012.
26. Hamidov A, Khaydarova Khamidov M, Neves MA, Beltrao J. *Apocynum lancifolium* and *Chenopodium album*—potencial species to remediate saline soils. WSEAS Trans Environ Dev. 2007;7(3):123–8.
27. Hassanzadehdelouei M, Vazin F, Nadaf J. Effect of salt stress in different stages of growth on qualitative and quantitative characteristics of cumin (*Cuminum cyminum* L.). Cercet Agron Mold. 2013;46(1):89–97.
28. Hirpara KD, Ramoliya PJ, Patel AD, Pandey AN. Effect of salinisation of soil on growth and macro- and micro-nutrient accumulation in seedlings of *Butea monosperma* (Fabaceae). An Biol. 2005;27:3–14.
29. Hixson AC, Crow WT, McSorley R, Trenholm LE. Saline irrigation affects belonolaimus longicaudatus and hoplolaimus galeatus on seashore paspalum. J Nematol. 2004;37:37–44.
30. Jampeetong A, Brix H. Effects of NaCl salinity on growth, morphology, photosynthesis and prolineaccumulation of *Salvinia natans*. Aquat Bot. 2009;91:181–6.
31. Jones WG, Gorham J. Intra- and inter-cellular compartments of ions. In: Lauchli A, Luttge U, editors. Salinity; environment–plant–molecules. Dordrecht: Kluwer; 2002. p. 159–80.
32. Kafi M, Rahimi Z. Effect of salinity and silicon on root characteristics, growth, water status, proline content and ion accumulation of purslane (*Portulaca oleracea* L.). Soil Sci Plant Nutr. 2011;57(2):341–7.
33. Ma T, Zuazaga G. Micro-Kjeldahl determination of nitrogen. A new indicator and an improved rapid method. Ind Eng Chem Anal Ed. 1942;14:280–2.
34. Mane AV, Karadge BA, Samant JS. Salt stress induced alteration in growth characteristics of a grass *Pennisetum alopecuroides*. J Environ Biol. 2011;32:753–8.
35. Marschner H. Adaptation of plants to adverse chemical soil conditions. In: Mineral nutrition of higher plants. 2nd edn. London: Academic Press; 1995. p. 596–80.
36. Mohammadzadeh M, Arouee H, Neamati SH, Shoor M. Effect of different levels of salt stress and salicylic acid on morphological characteristics of four mass native Basils (*Ocimum basilcum*). Int J Agron Plant Prod. 2013;4(S):3590–6.
37. Mozafariyan M, Bayat KSAE, Bakhtiari S. The effects of different sodium chloride concentrations on the growth and photosynthesis parameters of tomato (*Lycopersicum esculentum* cv. Foria). Int J Agric Crop Sci. 2013;6(4):203–7.
38. Munns R. Comparative physiology of salt and water stress. Plant Cell Environ. 2002;25:239–50.
39. Munns R, James RA. Screening methods for salinity tolerance: a case study with tetraploid wheat. Plant Soil. 2003;253:201–18.
40. Munns R, James RA, Läuchli A. Approaches to increasing the salt tolerance of wheat and other cereals. J Exp Bot. 2006;57:1025–43.
41. Munns R, Tester M. Mechanisms of salinity tolerance. Ann Rev Plant Biol. 2008;59:651–81.
42. Parida AK, Das AB. Salt tolerance and salinity effects on plants: a review. Ecotox Environ Saf. 2005;60:324–49.
43. Ramoliya PJ, Patel HM, Pandey AN. Effect of salinization of soil on growth and macro- and micro-nutrient accumulation in seedlings of *Salvadora persica* (Salvadoraceae). For Ecol Manag. 2004;202:181–93.
44. Ranjbar GH. Salt sensitivity of two wheat cultivars at different growth stages. World Appl Sci J. 2010;11(3):309–14.
45. Rohlf FJ. NTSYS-pc: numerical taxonomy system ver.2.1. Setauket, NY: Exeter Publishing Ltd.; 2002.
46. Saini HS. Effects of water stress on male gametophyte development in plants. Sex Plant Reprod. 1997;10:67–73.
47. SAS. The SAS system for Windows, version 9.3 (TS1M2). Cary, NC: SAS Institute Inc; 2013.
48. Schwabele KA, Iddo K, Knap KC. Drain water management for salinity mitigation inirrigated agriculture. Am J Agric Ecol. 2006;88:133–40.
49. Simopoulos AP. The importance of the omega-6/omega-3 fatty acid ratio in cardiovascular disease and other chronic diseases. Exp Biol Med. 2008;233:674–88.
50. Song J, Feng G, Zhang F. Salinity and temperature effect on three salt resistant euhalophytes, *Halostachys capsica* and *Halocnemum strobilaceum*. Plant Sci. 2006;279:201–7.
51. Takemura T, Hanagata N, Dubinsky Z, Karube I. Molecular characterization and response to salt stress of mRNAs encoding cytosolic Cu/Zn superoxide dismutase and catalase from *Bruguiera gymnorrhiza*. Tree. 2002;16:94–9.
52. Talei D, Kadir MA, Yusop MK, Valdiani A, Abdullah MP. Salinity effects on macro and micronutrients uptake in medicinal plant King of Bitters (*Andrographis paniculata* Nees.). Plant Omics J. 2012;5(3):271–8.
53. Tozlu I, Moore GA, Guy CL. Effect of increasing NaCl concentration on stem elongation, dry mass production, and macro-and micro-nutrient accumulation in Poncirus trifoliate. Austr J Plant Physiol. 2000;27:35–42.
54. Uddin MK, Juraimi AS. Salinity tolerance turfgrass: history and prospects. World Sci J. 2013;2013:1–6.
55. Uddin MK, Juraimi AS, Ismail MR, Alam MA. The effect of salinity on growth and ion accumulation in six turfgrass species. Plant Omics J. 2012;5(3):244–52.
56. Uddin MK, Juraimi AS, Ismail MR, Othman R, Rahim AA. Relative salinity tolerance of warm season turf grass species. J Environ Biol. 2011;32:309–12.
57. Volkmar KM, Hu Y, Steppuhn H. Physiological responses of plants to salinity: a review. Can J Plant Sci. 1998;78:19–27.
58. Yazici I, Turkan I, Sekmen AH, Demiral T. Salinity tolerance of purslane (*Portulaca oleracea* L.) is achieved by enhanced antioxidative system, lower level of lipid peroxidation and proline accumulation. Environ Exp Bot. 2007;61:49–57.
59. Zuazo VHD, Martínez-Raya A, Ruiz JA, Tarifa DF. Impact of salinity on macro- and micronutrient uptake in mango (*Mangifera indica* L. cv. Osteen) with different rootstocks. Span J Agric Res. 2004;2(1):121–33.

18

Essential oils from two *Eucalyptus* from Tunisia and their insecticidal action on *Orgyia trigotephras* (Lepidotera, Lymantriidae)

Badreddine Ben Slimane[1*†], Olfa Ezzine[2†], Samir Dhahri[2] and Mohamed Lahbib Ben Jamaa[2]

Abstract

Background: Essential oils extracted from aromatic and medicinal plants have many biological properties and are therefore an alternative to the use of synthetic products. The chemical composition of essential oils from two medicinal plants (*Eucalyptus globulus* and *E. lehmannii*) was determined and, their insecticidal effects on the third and fourth larval stages of *Orgyia trigotephras* were assessed.

Results: Larvae were collected from Jebel Abderrahmane (North-East of Tunisia), conserved in groups of 50/box (21 × 10 × 10 cm) at a temperature of 25°C. Larvae were tested for larvicidal activities of essential oils. Each oil was diluted in ethanol (96%) to prepare 3 test solutions (S1 = 0.05%, S2 = 0.10% and S3 = 0.50%). Essential oils were used for contact, ingestion and Olfactory actions and compared to reference products (*Bacillus thuringiensis* and Decis). Olfactory action of essential oils shows that larvae mortality is higher than contact action, lower than ingestion action. MTM and FTM of S3 of *E. lehmannii* were respectively 1 h 32 min and 1 h 39 min are higher than those of *E. globulus* (MTM = 51 min and FTM = 1 h 22 min 34 sec). Contact action of *E. lehmannii* oil shows low insecticidal activity compared to *E. globulus*. MTM are respectively (1 min 52 sec and 1 min 7 sec), FTM are (2 min 38 sec, 1 min 39 sec), are the shortest recorded for S3, on the third stage of larvae. The fourth stage of larvae, MTM are (2 min 20 sec and 2 min 9 sec), FTM are (3 min 25 sec, 3 min 19 sec). Ingestion action of essential oils is longer than the contact action, since the time of death exceeds 60 minutes for all species.

Conclusion: Results shows that essential oils have a toxic action on nerves leading to a disruption of vital system of insects. High toxic properties make these plant-derived compounds suitable for incorporation in integrated pest management programs.

Keywords: *Eucalyptus*, *Orgyia trigotephras*, Essential oils, Insecticides, Insect control

Background

Acquired resistance and environmental pollution due to repeated applications of persistent synthetic insecticides have created interest in discovering new natural insecticide products [1]. The use of plants with insecticidal activity has several advantages over the use of synthetic products, natural insecticides are obtained from renewable resources and quickly degradable, the development of insect resistance to these substances is slow, the substances do not leave residues in the environment, they are easily obtained by farmers and they cost less to produce [2]. The effects of essential oils on insects have been the subject of several studies. These oils are formed by a complex mixture of volatile constituents originating from the secondary metabolism of plants and are characterized by a strong scent [3]. The components in essential oils vary not only with plant species but also in relation to climate, soil composition, part of the plant and age of the plant. Since many trees were damaged by insects, the search for insecticides and repellents of botanical origin has been driven by the need to find new products that are effective, furthermore safer and cheaper than current products [4]. Additionally, people prefer natural products than synthetics [5]. Many secondary plant metabolites are known for their insecticidal properties, and in

* Correspondence: Badreddine.Benslimane@isste.rnu.tn
†Equal contributors
1Institut Supérieur des Sciences et Technologies de l'Environnement de Borj-Cédria, B.P. 1003, Hammam-Lif 2050, Tunisia
Full list of author information is available at the end of the article

many cases plants have a history of use as home remedies to kill or repel insects [6]. In recent decades, research on the interactions between plants and insects has revealed the potential use of plant metabolites or allelochemicals for this purpose [7]. It is known that some chemical constituents of essential oils have insecticidal properties [8]. In some studies, essential oils obtained from commercial sources were used. Specific compounds isolated from plant extracts or essential oils were tested for fumigation purposes [9].

Among essential oils, Eucalyptus oil, in particular, is more useful as it is easily extractable commercially (industrial value) and possesses a wide range of desirable properties worth exploiting for pest management [10,11]. Previous studies reported the fumigant toxicity of essential oils from various Eucalyptus species against different developmental stages [12]; furthermore the presence of volatile monoterpenes provides an important defense strategy to the plants, particularly against herbivorous insect pests and pathogenic fungi [13].

This study aims to evaluate toxic activities of essential oils obtained from two *Eucalyptus* species: *Eucalyptus lehmannii* and *Eucalyptus globulus* against third and fourth larval stage of *Orgyia trigotephras.*

Results

Essential oils composition

Essential oils efficiency from *E. globulus* and *E. lehmannii* leaves is above 1%. R = 1.25% for *E. globulus* and R = 1.05% for *E. lehmannii.*

As for essential identification, GC and GC/MS analysis of *E. globulus and E. lehmannii* essential oils led to the identification of 32 compounds. The *E. globulus* essential oil profile is characterized by α-pinene (13.61%) and 1.8-cineole (43.18%) as major compounds. Furthermore, *E. lehmannii* is characterized by 1.8-cineole (50.20%) and α-pinene (18.71%) as major compounds. Among other components, the majority belongs to sesquiterpenoïd hydrocarbon volatile compounds (Table 1).

Insecticidal activity

The evaluation of the contact action of essential oils on larvae of *O. trigotephras* showed a similar effect for the two tested oils. For all concentrations, the MTM and the FTM of larvae treated with essential oils were very short compared to the time of death of larvae treated with Decis.

Ethanol used as a solvent for essential oils, produce no toxic effect on larvae. Oils are revealed to be highly toxic on the third instar larvae. The MTM and FTM are the shortest recorded for a concentration of 0.5 ml. However, *E. lehmannii* oil shows low insecticidal activity compared to the oil of *E. globulus.*

Third instar larvae treated by *E. lehmannii* present a MTM = 11 min 22 sec and FTM = 16 min 55 sec, higher than *E. globulus* (MTM = 2 min and FTM = 5 min 20 sec) for S1 = 0.05%. MTM and FTM, obtained after treatment with 0.5% of *E. lehmannii* were respectively 1 min 52 sec and 2 min 38 sec are higher than those of *E. globulus* (with MTM = 1 min 7 sec and FTM = 1 min 39 sec) (Figure 1).

Fourth instar larvae treated by *E. lehmannii* present a MTM = 40 min and FTM = 54 min 49 sec, higher than *E. globulus* (MTM = 16 min 49 sec min and FTM = 32 min) for S1 = 0.05%. MTM and FTM, obtained after treatment with 0.5 ml of *E. lehmannii* were respectively 2 min 20 sec and 3 min 25 sec are higher than those of *E. globulus* (with MTM = 2 min 9 sec and FTM = 3 min 19 sec) (Figure 2).

Table 1 Chemical composition (%) of the essential oils of the analyzed *Eucalyptus sp.*

RT (min)	Compounds	*E. globulus*	*E. lehmani*
7.40	α-Pinene	13.61	18.71
8.94	β-Pinene	0.74	0.25
9.50	β-Myrcene	0.00	0.21
10.02	α-Phellandrene	0.22	0.22
10.87	β-Cymene	3.95	0.00
11.21	1. 8-Cineol	43.18	50.20
12.16	γ-Terpinene	0.36	2.49
13.31	α-Terpinolene	0.50	0.47
14.29	*endo*-Fenchol	0.00	0.41
14.77	α-Campholenal	0.20	0.19
15.28	*trans*-Pinocarveol	3.76	2.46
15.45	Camphor	0.00	1.68
16.16	Pinocarvone	2.99	0.60
16.29	Borneol	0.00	0.76
16.71	1-Terpinen-4-ol	0.40	0.57
17.05	*p*-Cymen-8-ol	0.23	0.00
17.24	α-Terpineol	1.65	0.00
18.27	Carveol	0.44	0.00
19.14	Carvone	0.22	0.00
22.55	Pulegone	0.15	0.00
17.25	β-Fenchol	0.00	2.86
20.63	Bornyl acetate	0.00	0.16
22.83	4-Carene	6.90	9.49
24.80	α-Gurjunene	1.33	0.20
25.12	β-Caryophyllene	0.81	0.84
25.52	Aristolene	0.70	0.00
25.79	Aromadendrene	10.09	4.26
25.88	β-Selinene	0.30	0.00
26.20	α-Humulene	0.52	0.00
26.44	Allo-Aromandrene	2.23	1.04
27.22	α-Selinene	0.00	0.32
27.50	Ledene	1.06	0.00
Total identified (%)		96.54	98.39

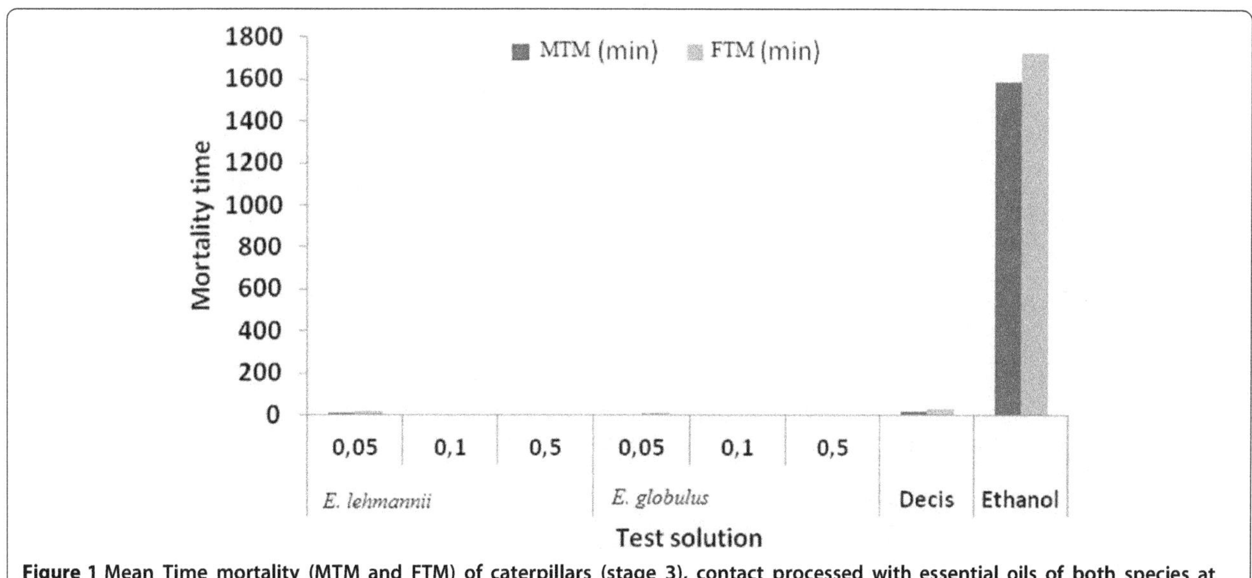

Figure 1 Mean Time mortality (MTM and FTM) of caterpillars (stage 3), contact processed with essential oils of both species at different concentrations.

Ingestion action of essential oils is longer than the contact action, since the time of death exceeds 60 minutes for all species. *E. globules* present the best insecticidal effect. Toxicity of *E. globulus* observed for 3 tested concentrations was particularly important when the concentration is high (S3). MTM = 1 h 40 min and FTM = 3 h 4 min for the third instar larvae and MTM = 1 h 37 min and FTM = 3 h 3 min (Figures 3 and 4).

Thus, it is necessary to do insect histology after treatment with oils to detect tissue target and to identify alteration type. Moreover, essential oils can cause cytoplasm coagulation, damage lipids and proteins or cause cell lysis [3]. A similar phenomenon was observed with *B. thurengiensis* treatment.

As for olfactory action, the third stage of larvae of *O. trigotephras* treated by *E. lehmannii* present a TMM = 4 h 1 min and TFM = 8 h 2 min, higher than *E. globulus* (MTM = 1 h 10 min and FTM = 2 h 11 min) for S1 = 0.05%. MTM and FTM, obtained after treatment with 0.5 ml of *E. lehmannii* were respectively 1 h 32 min and 1 h 39 min

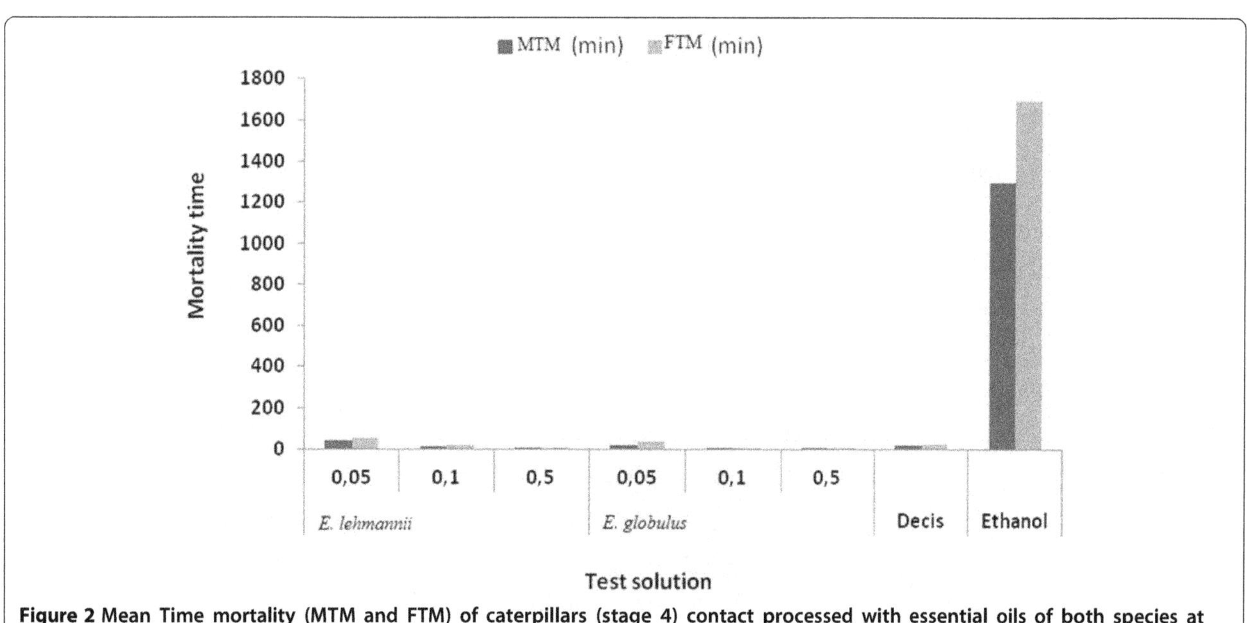

Figure 2 Mean Time mortality (MTM and FTM) of caterpillars (stage 4) contact processed with essential oils of both species at different concentrations.

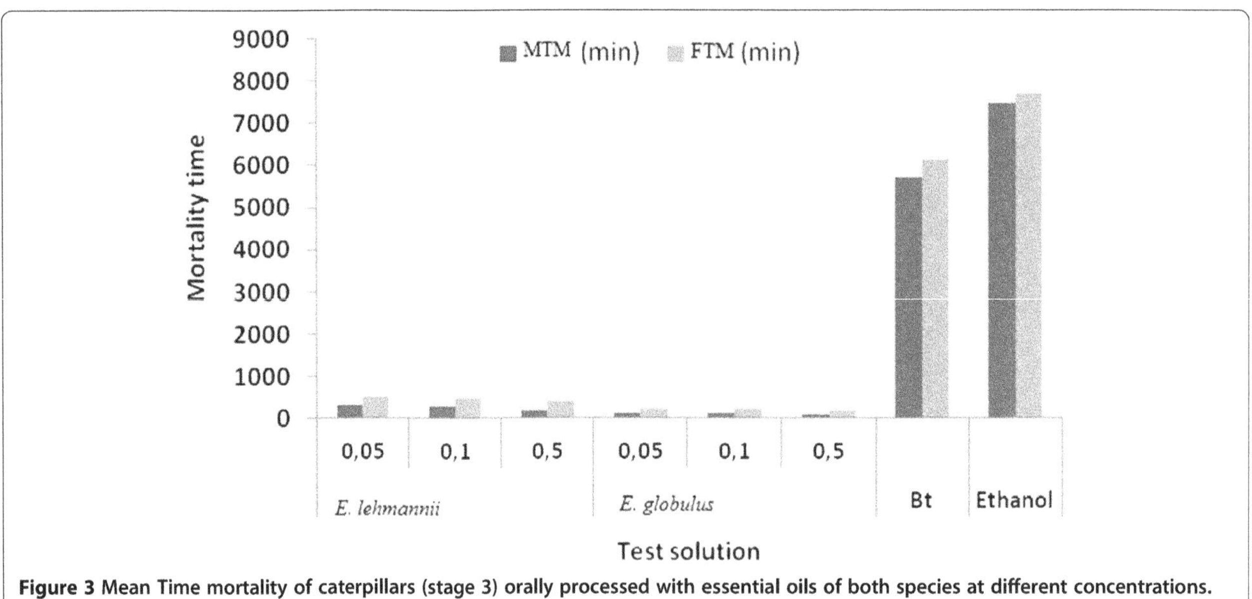

Figure 3 Mean Time mortality of caterpillars (stage 3) orally processed with essential oils of both species at different concentrations.

are higher than those of *E. globulus* (with MTM = 51 min and FTM = 1 h 22 min 34 sec) (Figure 5).

The fourth stage of larvae of *O. trigotephras* treated by *E. lehmannii* present a MTM = 1 h 40 min and FTM = 2 h 33 min, higher than *E. globulus* (MTM = 48 min min and FTM = 1 h 56 min) for S1 = 0.05%. MTM and FTM, obtained after treatment with 0.5 ml of *E. lehmannii* were respectively 1 h 50 min and 2 h 54 min are higher than those of *E. globulus* (with MTM = 35 min and FTM = 1 h 2 min) (Figure 6).

Olfactory action of essential oils shows that larvae mortality is higher than contact action, lower than ingestion action. MTM and FTM of *E. lehmannii* are lowest for S3. Larvae mortality is highest for S1 and S2 (Figures 5 and 6). Ingestion action is more effective and contact action is the less effective. It seems that contact action reduce processing times. However, the ingestion action is the most solicit because it indicates the specificity of the product to the insect. Although, more the necessary time for insect mortality is long, it is sure that the product will be toxic to the pest.

Discussion

Contact action of essential oils is comparable to chemical insecticide that affects the nervous system of larvae as cited by Enan [14] and Cetin et al. [15]. Essential oils

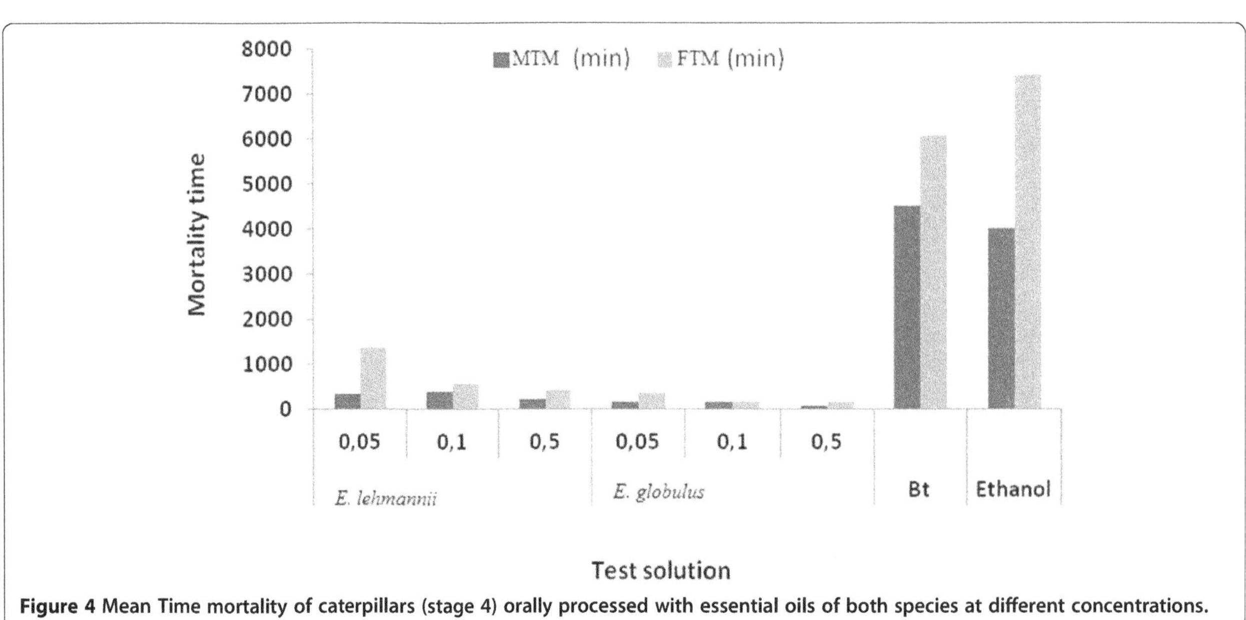

Figure 4 Mean Time mortality of caterpillars (stage 4) orally processed with essential oils of both species at different concentrations.

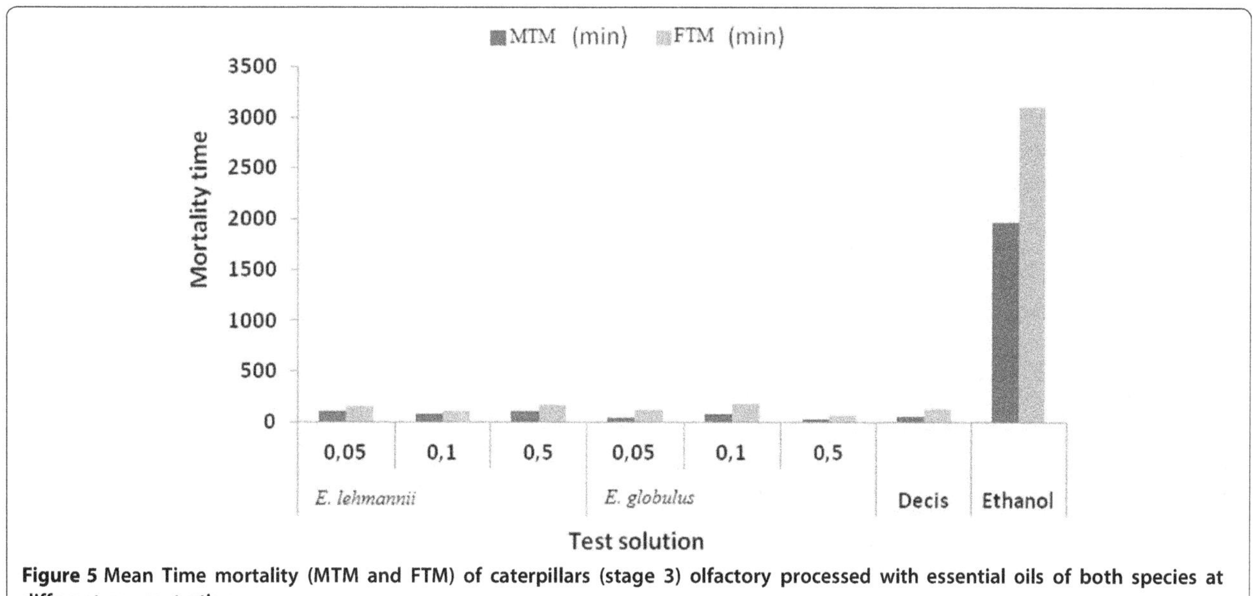

Figure 5 Mean Time mortality (MTM and FTM) of caterpillars (stage 3) olfactory processed with essential oils of both species at different concentrations.

have a toxic action on nerves leading to a disruption of vital system of insects [14,15]. However the highest MTM and FTM of Decis compared to essential oils may be attributed to the limited distribution on the body of larvae unlike essential oils that spread quickly and easily on the back of the insect. The variations between death times resulted from the change in percentages of essential oils components as elucidated by Aslan et al. [16]. The toxicity of certain compounds of essential oils on the fourth instars larvae of *Thaumetopoea pityocampa* has been carried out by Kanat and Alma [17], he showed that turpentine of *Pinus brutia* had the best insecticidal activity (MTM = 0.51 min) due to camphene presence. Kanat and Alma [18] revealed that essential oils of *Thymus vulgaris* composed by carvacrol, *p*-cymene; thymol and *Juniperus communis* composed by camphene and α-pinene have a better insecticide effect than *Lavandula angustifolia* composed by linalool acetate, linalyl, 1,8-cineole and borneol.

Cymbopogon citratus, Lippia sidoides, Ocimum americanum and *Ocimum gratissimum* essential oils showed good insecticidal activity against *Aedes aegypti*. Constituents of these oils are the monoterpenoids geranial: citral for *C. citratus*, thymol for *L. sidoides*, E-methylcinnamate

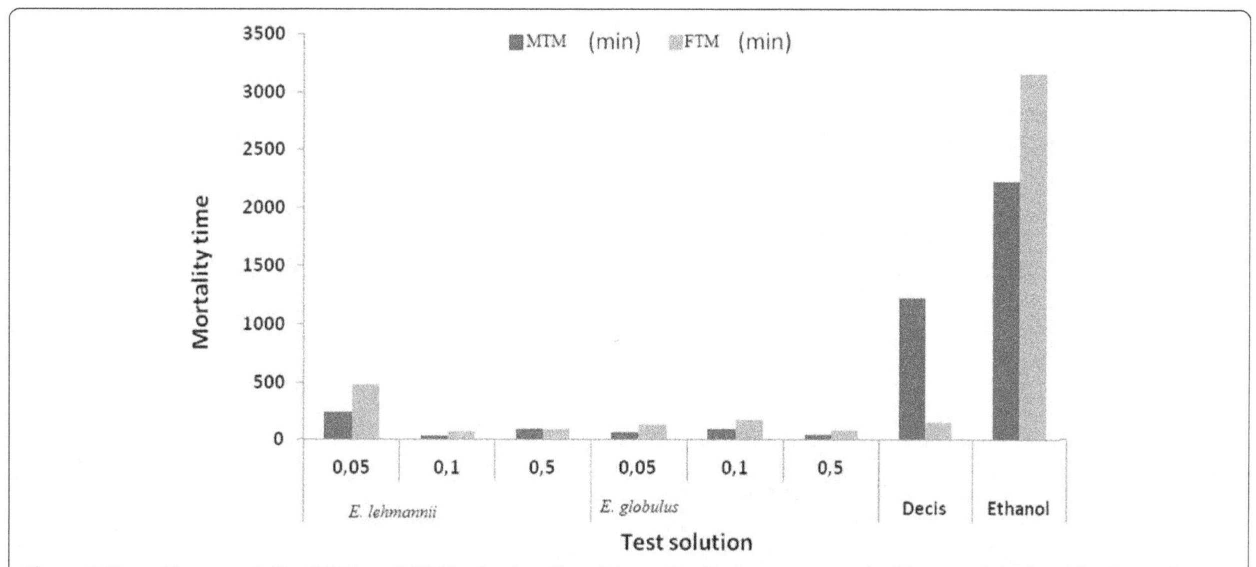

Figure 6 Mean Time mortality (MTM and FTM) of caterpillars (stage 4) olfactory processed with essential oils of both species at different concentrations.

for *O. americanum* and eugenol, 1,8-cineole for *O. gratissimum* [19]. *Myroxylum balsamum* essential oil presented good larvicidal activity against *A. aegypti* larvae, the monoterpenes α-pinene and β-pinene were the main constituents [17]. All these constituents are similar as *E. lehmannii* and *E. globules* essential oils.

Treatment with *Bacillus thurengiensis* is long since the action occurs after being gulped, release of toxin and its binding to specific receptors in the midgut of the insect. Ingestion action of *E. globulus* and *E. lehmannii* oils is faster for S1 for the third instar larvae (MTM = 5 h 3 min and FTM = 8 h 33 min) and for the fourth instar larvae (MTM = 5 h 37 min and FTM = 23 h 8 min), than *B. thurengiensis* (MTM = 1j 34 h 57 sec, FTM = 1j 42 h 3 min for the 3rd instar larvae and MTM = 1j 25 h 17 min, FTM = 1j 41 h 14 min for the 4th instar). It is well known that the biopesticide *B. thurengiensis* acts only by ingestion. This bacterium may have a different effect due to the diversity of toxins that can produce [20].

Previous study on monoterpenes' action on third instar larvae of *Anisakis simplex* Aslan et al. [16] showed that carvacrol was responsible for cells lysis, alteration of the membrane and perforation of the medgut. Essential oil composition varies not only with plant species but also in relation to climate, soil composition, part of the plant and age of the plant [4]. These substances are usually volatile and can be detected by the antennae or tarses of insects. The great majority of the literature on the terpenoids effects on insects has reported growth inhibition, impaired maturation, reduced reproductive capacity, appetite suppression and death of predator insects by starvation or direct toxicity [1]. Monoterpene limonene demonstrated insecticidal activity by penetrating the cuticle of the insect "contact effect", by respiration "fumigant effect" and through the digestive system "ingestion effect" [21].

Our results clearly demonstrated a high toxic effectiveness relatively to *B. thurengiensis* treatment (ingestion action) and Decis treatement (fumigant and contact action) against both larval phases of this pest.

Conclusion

To conclude, our study showed that *E. globulus* and *E. lehmannii* essential oils compositions were characterized by the presence of 1.8-cineole (43.18%; 50.20%), α-pinene (13.61%; 18.71%) respectively as major compounds. It is clear that essential oils from *Eucalyptus spp* are rich of monoterpenoid, compounds that possess insecticidal activity against various insect species. High larvicidal properties make derived compounds suitable for incorporation of integrated pest management program. These results show that application of natural plant products as *E. lehmannii* and *E. globules* which have a toxic effect against larvae of *Orgyia trigotephras* can be a potential method in environmental- friendly control management.

Methods

Plant material and larvae collect

Our study was carried out in the arboretum of Jebel Abderrahmaen (North-east of Tunisia). Four branches from the quadrant (N, S, E and W) were cut off from five trees of the two *Eucalyptus* species; *E. lehmannii* and *E. globules* using a telescopic tree pruner. Branches were separately placed in plastic bags. In the lab, leaves were carried out and air-dried at room temperature (20-25°C) for one week and stored for essential oil extraction. Larvae were collected from Jebel Abderrahmane, conserved in groups of 50 per box (21 × 10 × 10 cm) at a temperature of 25°C and fed every two days on fresh leaves of *Erica multiflora* Third and fourth stage larvae of *Orgyia trigotephras* were tested for larvicidal activities of essential oils.

Essential oil extraction and volatile compounds identification

100 g of dry matter of leaves were used for Essential oils extraction by hydro-distillation method during 90 min using a modified Clevenger-type apparatus. Anhydrous sodium sulphate was used to remove water after extraction. The extracted oils were stored in Eppendorf safe-lock tubes and stored at –4°C.

Essential oils were analyzed by gas chromatography (GC) using a Hewlett-Packard 6890 gas chromatograph (Agilent Technologies, Palo Alto, California, USA) equipped with a flame ionization detector (FID) and an electronic pressure control (EPC) injector. A polar HP-Innowax (PEG) column (30 m × 0.25 mm, 0.25 mm film thickness) and an apolar HP-5 column (30 m × 0.25 mm coated with 5% phenyl methyl silicone, and 95% dimethyl polysiloxane, 0.25 mm film thickness) from Agilent were used. Carrier gas flow (N2) was 1.6 ml/min and the split ratio 60:1. Analyses were performed using the following temperature program: oven kept isothermally at 35°C for 10 min, increased from 35 to 205°C at the rate of 3°C/min and kept isothermally at 205°C for 10 min. Injector and detector of temperatures were held, respectively, at 250 and 300°C. The GC/MS analyses were made using an HP-5972 mass spectrometer with electron impact ionization (70 eV) coupled with an HP-5890 series II gas chromatograph. An HP-5MS capillary column (30 m × 0.25 mm coated with 5% phenyl methyl silicone, and 95% dimethyl polysiloxane, 0.25 μm film thicknesses) was used. The oven temperature was programmed to rise from 50 to 240°C at a rate of 5°C/min. The transfer line temperature was 250°C. Helium was used as carrier gas with a flow rate of 1.2 ml/min and a split ratio of 60:1. Scan time and mass range were 1 s and 40e300 m/z respectively.

Essential oil volatile compounds were identified by calculating their retention index (RI) relative to (C9-C18) n-alkanes (Analytical reagents, Labscan, Ltd, Dublin, Ireland) and data for authentic compounds available in the literature and in our data bank, and also by matching their mass spectrum fragmentation patterns with corresponding data stored in the mass spectra library of the GC-MS data system (NIST) and other published mass spectra [22]. The relative percentage amount of each identified compound was obtained from the electronic integration of its FID peak area.

Essential oils efficiency

Essential oils efficiency (R) is expressed by the ratio between the amount of oil extracted and the amount of plant material used for extraction. R (%) = (mass of the essential oil obtained per mass of plant material used)*100.

Preparation of test solutions and chemical insecticide

Each oil was diluted in ethanol (96%) to prepare 3 test solutions (S1 = 0.05%, S2 = 0.10% and S3 = 0.50%). The essential oils were tested by contact action, ingestion action and olfactory action. The larvicidal effect of essential oils by contact is appreciated by comparison to a chemical insecticide Delta-metrine "Decis" (Atlas Agro-Tunisia). Ethanol used for dilutions was already used as control. The larvicidal effect by ingestion of essential oils is assessed by comparison to a biological insecticide *Bacillus thuringiensis* (reference product, provided by Atlas Agro-Tunisia).

Larvae preparation

Ten larvae were placed in Petri dishes (R = 9 cm). This experiment was replicated 6 times for each test. The rest of the larvae were placed in plastic boxes.

Contact and ingestion action of essential oils

Firstly, 10 µl of each oil solution prepared was deposited on the back of each larva; a total of 60 larvae from 3rd and 4th stage were used and secondary 100 µl from each oil concentration are spread over *Erica multiflora* leaves. Leaves are left in open air until total absorption, than are placed in Petri dishes with 10 fasted larvae to test ingestion action.

Residual toxicity test and evaluation of the insecticidal

100 µl of each test solution were deposited on the bottom of Petri dishes (R = 9 cm) and dried for 20 min at 21°C, 10 larvae per replication were placed to test olfactory action.

The larvicidal activity of essential oils, reference products (*Bacillus thuringiensis*, Decis) and ethanol were determined by measuring the average time of mortality rate (MTM) corresponding to the time required to kill 50% of larvae and the final time of mortality (FTM) corresponding to the death of the total larvae.

Statistical analysis

The statistical treatment of data is performed using SPSS (Version 10.0). MTM and FTM were analyzed for variance by the Fisher test to test the hypothesis of equality of means at the threshold 5%. It is complemented by multiple comparisons of means by the LSD test (Least Significant Difference).

Competing interests

The authors declare that they have no competing interests.

Authors' contributions

BB and OE contributed equally to the realization of this work. SD and MB supervised this scientific study. All authors read and approved the final manuscript.

Acknowledgments

Authors are grateful to Mohamed Laarbi KHOUJA, to Henia CHOGRANI for their help and to the "Institut National de Recherches en Génie Rural, Eaux et Forêts" (Tunisia) for providing assistance to undertake this work.

Author details

[1]Institut Supérieur des Sciences et Technologies de l'Environnement de Borj-Cédria, B.P. 1003, Hammam-Lif 2050, Tunisia. [2]Institut National de Recherches en Génie Rural, Eaux et Forêts, Tunis, Tunisia.

References

1. Viegas-Junior C: **Terpenos com atividade inseticida: uma alternativa para o controle quı'mico de insetos.** *Quim Nova* 2003, **26**:390–400.
2. Roel AR: **Utilização de plantas com propriedades inseticidas: uma contribuição para o Desenvolvimento Rural Sustentável.** *Revista Internacional do Desenvolvimento Local* 2001, **1**:43–50.
3. Bakkali F, Averbeck S, Averbeck D, Idaomar M: **Biological effects of essentials oils- a review.** *Food Chem Toxicol* 2008, **46**:446–475.
4. de Paula JP, Farago PV, Checchia LEM, Hirose KM, Ribas JLC: **Atividade repelente do oleo essencial de Ocimum selloi Benth (variedade eugenol) contra o Anopheles braziliensis Chagas.** *Acta Farm Bonaer* 2004, **23**:376–378.
5. Yaghoobi-Ershadi MR, Akhavan AA, Jahanifard E, Vantandoost H, Amin GH, Moosavi L, Ramazani ARZ, Abdoli H, Arandian MH: **Repellency effect of Myrtle essential oil and DEET against *Phlebotomus papatasi*, under laboratory conditions Iranian.** *J Public Health* 2006, **35**:7–13.
6. Broussalis AM, Ferraro GE, Martino VS, Pinzon R, Coussio JD, Alvarez JC: **Argentine plants as potential source of insecticidal compounds.** *J Ethnopharmacol* 1999, **67**:219–223.
7. Pavela R: **Insecticidal activity of some essential oils against larvae of *Spodoptera littoralis*.** *Fitoterapia* 2005, **76**:691–696.
8. Spitzer C: **Oleos volateis.** In *Farmacognosia: da planta ao medicamento Porto Alegre.* Edited by Simoes CMO, Schenkel EP, Gosmann G, Mello JCP, Mentz LA, Petrovick PR; 2004:467–495.
9. Rajendran S, Sriranjini V: **Plant products as fumigants for stored productinsect control.** *J Stored Prod Res* 2008, **44**:126–135.
10. Boland DJ, Brophy JJ, HOUSE, A.P.N (Eds): *Eucalyptus Leaf Oils. Use, Chemistry, Distillation and Marketing.* Melbourne/Sydney: Inkata Press; 1991.
11. Barton AFM: **The oil mallee project, a multifaceted industrial ecology case study.** *J Ind Ecol* 2000, **3**:161–176.
12. Ben Jemâa JM, Haouel S, Bouaziz M, Khouja ML: **Seasonal variations in chemical composition and fumigant activity of five *Eucalyptus* essential oils against three moth pests of stored dates in Tunisia.** *J Stored Prod Res* 2012, **48**:61–67.
13. Langenheim JH: **Higher plant terpenoids: a phytocentric overview of their ecological roles.** *J Chem Ecol* 1994, **20**:1223–1280.
14. Enan E: **Insecticidal activity of essential oils: octopaminergic sites of action.** *Comp Biochem Physiol* 2001, **130**:325–337.

15. Cetin H, Erler F, Yanikoglu A: **Toxicity of essential oils extracted from *Origanum onites* L and *Citrus aurentium* L against the pine processionary moth, Thaumetopoea wilkinsoni *Tams*.** *Folia Biol* 2006, **54:**153–157.
16. Aslan I, Özbek H, Çalma Ö, Fikrettin S: **Toxicity of essential oil vapours to two greenhouse pests, *Tetranychus urticae Koch* and *Bemisia tabaci* Genn.** *Industrial Crops and Products.* 2004, **19:**167–173.
17. Simas NK, Lima EC, Conceição SR, Kuster RM, Oliveira Filho AM: **Produtos naturais para o controle da transmissao da dengue– atividade larvicida de Myroxylon balsamum (oleo vermelho) e de terpenoides e fenilpropanoides.** *Quim Nova* 2004, **27:**46–49.
18. Kanat M, Alma MH: **Insecticidal effects of essential oils from various plants against larvae of pine processionary moth (*Thaumetopoea pityocampa* Schiff) (Lepidoptera: Thaumetopoeidae*)*.** *Pest Manag Sci* 2003, **60:**173–177.
19. Cavalcante GM, Moreira AFC, Vasconcelos SD: **Potencialidade inseticida de extratos aquosos de essências florestais sobre moscabranca.** *Presquisa Agropecuària Brasileira* 2006, **41:**9–14.
20. Raussel C, Martinez-Ramirez AC, Garcia-Robles I, Real MD: **The toxicity and physiological effects of *Bacillus thuringiensis* toxins and formulations on *Thaumetopoea pityocampa*, the pine processionary caterpillar.** *Pestic Biochem Physiol* 1999, **65:**44–54.
21. Prates HT, Santos JP, Waquil JM, Fabris JD, Oliveira AB, Foster JE: **Insecticidal activity of monoterpenes against *Ryzopertha dominica* (F) and *Tribolium castaneum* (Herbst).** *J Stored Prod Res* 1998, **34:**243–249.
22. Adams RP: *Identification of Essential Oil Components by Gas Chromatography/ Quadrupole Mass Spectroscopy Allured.* USA: Carol Stream Il; 2001.

19

In vitro response of date palm (*Phoenix dactylifera* L.) to K/Na ratio under saline conditions

Suliman A. Alkhateeb[1*], Abdullatif A. Alkhateeb[2] and Mohei EL-Din Solliman[2,3]

Abstract

Background: Salinity is a serious factor limiting the productivity of agricultural plants. One of the potential problems for plants growing under saline conditions is the inability to up take enough K^+. The addition of K^+ may considerably improve the salt tolerance of plants grown under salinity. It is assumed that increasing the K^+ supply at the root zone can ameliorate the reduction in growth imposed by high salinity. The present study aims to determine whether an increase in the K/Na ratio in the external media would enhance the growth of date palm seedlings under in vitro saline conditions.

Methods: Date palm plants were grown at four concentrations of Na + K/Cl (mol/m^3) with three different K/Na ratios. The 12 salt treatments were added to modified MS medium. The modified MS medium was further supplemented with sucrose at 30 g/l.

Results: Growth decreased substantially with increasing salinity. Growth expressed as shoot and root weight, enhanced significantly with certain K/Na ratios, and higher weight was maintained in the presence of equal K and Na. It is the leaf length, leaf thickness and root thickness that had significant contribution on total dry weight. Na^+ contents in leaf and root increased significantly increased with increasing salinity but substantial decreases in Na^+ contents were observed in the leaf and root with certain K/Na ratios. This could be attributed to the presence of a high K^+ concentration in the media. The internal Na^+ concentration was higher in the roots in all treatments, which might indicate a mechanism excluding Na^+ from the leaves and its retention in the roots. K/Na ratios up to one significantly increased the leaf and root K^+ concentration, and it was most pronounced in leaves. The K^+ contents in leaf and root was not proportional to the K^+ increase in the media, showing a high affinity for K^+ uptake at lower external K^+ concentrations, but this mechanism continues to operate even with high external Na^+ concentrations.

Conclusion: Increasing K/Na ratios in the growing media of date plam significantly reduced the absorption of Na^+ less than 200 mM and also balance ions compartmentalization.

Keywords: Date palm, In vitro, Ion relations, K/Na ratio, Salinity

Background

Salinity is a serious factor limiting the productivity of agricultural crops [20]. Although drainage and the supply of high-quality water can solve this problem, these measures are extremely costly and not feasible for extensive application to agriculture [25]. High salinity adversely affects plants due to water stress, ion toxicity, nutritional disorders, membrane disorganization, reduction of cell division and expansion [19, 28].

One of the potential problem for plants growing under saline conditions is the inability to uptake enough K^+, thus creating K^+ deficiency as a result of high concentration of Na^+ and its competition with K^+ [28]. Low uptake of K^+ may occurs because, in saline soils, K^+ is usually present at lower concentrations than Na^+. Under

*Correspondence: skhateeb@kfu.edu.sa
[1] Environment and Natural Resources Department, College of Agriculture and Food Sciences, King Faisal University, P.O. Box 400, Hofuf, Alhassa 31982, Kingdom of Saudi Arabia
Full list of author information is available at the end of the article

non-saline conditions, plants are able to limit their Na^+ uptake. However, with increasing salinity, Na^+ is abundantly absorbed by most plants, even to lethal levels [20, 19]. The addition of K^+ may considerably improve the salt tolerance of many crops [15, 32] by maintaining the ion transport balance across the plasma- and intra-organelle membrane [10, 27]. Application of Multi-K (potassium nitrate) is a very efficient method of combating stresses and enhancing crop performance under saline conditions. This concept has been validated for five crops [1].

It is generally assumed that Na^+ is compartmentalized into vacuole, unlike K^+, which is sequestered in the cytoplasm, resulting in maintenance of a high K/Na ratio in cytoplasm [17, 24]. However, Cuin et al. [8] in quantitative measurements of cytosolic potassium activity in leaf cell compartments of plants subjected to salt stress, reported reduced cytosolic potassium activity to 15 mM despite the vacuole still contain 47 mM. Potassium is ultimately involved in mitigating the detrimental effects of salinity to plant metabolism [29]. Very recently Anschutz et al. [6] in their review showed that regulation of intercellular potassium homeostasis is essential to mediate plant response to a broad range of biotic stresses including drought, salinity and oxidative stress.

An inadequate rate of accumulation of osmotic solute in growing tissues of roots and leaves may limit the growth of many crop plants. High Na^+ concentrations in the leaves may help to maintain turgor, which drives growth in the growing zone [31], but it cannot substitute for adequate K^+ concentration, presumably because K^+ plays essential roles in energy transfer and utilization, protein synthesis, carbohydrate metabolism, transport of sugars from leaves to fruits, and production and accumulation of oils [23]. However, high accumulation of Na^+ in plant tissue may cause damaging effects due to ion excess. Moreover, the ability of the plant to maintain high cytosolic K/Na ratio has been named as a key determinator of a plant salt tolerance [18, 28].

Date palm (*Phoenix dactylifera* L.), being cultivated mostly in arid and saline conditions, and affected by excess salts, has not been previously explored for most of the physiological responses such as K^+/Na^+ ratio. The present studies were therefore, aimed to determine whether an increase in the K/Na ratios in the external media would enhance the growth of date palm seedlings under in vitro saline conditions.

Results

Growth expressed as root, shoot and total dry weights reduced substantially with 200 mol/m^3 (Na + K)/Cl compared to 10 mol/m^3 (Na + K)/Cl (Table 1). Adverse effects of increasing (Na + K)/Cl concentration were more pronounced on shoots than on roots. Root and shoot dry weights of date palm seedlings were drastically increased in the presence of equal concentrations of K^+ and Na^+, even at 200 mol/m^3 (Na + K)/Cl (Table 1). The shoot/root ratio significantly reduced at

Table 1 Effects of Na + K/Cl concentration and K/Na ratio on leaf and root dry weight (mg/plant), total dry weight (mg/plant) and leaf dry weight/root dry weight ratio

Interaction (salinity x K/Na ratio)	Leaf dry weight	Root dry weight	Total dry weight	Ratio of leaf d.w./root d.w.
10 mol/m^3				
0	268	125	393	2.329
0.5	428	134	562	3.953
1	359	208	567	1.845
50 mol/m^3				
0	267	132	399	2.923
0.5	262	165	427	1.742
1	338	133	471	3.107
100 mol/m^3				
0	408	143	551	2.875
0.5	351	113	465	3.821
1	393	152	545	2.676
200 mol/m^3				
0	180	135	315	1.591
0.5	224	179	403	1.272
1	281	137	418	2.585
SEDs (d.f.)	51 (6)	11.5 (6)	48.3 (6)	0.46 (6)

Stander error deviation (SEDs) at $P < 0.05$

the highest salinity level (Table 1). However, it was significantly increased with increasing K/Na ratios from 0–1.

Regression between root thickness and total dry weight gave nonlinear correlations but with relatively good to fair coefficient of determination $r^2 = 0.45$ while coefficient of determinations r^2 of root number and root length were very poor (Table 2; Fig. 1). On the other hand Regression between leaf length, and leaf thickness and total dry weight gave nonlinear correlations but with relatively good to fair coefficient of determinations $r^2 = 0.55$ and 0.37 respectively, while coefficient of determination r^2 of leaf no. was very poor (Table 2; Fig. 2).

The Na^+ content increased significantly as the (Na + K)/Cl concentration increased, particularly at the 0 K/Na ratio (Fig. 3). However, at 200 mol/m^3 (Na + K)/Cl, the leaf Na^+ concentration decreased significantly at both the 0.5 and 1 K/Na ratios (Fig. 3). Varying K/Na ratios from 0–1 consistently decreased the Na^+ concentration in roots and leaves at all (Na + K)/Cl concentrations. The only exception was observed in leaves at 10 mol/m3 (Na + K)/Cl, for which K/Na at the 0.5 ratio had the highest Na^+ concentration (Fig. 3). The K^+ contents in leaves and roots decreased significantly as the (Na + K)/Cl concentration increased (Fig. 4). K/Na ratios up to 1 increased significantly the K^+ concentration in leaves and roots, and it was especially pronounced in roots at higher K/Na ratios (Fig. 4).

The root Ca^{2+} concentration of 10 mol/m^3 (Na + K)/Cl was highest at the K/Na ratios of 0.5 and 1. The root Ca^{2+} concentration was significantly reduced at the K/Na ratio of 1.0 at the highest (Na + K)/Cl concentrations (Fig. 4). At 10 mol/m^3 (Na + K)/Cl, the Ca^{2+} concentration in the leaves was significantly increased with increasing K/Na ratios. However, at 100 mol/m^3 (Na + K)/Cl, the Ca^{2+} concentration in the leaves remained relatively unchanged with varying K/Na ratios, and it was lower than in the 10 and 50 mol/m^3 (Na + K)/Cl conditions (Fig. 3).

Root Cl^- concentration was significantly increased as Cl^- concentration increased in the root media (Fig. 4). Varying K/Na ratios had no significant effects on root Cl^- concentrations. The Cl^- concentration followed approximately the same patterns in leaves and in roots (Figs. 3, 4).

Discussion

Growth expressed as root, leaf and total dry weights reduced substantially in the presence of 200 mol/m^3 (Na + K)/Cl (Table 1). This reduction under salinity is consistent with the results of Aljuburi et al. [3], Al-Abdoulhadi et al. [2], Darwesh [9] and Sperling et al. [30] in date palm. High NaCl levels inhibited leaf expansion, largely due to an inhibition of cell division rather than of cell expansion [7]. Adverse effects of increasing (Na + K)/Cl concentration were more pronounced in leaves than in roots, indicating that root growth was less affected by salinity; this was supported by the results for root and leaf length (Table 2). Consequently, the leaf/root

Table 2 Effects of Na + K/Cl concentration and K/Na ratio on root and leaf number, root and leaf length (mM), root and leaf thickness (μM)

Interaction (salinity x K/Na ratio)	Root no.	Leaf no.	Root length	Leaf length	Root thickness	Leaf thickness
10 mol/m^3						
0	6.00	2.50	9.96	26.22	58.26	7.26
0.5	1.76	3.00	13.58	27.32	79.00	17.26
1	3.26	3.26	8.500	34.88	69.50	13.50
50 mol/m^3						
0	3.26	3.76	8.56	19.62	55.26	7.48
0.5	6.50	4.00	7.62	27.12	47.76	7.76
1	3.76	3.00	7.92	21.50	62.40	17.50
100 mol/m^3						
0	3.50	3.50	7.48	25.56	76.00	12.26
0.5	4.00	3.80	8.16	28.08	60.80	11.00
1	5.50	3.00	6.68	28.50	70.50	23.50
200 mol/m^3						
0	3.00	3.26	8.10	16.72	70.26	12.00
0.5	3.68	3.32	9.00	25.10	75.32	10.00
1	4.00	3.00	8.38	23.26	67.26	14.26
SEDs (d.f)	0.365 (6)	0.352 (6)	1.21 (6)	2.89 (6)	10.5 (6)	1.76 (6)

Stander error deviation (SEDs) at $P < 0.05$

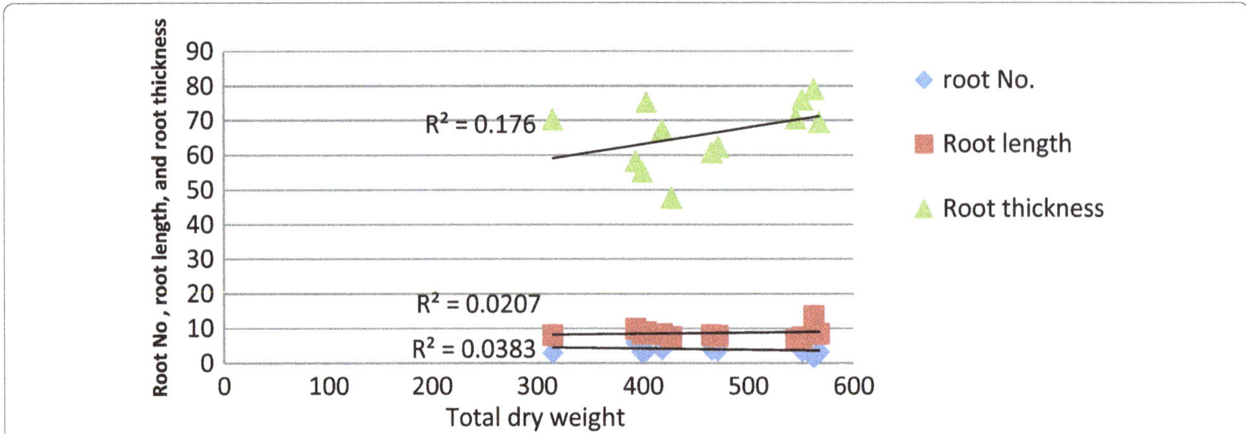

Fig. 1 Relationship between root no., root length (mM) and root thickness (μM) and total dry weight (mg/plant) of date palm as affected by salinity and K/Na ratio

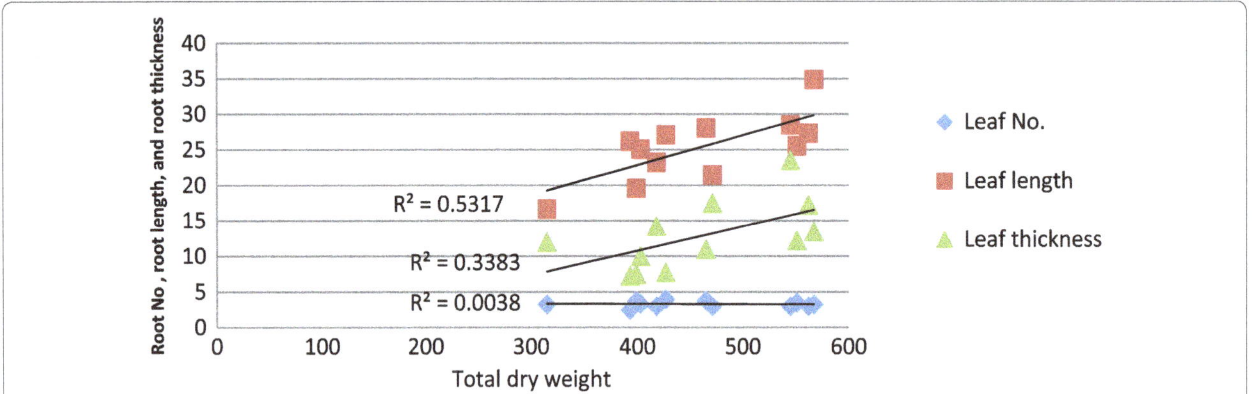

Fig. 2 Relationship between root no., root length (mM) and root thickness (μM) and total dry weight (mg/plant) of date palm as affected by salinity and K/Na ratio

ratio was expected to decrease with increasing (Na + K)/Cl concentrations. Moreover, visual observation (Fig. 5) of date palm offshoots grown under the highest salinity concentration showed necrosis at the tips and margins of leaves, which could be attributed to salt toxicity.

The increase in root and shoot dry weights of seedlings at K/Na ratio of 1, even at the highest salinity level (Table 1) is consistent with the results of Achilea [1] in five crop species. Concerning the relationship of total dry weight and growth parameters, the relatively good to fair correlations indicated that leaf length, leaf thickness, and root thickness had significant contribution in total dry weight (Figs. 1, 2). Shoot/root ratio reduced significantly at high salinity, but increased substantially with increasing K/Na ratios from 0–1. This might be due to dry weight partitioning between shoot and root, being significantly affected by varying K/Na ratios, in addition, root growth was always less affected by salinity than shoot [24].

Increasing leaf and root internal Na^+ concentrations with increasing salinity levels, particularly at lower K/Na ratio (Figs. 3, 4), were reported in date palm [2, 3, 9, 30]. However, at the highest (Na + K)/Cl concentration, the leaf Na^+ concentration decreased significantly at both 0.5 and 1 K/Na ratios (Fig. 3). This could be attributed to the presence of a high concentration of K^+ in the media. The internal Na^+ concentration was higher in the roots than leaves with all treatments, which might indicate a mechanism that excluded Na^+ from the leaves and caused its retention in the roots. Such a mechanism helps to maintain the level of Na^+ in leaves at low concentration. However, to protect metabolism from adverse effects of high Na^+ concentrations in roots or leaves, ion compartmentation is required to take place in the different cell components. Leigh and Wyn Jones [16] reported that Na^+ ions often excluded from the cytoplasm and accumulated in vacuoles. Varying K/Na ratios decreased Na^+ concentrations in roots and leaves at all (Na + K)/Cl

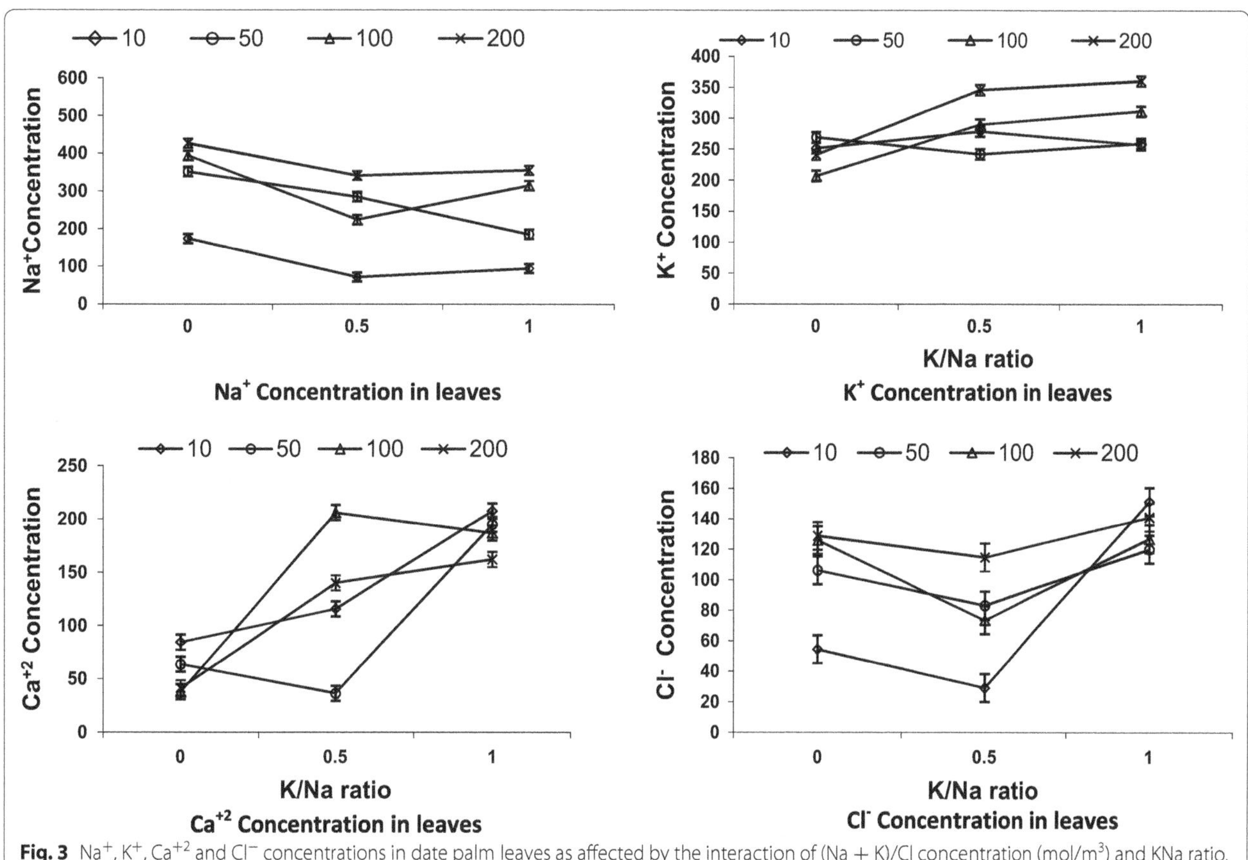

Fig. 3 Na^+, K^+, Ca^{+2} and Cl^- concentrations in date palm leaves as affected by the interaction of (Na + K)/Cl concentration (mol/m^3) and KNa ratio. *Error bars* are SEDs, d.f. = 6

concentrations. The lower Na^+ concentrations in leaves and roots of plants grown with K/Na ratios of 1 could be attributed to the slower growth of these plants, particularly for the leaves. Alternatively, the rate of exporting Na^+ from roots to leaves may have been lower in plants grown with a K/Na ratio of 1 than in those grown with K/Na ratios lower than 1.

The K^+ contents in leaves and roots decreased significantly with increase in Na + K/Cl concentration in the growing medium (Figs. 3, 4). The lower internal K^+ contents with an increase in external Na^+ concentration in the absence of K^+ in date palm might be due to the tendency of the Na^+ to compete with K^+ for major binding sites including control of enzymatic activity or a direct competition between K^+ and Na^+ [29]. K/Na ratios up to 1 significantly increased leaf and root K^+ contents with substantial increase in leaves particularly at higher K/Na ratios (Figs. 3, 4). However, there was no proportionality of leaf and root K^+ contents to K^+ increase in the external media, showing a high affinity for K^+ uptake at low external K^+ concentration, but this mechanism is still operative even with high external Na^+ concentrations [11]. K^+ contents were higher in roots than in leaves with the absence of salinity. This was expected because K^+ and Na^+ contents in the roots depend on active and passive fluxes of Na^+ and K^+, respectively, into and out of the root.

The date palm seedlings grown at 100 mol/m^3 Na + K/Cl with a K/Na ratio of 0 and 1, had 206 and 210, 290 and 310 mol/m^3 K^+ in roots and leaves, respectively (Figs. 3, 4). However, to protect the plant metabolism from excess ions, compartmentalization is required. [8] reported unequal compartmentation of K^+ between cytosol and vacuole, with a threefold more K^+ in vacuole than cytosol. With this assumption, cytoplasm of root and leaf will have around 70 and 100 mol/m^3 K^+ in a K/Na ratio of 0 and 1, respectively. According to He and Wang [14], seedlings grown under saline conditions accumulated more than 60 % of Na^+ in vacuoles. Assuming this, Na^+ contents in the cytoplasm of roots and leaves will be approximately 178 and 157 mol/m^3, 122, and 125 with 0 and 1 K/Na ratios, respectively. Groham et al. [13] reported that the concentration of inorganic ions in the cytoplasm (especially of meristematic cells) is in the range of 100–200 mol/m^3. This situation reflects a low internal K/Na contents in cytoplasm particularly under high salinity

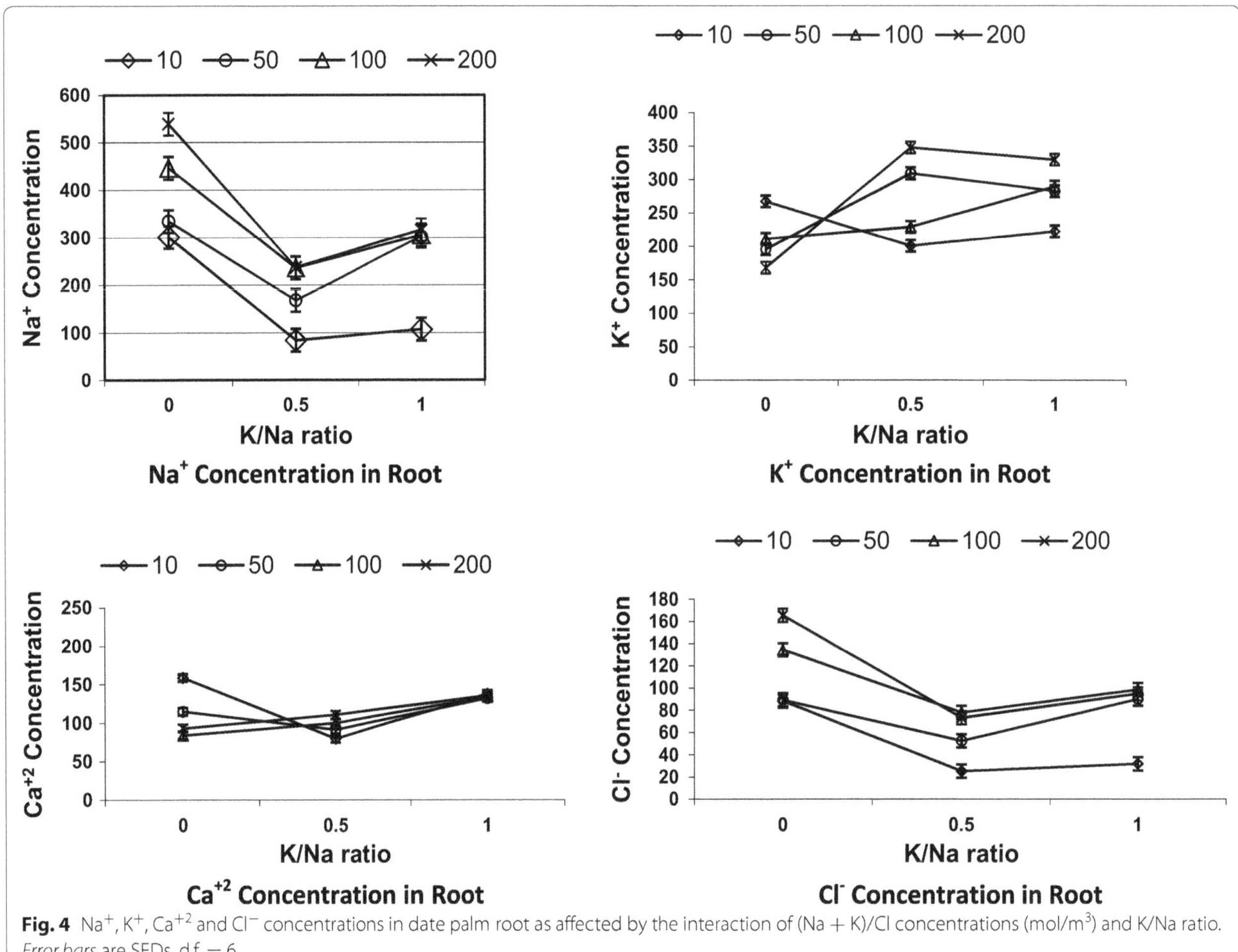

Fig. 4 Na^+, K^+, Ca^{+2} and Cl^- concentrations in date palm root as affected by the interaction of (Na + K)/Cl concentrations (mol/m^3) and K/Na ratio. *Error bars* are SEDs, d.f. = 6

with the absence of K^+ supply. Alternatively, ions may accumulate in the cell wall reducing turgor pressure [22], which is the driving force of plant growth. Moreover, the regulation of intercellular potassium homeostasis is also essential to mediate plant response to a broad range of biotic stresses including drought, salinity and oxidative stress [6]. Moreover it is suggested that not only cytosolic K/Na ratios but also absolute concentrations of K^+ are essential for conferring salinity stress tolerance [29].

Decrease in Ca^{2+} concentration in roots and leaves of date palm with increasing (Na + K)/Cl concentrations might have induced displacement of Ca^{2+} by Na^+ in cell membrane [11]. However, Darwesh [9] elucidated that date palm plantlets clarified high calcium contents under salinity applications. The removal of Ca^{2+} from the membrane affects adversely the mechanism of selective ion transport and increase membrane permeability [5]. However, it appears that the presence of K^+ in the external media positively affects Na^+ displacement of Ca^{2+}, particularly with the highest (Na + K)/Cl concentrations.

Root Cl^- concentration increased significantly as Cl^- concentration increased in the root media (Fig. 4). Darwesh [9] obtained similar results growing date palm seedlings under salinity and amino acid treatment. It appears that varying K/Na ratios had no significant effects on root Cl^- concentrations. Roots had much higher Cl^- concentrations than leaves. Leaf Cl^- concentrations followed approximately the same pattern as in root (Figs. 3, 4). Groham et al. [13] showed that ion toxicity was usually associated with either excessive chloride or sodium intake. However, there was a slight increase in the leaf Cl^- concentration with increasing K/Na ratios. Moreover, Cl^- is a prevalent anion accompanying K^+ or Na^+; therefore, its concentration is expected to be equivalent to the sum of Na + K. This concurrence of Na + K complicates the evaluation of Cl^- specific toxicity. At 200 mol/m^3 (Na + K)/Cl, the internal Na + K concentration in roots and leaves were much larger than Cl^- concentration. Although Na^+ appears to reach a toxic concentration before Cl^- for most species, while for other species such as soybean,

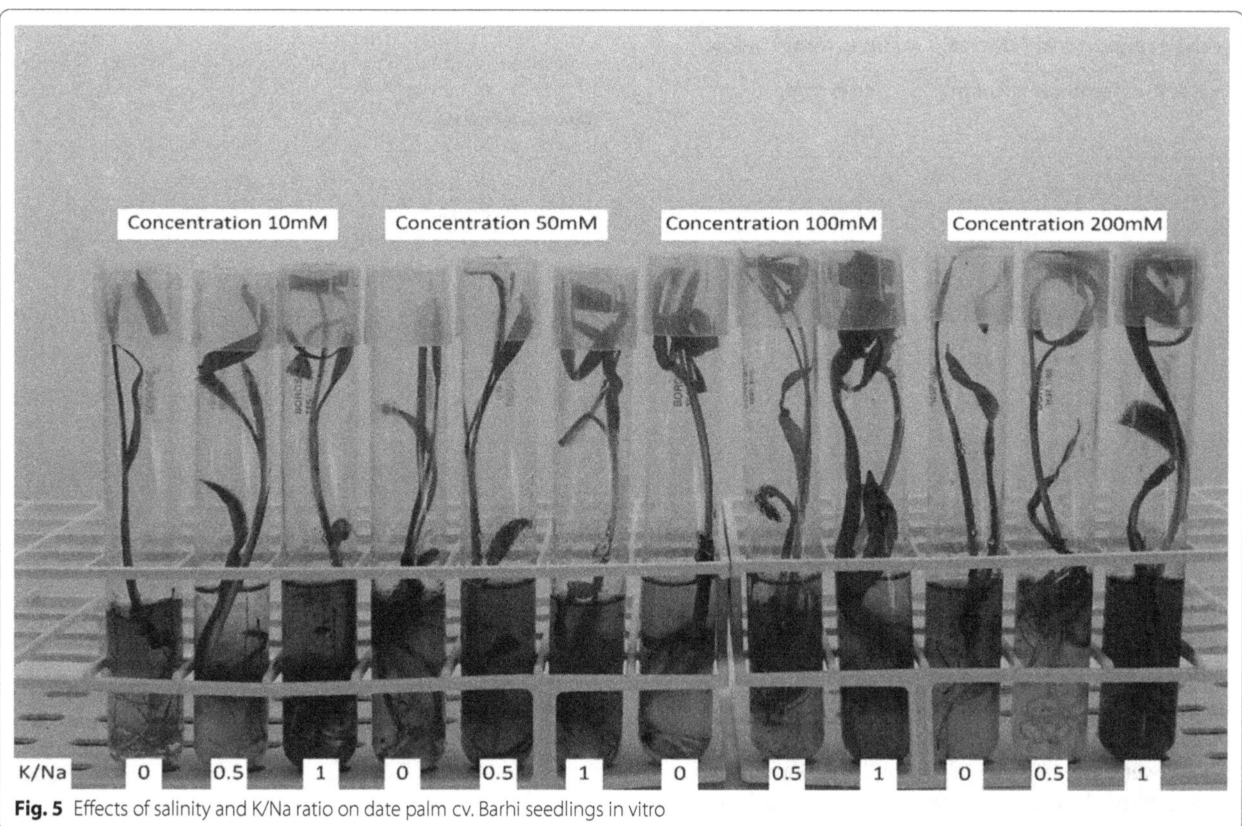

Fig. 5 Effects of salinity and K/Na ratio on date palm cv. Barhi seedlings in vitro

citrus and grapevine, Cl^- is regarded as more toxic than Na^+ [24].

Conclusion

The growth of date palm reduced substantially with increasing salinity. Increasing K/Na ratio on the growing media enhanced date palm seedling growth. This improvement in growth was accompanied by a decrease in Na^+ concentration and an increase in K^+ concentration in the plant tissue with lower Cl^- concentrations in leaves and roots of date palm. Adding K^+ to the salt containing media of date palm reduced the absorption of Na^+ less than 200 mM and also balances ions compartmentalization.

Methods

This experiment was conducted in the Tissue Culture Laboratory of the College of Agriculture and Food Sciences, King Faisal University, Kingdom of Saudi Arabia. Date palm offshoots cv. Barhi of approximately 3 to 4 years old and weighing 5–7 kg were separated from healthy mother trees. Offshoots were cleaned thoroughly and the outer leaves were carefully removed to expose the region of the shoot tip and lateral buds. The exposed region was excised and placed immediately in antioxidant solution containing 15 mg/l ascorbic acid and 100 mg/l citric acid. The shoot tip and lateral buds were sterilized in 20 % v/v Clorox solution for 15 min, followed by rinsing 3 times with distilled water. The tissues were kept in the previous antioxidant solution until explant excision for culturing. The shoot tip and lateral buds were sectioned into explants of approximately 1 cm, which were used for culture initiation as described by Alkhateeb and Ali-Dinar [4]. One rooted plant resulted from rooting media was transferred to a test tube of 20 mm in diameter and 200 mm in length filled with 15 ml of modified MS salts medium [21] supplemented with 125 mg/l inositol; 200 mg/l glutamine; 1 mg/l thiamine HCl; 1 mg/l pyridoxine HCl; 1 mg/l nicotinic acid; 1 mg/l calcium pantothenate; 1 mg/l biotin; 7 g/l purified agar, and 30 g/l sucrose. Potassium phosphate, potassium nitrate, potassium iodide, sodium molybdate, and $Na2EDTA{\cdot}2H_2O$ were eliminated from the MS media to avoid any interference with the treatment concentrations of Na and K. The modified MS media was supplemented with the 12 salt treatments (Table 3). Cultures were incubated at 25 ± 2 °C with 16 h of light daily supplied by 65/80 Warm White Weisse 3500 fluorescent tubes. Each treatment was represented by 10 replicates (tubes) in a factorial, completely randomized design.

Table 3 Concentration of K^+ and Na^+ (mol/m^3) required for the evaluation of K/Na ratio in the external media

(Na + K) Cl concentration (mol/m^3)	K/Na ratio		
	0.0	0.5	1.0
10			
K^+	0	3.33	5
Na^+	10	6.67	5
50			
K^+	0	16.7	25
Na^+	50	33.3	25
100			
K^+	0	33.3	50
Na^+	100	66.7	50
200			
K^+	0	66.7	100
Na^+	200	133.3	100

(K and N) were supplied as KCl and NaCl, respectively

Plants were harvested 3 months after the treatments were applied. Plants were separated into shoot and roots, and their fresh weights were determined. The shoots were washed twice in distilled water, and ions were removed from the free spaces around roots by washing for 2 min in sorbitol solutions isotonic with the treatment concentration in which the plants had grown. To determine the dry weight, shoots and roots were dried at 85 °C for 48 h. For the analysis of K^+, Na^+, Ca^{2+}, and Cl^-, samples of 500 mg of fresh material of leaves or roots were homogenized using a mortar and pestle and were extracted in 25 ml of distilled deionized water at 90 °C for 4 h. The Na^+ and K^+ contents were determined with a flame photometer (Jenway, PFP7). Ca^{2+} was measured with a GBS 905 atomic absorption spectrophotometer. Cl^- was determined using a chloride meter (Jenway, PCLLM3).

Data was subjected to statistical analysis as a factorial design according to Gomez and Gomez [12]. Statistical analyses were performed using SAS software [26]. Means were separated by standard error deviation with their corresponding degrees of freedom.

Authors' contributions

SK and AK conceived and designed research. AK and MS conducted experiment. SK contributed in ion relations. SK and AK analyzed data. SK wrote the manuscript. All authors read and approved the final manuscript.

Author details

[1] Environment and Natural Resources Department, College of Agriculture and Food Sciences, King Faisal University, P.O. Box 400, Hofuf, Alhassa 31982, Kingdom of Saudi Arabia. [2] Agriculture Biotechnology Department, College of Agriculture and Food Sciences, King Faisal University, P.O. Box 400, Hofuf, Alhassa 31982, Kingdom of Saudi Arabia. [3] Plant Biotechnology Department, National Research Centre, Dokki, 12622 Cairo, Arab Republic of Egypt.

Acknowledgements

This work was financed by a Grant No. 130062 from Deanship of Research at King Faisal University.

Competing interests

The authors declare that they have no competing interests.

References

1. Achilea O. Alleviation of salinity-induced stress in cash crops by multi-K (potassium nitrate), five cases typifying the underlying pattern. Acta Hortic. 2002;573:43–8.
2. Al-Abdoulhadi IA, Dinar HA, Ebert G, Büttner C. Effect of salinity on leaf growth, leaf injury and biomass production in date palm (*Phoenix dactylifera* L.) Cultivars. Indian J Sci Technol. 2011;4:1542–6.
3. Aljuburi HJ, Maroff A, Wafi M. The growth and mineral composition of Hatamy date palm seedlings as affected by sea water and growth regulators. Acta Hortic. 2007;736:161–75.
4. Alkhateeb AA, Ali-Dinar HM. Date palm in Kingdom of Saudi Arabia: cultivation, production and processing. Hofuf: Translation, authorship and publishing Center, King Faisal University; 2002.
5. Alkhateeb SA. Effect of calcium/sodium ratio on growth and ion relations of alfalfa (*Medicago sativa* L.) seedling grown under saline condition. J Agron. 2006;5:175–81.
6. Anschutz U, Becker D, Shabala S. Going beyond nutrition: regulation of potassium in homoeostasis as a common denominator of plant adaptive responses to environment. J Plant Physiol. 2014;171:670–787.
7. Chartzoulakis K, Klapaki G. Response of two greenhouse pepper hybrids to NaCl salinity during different growth stages. Sci Hortic. 2000;86:247–60.
8. Cuin TA, Miller AJ, Laurie SA, Leigh RA. Potassium activities in cell components of salt-grown barley leaves. J Exp Bot. 2003;54(383):657–61.
9. Darwesh RSS. Improving growth of date palm plantlets grown under salt stress with yeast and amino acids applications. Ann Agric Sci. 2013;58:247–56.
10. Dreyer I, Uozumi N. Potassium channels in plant cells. FEBS J. 2011;278:4293–303.
11. Epstein E. Mineral nutrition of plants: principles and perspectives. New York: Wiley; 1972.
12. Gomez KA, Gomez AA. Statistical procedures for agricultural research. 2nd ed. New York: Wiley; 1984.
13. Groham J, Wyn Jones RG, McDownell E. Some mechanisms of salt tolerance in crop plants. Plant Soil. 1985;89:15–40.
14. He ZL, Wang HC. Effect of NaCl pretreatment on the accumulation and distribution of Na^+, Cl^- and proline in alfalfa under salt stress. Plant Physiol Commun. 1992;28:330–4.
15. Jin SH, Huang JQ, Li XQ, Zheng BS, Wu JS, Wang ZJ, Liu GH, Chen M. Effects of potassium supply on limitations of photosynthesis by mesophyll diffusion conductance in Caryacathayensis. Tree Physiol. 2011;31:1142–51.
16. Leigh RA, Wyn Jones RG. Cellular compartmentation in plant nutrition: the selective cytoplasm and the promiscuous vacuole. In: Tinker B, Lauchi A, editors. Advances in plant nutrition, vol. 2. Santa Barbara: Praeger Publishers; 1986. p. 249–79.
17. Marschner H. Mineral nutrition of higher plants. London: Elsevier Academic Press; 2012.
18. Maathius FJM, Amtmann A. K^+ nutrition and Na^+ toxicity: the basis of cellular K^+/Na^+ ratios. Ann Bot. 1999;84:123–33.
19. Munns R. Comparative physiology of salt and water stress. Plant Cell Environ. 2002;25:239–50.
20. Munns R, Tester M. Mechanisms of salinity tolerance. Annu Rev Plant Biol. 2008;59:651–81.
21. Murashige T, Skoog F. A revised medium for rapid growth and bioassays with tobacco tissue cultures. Physiol Plant. 1962;15:473–97.
22. Oertiel JJ. Extracellular salt accumulation. Agrochimica. 1968;12:461–9.
23. Römheld V, Kirkby EA. Research on potassium in agriculture: needs and prospects. Plant Soil. 2010;335:155–80.

24. Roychoudhury A, Chakraborty M. Biochemical and molecular basis of varietal difference in plant salt tolerance. Ann Rev Res Biol. 2013;3:422–54.
25. Ruiz JR. Engineering salt tolerance in crop plants. Trends Plant Sci. 2001;6:451.
26. SAS Institute. SAS for windows, SAS users guide: statistics version 8.0 e. Cary: SAS Institute; 2001.
27. Shabala S. Regulation of potassium transport in leaves:from molecular to tissue level. Ann Bot. 2003;92:627–34.
28. Shabala S, Cuin TA. Potassium transport and plant salt tolerance. Physiol Plant. 2008;133:651–69.
29. Shabala S, Pottosin I. Regulation of potassium transport in plants under hostile conditions: implications for abiotic and biotic stress. Physiol Plant. 2014;151:257–79.
30. Sperling O, Lazarovitch N, Schwartz A, Shapira O. Effects of high salinity irrigation on growth, gas-exchange, and photoprotection in date palms (*Phoenix dactylifera* L., cv. Medjool). Environ Exp Bot. 2014;99:100–9.
31. Tomos AD. The physical limitations of leaf cell expansion. In: Baker NR, Davies WD, Ong C, editors. Society of experimental biology symposium. Cambridge: Cambridge University Press; 1985. p. 1–33.
32. Tzortzakis NG. Potassium and calcium enrichment alleviate salinity-induced stress in hydroponically grown endives. Horticultural Science (Prague). 2010;37:155–62.

PERMISSIONS

All chapters in this book were first published in BR, by BioMed Central; hereby published with permission under the Creative Commons Attribution License or equivalent. Every chapter published in this book has been scrutinized by our experts. Their significance has been extensively debated. The topics covered herein carry significant findings which will fuel the growth of the discipline. They may even be implemented as practical applications or may be referred to as a beginning point for another development.

The contributors of this book come from diverse backgrounds, making this book a truly international effort. This book will bring forth new frontiers with its revolutionizing research information and detailed analysis of the nascent developments around the world.

We would like to thank all the contributing authors for lending their expertise to make the book truly unique. They have played a crucial role in the development of this book. Without their invaluable contributions this book wouldn't have been possible. They have made vital efforts to compile up to date information on the varied aspects of this subject to make this book a valuable addition to the collection of many professionals and students.

This book was conceptualized with the vision of imparting up-to-date information and advanced data in this field. To ensure the same, a matchless editorial board was set up. Every individual on the board went through rigorous rounds of assessment to prove their worth. After which they invested a large part of their time researching and compiling the most relevant data for our readers.

The editorial board has been involved in producing this book since its inception. They have spent rigorous hours researching and exploring the diverse topics which have resulted in the successful publishing of this book. They have passed on their knowledge of decades through this book. To expedite this challenging task, the publisher supported the team at every step. A small team of assistant editors was also appointed to further simplify the editing procedure and attain best results for the readers.

Apart from the editorial board, the designing team has also invested a significant amount of their time in understanding the subject and creating the most relevant covers. They scrutinized every image to scout for the most suitable representation of the subject and create an appropriate cover for the book.

The publishing team has been an ardent support to the editorial, designing and production team. Their endless efforts to recruit the best for this project, has resulted in the accomplishment of this book. They are a veteran in the field of academics and their pool of knowledge is as vast as their experience in printing. Their expertise and guidance has proved useful at every step. Their uncompromising quality standards have made this book an exceptional effort. Their encouragement from time to time has been an inspiration for everyone.

The publisher and the editorial board hope that this book will prove to be a valuable piece of knowledge for researchers, students, practitioners and scholars across the globe.

LIST OF CONTRIBUTORS

Aysel Sivaci
Department of Biology, Art and Science Faculty, Adiyaman University, Adiyaman, Turkey

Sevcan Duman
Graduate School of Sciences, Adiyaman University,Adiyaman, Turkey

Sadiye Peral Eyduran
Agricultural Faculty, Department of Horticulture, Igdır University, Igdir,Turkey

Meleksen Akin
Agricultural Faculty, Department of Horticulture, Oregon State University, Corvallis, Oregon, USA

Sezai Ercisli
Agricultural Faculty, Department of Horticulture, Atatürk University, Erzurum,Turkey

Ecevit Eyduran
Agricultural Faculty, Department of Animal Science, Biometry Genetics Unit, Igdır University, Igdır, Turkey

David Maghradze
Scientific-Reaserch Center of Agriculture, Tbilisi, Georgia

Muhammad Zia-Ul-Haq
The Patent Office, Karachi, Pakistan

Shakeel Ahmad
Department of Agronomy, Bahauddin Zakariya University, Multan 60800, Pakistan

Shazia Anwer Bukhari
Department of Applied Chemistry and Biochemistry, Government College University, Faisalabad,Pakistan

Ryszard Amarowicz
Institute of Animal Reproduction and Food Research of the Polish Academy of Sciences, Tuwima Str. 10, 10-747 Olsztyn, Poland

Sezai Ercisli
Agricultural Faculty, Department of Horticulture, Ataturk University, Erzurum, Turkey

Hawa ZE Jaafar
Department of Crop Science, Faculty of Agriculture, 43400 UPM Serdang, Selangor, Malaysia

Emine Sema Cetin
Department of Horticulture, Faculty of Agriculture and Natural Science, Bozok University, 66200 Yozgat, Turkey

Zehra Babalik
Fruit Research Station, Republic of Turkey Ministry of Food, Agriculture and Livestock, Egirdir, Isparta, Turkey

Filiz Hallac-Turk
Department of Horticulture, Faculty of Agriculture, Suleyman Demirel University, Isparta, Turkey

Nilgun Gokturk-Baydar
Department of Agricultural Biotechnology, Faculty of Agriculture, Suleyman Demirel University, Isparta, Turkey

Yuhang Chen
Institute of Chinese Medicinal Materials, Nanjing Agricultural University, Nanjing
210095, People's Republic of China
College of Pharmaceutical Sciences,Chengdu Medical College, Chengdu 610083, People's Republic of China

Li Liu, Qiaosheng Guo, Zaibiao Zhu and Lixia Zhang
Institute of Chinese Medicinal Materials, Nanjing Agricultural University, Nanjing
210095, People's Republic of China

Sadaf Naz Ashraf, Muhammad Zubair, Komal Rizwan and Nasir Rasool
Department of Chemistry, Government College University, Faisalabad 38000, Pakistan

Rasool Bakhsh Tareen
Department of Botany, University of Balochistan, Quetta, Pakistan.

Muhammad Zia-Ul-Haq
The Patent Office, Karachi, Pakistan

Sezai Ercisli
Ataturk University Agricultural Facultu Department of Horticulture, 25240 Erzurum, Turkey

Ferhad Muradoglu and Muttalip Gundogdu
Department of Horticulture, Faculty of Agriculture and Natural Sciences, Abant Izzet Baysal University, Bolu, Turkey

Sezai Ercisli
Department of Horticulture,Faculty of Agriculture, Ataturk University, Erzurum, Turkey

Tarik Encu
Department of Horticulture, Faculty of Agriculture, Yuzuncu Yil University, Van, Turkey

Fikri Balta
Department of Horticulture, Faculty of Agriculture, Ordu University, Ordu, Turkey

Hawa ZE Jaafar
Department of Crop Science, Faculty of Agriculture, University Putra Malaysia, 43400 Selangor, Malaysia

Muhammad Zia-Ul-Haq
The Patent Office, Karachi, Pakistan

Yu Liu, Lu Wang, Heng Liu, Rongrong Zhao, Bin Liu and Yuanhu Zhang
State Key Laboratory of Crop Biology, College of Life Sciences, Shandong Agricultural University, 61 Dai Zong Street, Tai'an 271018, Shandong, People's Republic of China

Quanjuan Fu
Shandong Institute of Pomology, 66 Long Tan Road, Tai'an 271018, Shandong, People's Republic of China

Bingyu Ye
Beijing Institute of Biotechnology, No. 20, Dongdajie Street, Beijing, Fengtai District 100071, China
College of Life Science, Capital Normal University, 105 Xisihuanbei Road, Beijing, Haidian District 100048, China

Paz Zúñiga-González
Laboratorio de Micología y Micorrizas, Facultad de Ciencias Naturales y Oceanográficas and Laboratorio de Investigación en Agentes Antibacterianos, Facultad de Ciencias Biológicas, Universidad de Concepción, Barrio Universitario s/n, Concepción, Chile

Gustavo E. Zúñiga and Marisol Pizarro
Departamento de Biología, Facultad de Química y Biología, Universidad de Santiago, Alameda, 3363 Santiago, Chile.

Angélica Casanova-Katny
Núcleo de Estudios Ambientales, Universidad Católica de Temuco, Casilla 15-D, Temuco, Chile
Facultad de Química y Biología, Universidad de Santiago,Alameda, 3363 Santiago, Chile

Babita Paudel, Hari Datta Bhattarai, Il Chan Kim, Hyoungseok Lee and Joung Han Yim
Division of Life Sciences, Korea Polar Research Institute, KOPRI, Incheon 406-840, Republic of Korea

Roman Sofronov, Lena Ivanova and Lena Poryadina
Institute for Biological Problems of Cryolithozone, Siberian Branch of Russian Academy of Sciences, Moscow, Russia

Adnan Muzaffar
National Center of Excellence in Molecular Biology, University of the Punjab, Lahore 53700, Pakistan
Institute of Molecular Biology, Academia Sinica,Taipei 115, Taiwan

Sarfraz Kiani, Muhammad Azmat Ullah Khan, Abdul Qayyum Rao, Arfan Ali, Mudassar Fareed Awan, Adnan Iqbal, Idrees Ahmad Nasir, Ahmad Ali Shahid and Tayyab Husnain
National Center of Excellence in Molecular Biology, University of the Punjab, Lahore 53700, Pakistan

Evelyn Villagra, Carola Campos-Hernandez, Pablo Cáceres, Gustavo Cabrera, Yamilé Bernardo, Ariel Arencibia and Rolando García-Gonzales
Departamento de Ciencias Forestales, Centro de Biotecnología de los Recursos Naturales, Universidad Católica del Maule, Campus San Miguel. Av. San Miguel 3605, casilla 617, Talca, Maule Region, Chile

Basilio Carrasco
Facultad de Agronomía, e Ingeniería Forestal, Pontificia Universidad Católica de Chile,Vicuña Mackenna 4860, Macul, Santiago, Chile

Peter DS Caligari and José Pico
Instituto de Biología Vegetal y Biotecnología, Universidad de Talca, Avenida Lircay s/n, Talca, Chile

Ali Ikinci and Ibrahim Bolat
Horticulture Department, Harran University, Agriculture Faculty, 63330 Sanliurfa, Turkey

Sezai Ercisli
Horticulture Department, Ataturk University, Agriculture Faculty, 25240 Erzurum, Turkey

Ossama Kodad
Department of Pomology, National School of Agriculture, Meknes, Morocco

Corina Danciu
Department of Pharmacognosy, Faculty of Pharmacy, University of Medicine and Pharmacy Victor Babes", Eftimie Murgu Square, No. 2, Timisoara 300041, Romania

Lavinia Vlaia
Department of Pharmaceutical Technology, Faculty of Pharmacy, University of Medicine and Pharmacy Victor Babes", Eftimie Murgu Square, No. 2, Timisoara 300041, Romania

Florinela Fetea
Department of Chemistry and Biochemistry, University of Agricultural Sciences and Veterinary Medicine ofCluj-Napoca, Mănăştur Str.,No. 3-5, Cluj-Napoca 400372, Romania

Monica Hancianu and Cristina A Dehelean
Department of Pharmacognosy, Faculty of Pharmacy, University of Medicine and Pharmacy "Gr.T.Popa", Iasi, Romania

Dorina E Coricovac and Sorina A Ciurlea
Department of Toxicology, Faculty of Pharmacy, University of Medicine and Pharmacy "Victor Babes", Eftimie Murgu Square, No. 2, Timisoara 300041, Romania

Codruța M Şoica and Cristina Trandafirescu
Department of Pharmaceutical Chemistry, Faculty of Pharmacy, University of Medicine and Pharmacy "Victor Babes", Eftimie Murgu Square, No. 2, Timisoara 300041,Romania

Iosif Marincu
Faculty of Medicine, University of Medicine and Pharmacy "Victor Babes", Eftimie Murgu Square, No. 2, Timisoara 300041, Romania

Vicentiu Vlaia
Department of Organic Chemistry, Faculty of Pharmacy, University of Medicine and Pharmacy "Victor Babes", Eftimie Murgu Square, No. 2, Timisoara 300041, Romania

Corina Danciu
Department of Pharmacognosy, Faculty of Pharmacy, University of Medicine and Pharmacy Victor Babes", Eftimie Murgu Square, No. 2, Timisoara 300041, Romania

Lavinia Vlaia
Department of Pharmaceutical Technology, Faculty of Pharmacy, University of Medicine and Pharmacy Victor Babes", Eftimie Murgu Square,No. 2, Timisoara 300041, Romania

Florinela Fetea
Department of Chemistry and Biochemistry, University of Agricultural Sciences and Veterinary Medicine of Cluj-Napoca, Mănăştur Str.,No. 3-5, Cluj-Napoca 400372, Romania

Monica Hancianu and Cristina A Dehelean
Department of Pharmacognosy, Faculty of Pharmacy, University of Medicine and Pharmacy "Gr.T.Popa", Iasi, Romania

Dorina E Coricovac and Sorina A Ciurlea
Department of Toxicology, Faculty of Pharmacy, University of Medicine and Pharmacy "Victor Babes", Eftimie Murgu Square, No. 2, Timisoara 300041, Romania

Codruța M Şoica and Cristina Trandafirescu
Department of Pharmaceutical Chemistry, Faculty of Pharmacy, University of Medicine and Pharmacy "Victor Babes", Eftimie Murgu Square, No. 2, Timisoara 300041,Romania

Iosif Marincu
Faculty of Medicine, University of Medicine and Pharmacy "Victor Babes", Eftimie Murgu Square, No. 2, Timisoara 300041, Romania

Vicentiu Vlaia
Department of Organic Chemistry, Faculty of Pharmacy, University of Medicine and Pharmacy "Victor Babes", Eftimie Murgu Square, No. 2, Timisoara 300041, Romania

Oksana Sytar
Plant Physiology and Ecology Department, Taras Shevchenko National University of Kyiv, Institute of Biology, Volodymyrskya str., 64, Kyiv 01033, Ukraine

Asel Borankulova
Department of Technology of Food Products, Processing Industries and Biotechnology, Taraz State University named after MK Dulati, Suleimen Str., 7, Taraz 080012, Republic of Kazakhstan

Irene Hemmerich and Cornelia Rauh
Department of Methods of Food Biotechnology, Berlin University of Technology, Institute of FoodTechnology and Food Chemistry, Koenigin Luise Str. 22, Berlin D-14195, Germany

Iryna Smetanska
Department of Methods of Food Biotechnology, Berlin University of Technology, Institute of Food Technology and Food Chemistry, Koenigin Luise Str. 22, Berlin D-14195,Germany
Agricultural Faculty, Department of Plant Food Processing, University of Applied Science Weihenstephan-Triesdorf, Steingruberstr. 2, Weidenbach 91746, Germany

Josiah Bitrus Habu
Bioresources Development Centre Odi, Bayelsa, National Biotechnology Development Agency, Abuja, Nigeria

Bartholomew Okechukwu Ibeh
Department of Biochemistry, College of Natural and Applied Sciences, Michael Okpara University of AgricultureUmudike, Umudike, Nigeria
National Biotechnology Development Agency,Abuja, Nigeria

Md. Amirul Alam
School of Agriculture Science and Biotechnology, Faculty of Bioresources and Food Industry, Universiti Sultan Zainal Abidin, Tembila Campus,22200 Besut, Terengganu, Malaysia

Abdul Shukor Juraimi and Farzad Aslani
Department of Crop Science, Faculty of Agriculture, Universiti Putra Malaysia, UPM Serdang, 43400 Serdang, Selangor,DE, Malaysia

M. Y. Rafii
Department of Crop Science, Faculty of Agriculture, Universiti Putra Malaysia, UPM Serdang, 43400 Serdang, Selangor,DE, Malaysia
Institute of Tropical Agriculture, Universiti Putra Malaysia, UPM Serdang, 43400 Serdang, Selangor, DE, Malaysia

Azizah Abdul Hamid
Faculty of Food Science and Technology, Universiti Putra Malaysia, UPM Serdang, 43400 Serdang,Selangor, DE, Malaysia

M. A. Hakim
Institute of Tropical Agriculture, Universiti Putra Malaysia,UPM Serdang, 43400 Serdang, Selangor, DE, Malaysia

Badreddine Ben Slimane
Institut Supérieur des Sciences et Technologies de l'Environnement de Borj-Cédria, B.P. 1003, Hammam-Lif 2050, Tunisia

Olfa Ezzine, Samir Dhahri and Mohamed Lahbib Ben Jamaa
Institut National de Recherches en Génie Rural, Eaux et Forêts, Tunis, Tunisia

Suliman A. Alkhateeb
Environment and Natural Resources Department, College of Agriculture and Food Sciences, King Faisal University, Hofuf, Alhassa 31982, Kingdom of Saudi Arabia

Abdullatif A. Alkhateeb
Agriculture Biotechnology Department, College of Agriculture and Food Sciences, King Faisal University, Hofuf, Alhassa 31982 Kingdom of Saudi Arabia

Mohei EL-Din Solliman
Agriculture Biotechnology Department, College of Agriculture and Food Sciences, King Faisal University, Hofuf,Alhassa 31982, Kingdom of Saudi Arabia
Plant Biotechnology Department,National Research Centre, Dokki, 12622 Cairo, Arab Republic of Egypt

Index

www.ingramcontent.com/pod-product-compliance
Lightning Source LLC
Chambersburg PA
CBHW082011190326
41458CB00010B/3153

* 9 7 8 1 6 8 2 8 6 7 4 1 9 *